MICROPROCESSOR PROGRAMMING, TROUBLESHOOTING, AND INTERFACING
The Z80, 8080, and 8085

James W. Coffron

PRENTICE-HALL, Englewood Cliffs, New Jersey 07632

Library of Congress Cataloging-in-Publication Data

Coffron, James.
    Microprocessor programming, troubleshooting, and interfacing : the
Z80, 8080, and 8085 / James W. Coffron.
        p.   cm.
    Includes index.
    ISBN 0-13-581976-8
    1. Zilog Z-80 (Microprocessor) 2. Intel 8080 (Microprocessor)
3. Intel 8085 (Microprocessor)   I. Title.
QA76.8.Z54C64 1988
004.165--dc19   87-25820
                                                            CIP

Editorial/production supervision and
   interior design: *Ellen Denning*
Cover design: *Photo Plus Art*
Manufacturing buyer: *Peter Havens*

To my Mother and Father:
Two people who always give me their best.

Z80 is a registered trademark of ZILOG Corporation.

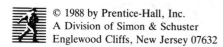 © 1988 by Prentice-Hall, Inc.
A Division of Simon & Schuster
Englewood Cliffs, New Jersey 07632

Printed in the United States of America

10 9 8 7 6 5 4 3 2 1

ISBN   0-13-581976-8

Prentice-Hall International (UK) Limited, *London*
Prentice-Hall of Australia Pty. Limited, *Sydney*
Prentice-Hall Canada Inc., *Toronto*
Prentice-Hall Hispanoamericana, S.A., *Mexico*
Prentice-Hall of India Private Limited, *New Delhi*
Prentice-Hall of Japan, Inc., *Tokyo*
Simon & Schuster Asia Pte. Ltd., *Singapore*
Editora Prentice-Hall do Brasil, Ltda., *Rio de Janeiro*

CONTENTS

PREFACE

Welcome to the world of microprocessors! For those of you just beginning the study of these powerful digital electronic devices, get ready for some fun and excitement. Microprocessors are finding their way into almost all aspects of electronic systems from spacecraft to children's toys. With this continued growth in the use of microprocessors comes the need for qualified people to design, build, test, install, operate, and maintain the equipment designed with these devices.

This text is the starting point in the study of microprocessors. It is written with the assumption that the reader has had a first course in digital electronics or its equivalent. Eight-bit microprocessors, including the 8080, 8085, and Z80, are used in this text. A firm understanding of these 8-bit devices puts you on very solid ground for understanding other 8-, 16-, and 32-bit microprocessors. Details may change from microprocessor to microprocessor, but the fundamental concepts given in this book do not.

Starting with Chapter 1 you learn about general microprocessor system architecture. This includes topics such as the data bus, address bus, control bus, I/O, and memory. Also in Chapter 1 you are shown how to convert between different number bases used in the study of microprocessors.

In Chapter 2 programming the 8080, 8085, and Z80 is introduced. These three microprocessors can use the same instructions. Assembly language programming is defined, flowcharting techniques are shown, and steps in assembling your own programs are discussed.

Chapter 3 continues the presentation of programming, covering topics such as addressing modes, flags, and arithmetic and conditional branching. Several programming examples are given that allow you to see exactly how the topics covered in the chapter operate.

In Chapter 4 more programming topics are given, such as stack area, subroutines, and special instructions for the Z80. This chapter concludes the formal study on programming, and the remaining chapters of the text build upon this knowledge by showing different hardware and explaining how to apply your programming knowledge for system control.

Hardware fundamentals of the microprocessor system are discussed in Chapter 5. Topics include the address bus for the 8080, 8085, and Z80. The data bus and control bus are also presented for these three microprocessors. This technique of presenting topics for all three microprocessors allows you to understand how each functions. You finish the text knowledgeable about the details of three microprocessors instead of just one.

The discussion of hardware continues in Chapter 6 with the ROM. You are first shown how a ROM operates. From this introduction you are shown how to design ROM into a microprocessor environment. This knowledge will allow you to understand how ROM is used in most microprocessor systems.

Chapters 7 and 8 show how to use RAM and I/O in a microprocessor-controlled system. A typical RAM device, the 6264, is discussed and you are then shown how to connect it to a microprocessor for reliable communication. Important timing considerations are covered for the RAM, ROM, and I/O. A general input and output port is designed and presented in detail.

In Chapter 9 you are introduced to troubleshooting a microprocessor system using static stimulus testing (SST). This is a hardware troubleshooting technique, which is completely independent of software. You are shown how to troubleshoot ROM, RAM, and I/O using SST. Further, you are given actual schematics of SST hardware that allow you construct your own SST from inexpensive, off-the-shelf hardware.

Chapter 10 presents the important concepts and details of interrupts. Both the software and hardware of the interrupt system is explained. You are shown how the system responds to an interrupt and what an interrupt service routine is and how to write one using assembly language.

In Chapter 11 two classic troubleshooting techniques, logic state analysis

(LSA) and digital signature analysis, are presented. You are shown how each may be used to troubleshoot a defective system.

Chapters 12, 13, and 14 are dedicated to three popular peripheral devices that are used with a microprocessor: (1) the timer chip, (2) programmable interface adapter (PIA), and (3) serial communication devices. The hardware connection of each type of device to the microprocessor system buses is given and discussed. Once the devices are connected to the system, application software is presented that shows you how to program the devices to perform the way you want them to for your desired application.

Also in these chapters discussions are given that show you how to troubleshoot the hardware and software of a system that uses this type of chip. Static stimulus testing and logic state analysis are valuable debugging aids in the development phase of the system, whereas DSA may be designed in a production troubleshooting problem.

Chapter 15 shows how to use dynamic RAMS. These devices have vastly different operating characteristics than static RAMs and this chapter presents the details of their use. A popular dynamic RAM is presented and then a complete dynamic RAM system is designed and discussed. At the conclusion of this chapter, you will have a good understanding of how a dynamic RAM operates and what to look for in systems which use them.

Chapters 16 and 17 cover the topics of analog-to-digital and digital-to-analog conversion. Both of these concepts are used in many microprocessor system applications where the digital machine must interface to the analog world.

At the conclusion of this text you will have been shown the details of three 8-bit microprocessors and the general concepts for almost any microprocessor system in use today. Software and hardware are given throughout the entire text showing how both are used to realize a particular application. Troubleshooting of systems is given in enough detail to allow you to understand how to approach and actually debug a malfunctioning microprocessor system.

Once you have completed this text, it will be an easy matter to learn another microprocessor's hardware and software by simply studying the data sheets and assembly language instructions of the device.

James W. Coffron

ACKNOWLEDGMENTS

The following figures are being reproduced with the permission of Prentice-Hall. (Figure numbers refer to *Microprocessor Programming, Troubleshooting, and Interfacing: The Z80, 8080, and 8085.*)

From Coffron, James W., *Using and Troubleshooting the MC68000,* © 1983.
Figure 1-1

From Coffron, James W., *Understanding and Troubleshooting the Microprocessor,* © 1980.
Figures 1-2, 1-3

From Coffron, James W., *Practical Hardware Details for the 8080, 8085, Z80, and 6800 Microprocessor Systems* © 1981.
Figures 5-6 through 5-10, 5-16, 5-17, 5-19, 5-20, 5-22 through 5-26, 9-6, 9-9(a), 10-3, 10-4, 10-5, 10-7 through 10-17, 10-19, 10-20, 10-22 through 10-26, 10-29, and Tables 5-1 and 5-2

From Coffron, James W., *Practical Troubleshooting Techniques for Microprocessor Systems,* © 1981.

Figures 9-2, 9-3, 11-2, 11-4 through 11-9, 11-20, 11-21, 11-23, 11-25 through 11-35

The following figures are being reproduced with the permission of SYBEX, Inc. (Figure numbers refer to *Microprocessor Programming, Troubleshooting, and Interfacing: The Z80, 8080, and 8085.*)

From Coffron, James W., *Z80 Applications,* © 1983.

Figures 6-1, 6-2, 6-3, 6-17, 6-22, 7-1 through 7-5, 7-8, 7-9, 10-28, 12-6, 12-8, 12-10 through 12-14, 12-16, 12-18, 12-19, 13-1 through 13-6, 13-9, 13-11, 13-12, 13-15, 13-17, 13-18, 13-19, 13-21, 13-22, 14-3, 14-8 through 14-11, 14-14, 14-16, 14-19, 14-20, 14-22, 14-24

From Coffron, James W., *The VIC 20™ Connection,* © 1983.

Figures 16-1 through 16-6 and 17-1 through 17-10

1

WHAT IS A
MICROPROCESSOR?

In this chapter we discuss the concept of a microprocessor and a microprocessor system without going into great technical detail about any single microprocessor available on the market today. Instead, we present a view of microprocessors that allows you to understand how these devices "fit" into systems designed to use them. The discussion is calculated to provide beginners with basic guidelines that may be followed in understanding microprocessor systems regardless of the type of microprocessor used.

Many microprocessor systems used in industry today have a number of points in common. This text will provide the background necessary to recognize these points. By understanding the general microprocessor system, you will be on much firmer ground when confronted with a new, unknown system.

After the microprocessor has been introduced, you will be shown how to work with different number systems, such as decimal, binary, octal, and hexadecimal. These number systems were chosen as they are the ones most often encountered when working with and programming microprocessors.

1-1 A GENERAL MICROPROCESSOR SYSTEM

The term microprocessor refers to a VLSI (very large scale integrated) circuit. Microprocessor circuits contain from 20,000 to 100,000 transisitors! Most microprocessors on the market today are fabricated using MOS (metal oxide semiconductor) technology. Microprocessor such as the Z80, 8080, 8085, 6800, 8086, 8088 are packaged in a 40-pin, DIP (dual in-line package), whereas the MC68000 microprocessor is packaged in a 64-pin DIP, as shown in Figure 1-1. A microprocessor device consists of many circuits but is very small physically. A microprocessor is used to process information, or it may be the CPU (central processing unit) in a system. Because of its small physical size (micro) and the ability to process information the name *microprocessor* was given.

One of the first goals for anyone beginning to study microprocessors is to understand the basic function of the microprocessor in a system. Once you understand this function, it will be much easier for you to see how the internal logic blocks contained inside the device help to accomplish its task.

To present the overall function of the microprocessor in a system, we first discuss another system that is more familiar. Let us examine the system of a person following a recipe or procedure. The essential parts of this "human" system are as follows:

1. The recipe or procedure (data input)
2. The person working or following the procedure (CPU)
3. The function performed by the person (output)

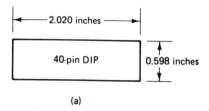

(a)

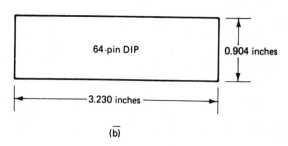

(b)

Figure 1-1 These figures show the relative sizes of the 40-pin dip package and the 64-pin dip package.

If you have ever cooked or followed some type of procedure, such as a recipe, you know the instructions must be exact. If they are not, the output is unpredictable. Further, these instructions must be written in a language that the person doing the baking or following the procedure can understand.

Next, the person following the procedure (CPU) reads the instructions and performs the action specified. If the instructions are followed in the order they are given, then the output operation will be correct.

Finally, the function is performed by the person mixing the ingredients, setting a temperature on the oven or putting pieces together in a certain order.

Each of these three sections in the example just described is found in a microprocessor-based system. The procedures or instructions are found in the memory section of the microprocessor system. The person reading the instructions, the CPU, reads the instructions in from the system memory and finally performs some action. The output or action performed by the microprocessor system will depend upon the sequence of instructions and the function of the system.

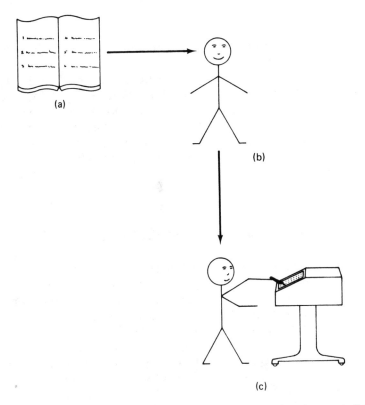

Figure 1-2   The "human" system consists of (a) list of instructions (memory), (b) person (CPU), and (c) action performed by the CPU (I/O).

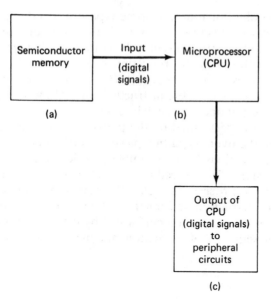

Figure 1-3 Block diagram of a typical microprocessor system relating the blocks to the human system shown in Figure 1-2.

It is important that the instructions in memory be written in a language that the CPU (microprocessor) can understand. Further, the instructions must be in an exact order. If the instructions are not, then the microprocessor can interpret them, but the output operations will not be correct. That is, the actions taken by the microprocessor may not be what is desired.

Figure 1-2 shows the "human" system just described, and Figure 1-3 shows the microprocessor-based system that relates to Figure 1-2.

Now that you have had an introduction to the function of the microprocessor in a system, your next step is to examine a block diagram of a typical microprocessor-based system.

1-2 BLOCK DIAGRAM OF A MICROPROCESSOR SYSTEM

Let us now drop the human system analogy and concentrate on the microprocessor system. In later chapters you will be shown how to write and run programs for an 8080, 8085, and Z80 microprocessor system. However, before you can do this, you should understand how most microprocessor systems are organized. This information will allow you to write your programs with the confidence of knowing how each instruction affects the system operation.

Figure 1-4 shows the block diagram of a microprocessor system. You see that the block diagram of a microprocessor system. You see that the block diagram of Figure 1-4 is very similar to the system shown in Figure 1-3. The system of Figure 1-4 is called the *3-bus microprocessor system*. With this definition comes the question, what is a bus? We will reserve the answer to this

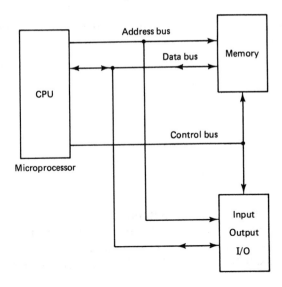

Figure 1-4 Block diagram of a microprocessor system showing the three buses that interconnect the functional blocks.

question until a little later in the discussion. Let us first examine the major hardware blocks shown in Figure 1-4.

The first block shown is the *memory*. This is the electrical means by which the microprocessor obtains its instructions and stores temporary answers. Later we will talk more about the memory. For now, you should understand that the microprocessor (CPU) will read information from the memory and write information to the memory.

The next block shown in Figure 1-4 is labeled CPU. This block is the microprocessor itself. All information in the system is communicated between the CPU and memory and the CPU and I/O (input/output). The information being transferred between the microprocessor and memory and the microprocessor and I/O gives the CPU its name. There are cases where information may be passed directly between the memory and I/O, as during a DMA (direct memory access). You need not consider this case at the present time.

The final block shown in Figure 1-4 is I/O. This is the section of the microprocessor system that gives the unit its personality. That is, the I/O devices connect the microprocessor to the "real world." Examples of devices that input information to the microprocessor are

1. Keyboard
2. Cassette tape
3. Floppy disc
4. Hard disc
5. Temperature sensor
6. Pressure sensor
7. Flow meters

Examples of devices that accept output information from the microprocessor are

1. CRT terminal
2. Printer
3. Floppy disc
4. Hard disc
5. Numeric readouts
6. Electronic voice
7. Relays and valves

These are just a few of the input and output devices that may be attached to a microprocessor system. As you become more and more familiar with microprocessor systems, you will see an enormous variety of input and output devices used. For the present time you may use the following definition of an input or output device:

> In general, any destination for data in the system that is not memory is ouput. Any source of data in the system that is not memory is input.

Almost any microprocessor system that you encounter may be organized in the manner shown in Figure 1-4. Keeping this figure in mind will help you understand how the system is being affected when a program is running. To help in the understanding, let us review the important points we have covered about Figure 1-4.

1. There are three major blocks in the system.
   a. Memory
   b. CPU (microprocessor)
   c. I/O devices
2. Memory has information written into it and read from it by the CPU.
3. Input devices send information from outside the system to the microprocessor.
4. Output devices send information from the microprocessor to a destination ouside the system.
5. Information does not usually transfer directly to or from memory to an I/O device. The information will pass through the CPU first.

Using this block diagram you can see that the microprocessor communicates with memory or I/O. Therefore, the microprocessor does only a few jobs. No matter how complex the system or how long the program, the microprocessor performs only a few functions:

1. Read from memory
2. Write to memory
3. Read from input
4. Write to output
5. An internal operation such as ADD

The first four events may be obtained by examining the block diagram of the system given in Figure 1-4. Event 5 is an operation that is internal to the microprocessor device, such as adding two numbers. You will become very familiar with the internal operations of the microprocessor when the topic of programming is introduced and discussed in Chapter 2.

Programming allows you to make the microprocessor perform any of these operations in the order you wish. We will be adding a few more operations to this list later in the text. However, many systems are designed with the microprocessor performing only the five events listed. This may sound unbelievable, but it is true. You will see for yourself when you start programming the microprocessor that only the five events listed will be used. The complexity of the programs is due to the many different internal operations you can instruct the microprocessor to perform. However, these will become very familiar to you by the end of this text. You will be writing many programs to enable the microprocessor to perform some very elegant functions using only the five operations shown.

In the next section you will learn about the jobs, or "functions," of the system buses in making these five events occur.

1-3 BUSES: THE ADDRESS BUS

In the preceeding section we discussed the block diagram of a general microprocessor system. The diagram had three major blocks connected with buses. In fact, the diagram was called the 3-bus system architecture. This section introduces the concept of a bus and shows the importance of buses in a microprocessor system.

The three buses of Figure 1-4 are the

1. Address bus
2. Data bus
3. Control bus

We discuss each of these buses and show how each bus operates within the microprocessor system to produce reliable communication. We do not spend much time discussing the actual hardware generation of these buses. This topic is covered in detail in Chapter 5. The goal of this presentation is to make you aware of how communication is accomplished in a typical microprocessor system. The buses are shown, and you will need to accept that they can be realized with hardware.

What is a Bus?

Let us first discuss what is meant by the term *bus* as it relates to microprocessor systems. The following definition applies to a bus in a microprocessor system.

Bus.   A *bus* is a collection of signal lines that originate at a common source. The signal lines are physically connected to many circuits in a parallel fashion within the system. The collection of signal lines will perform some common electrical function within the system.

Let us now show how each of the three buses fits this definition and explain exactly what every part of the definition just given means. First, we consider the address bus.

Address Bus

The *address bus* for the 8080, 8085 and Z80 microprocessors comprises 16 physical lines. An 8085 has some special requirements for generating the 16 bits of the address bus; these will be pointed out later in this text. For the present time it will be a valid model of all three microprocessors to represent them with 16 physical address lines. These lines are labeled A0–A15, as shown in Figure 1-5. When you describe a bus in a microprocessor system that is contains 16 physical lines, you say the bus is 16 bits wide. If the address bus contains only 12 lines, then the bus is 12 bits wide.

As a first approximation we will say that the address bus is unidirectional (one direction.) That is, the bus originates at the microprocessor and is "bused" out to the system. There are times when the microprocessor is not outputting the address bus, as during a DMA operation. However, this is a special case, and we ignore it for now. Ignoring this special case will not detract from your

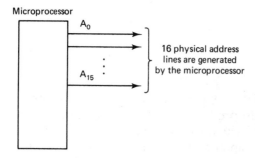

Figure 1-5 Diagram showing the address lines of a typical microprocessor system. There are 16 physical address lines for the 8080, 8085, and Z80 microprocessors.

understanding of microprocessor system; in fact many systems are designed in which the microprocessor is always in control of the address bus.

A point to note about all system buses is that each has a point of electrical origin and a point of destination. Remembering this fact will help you understand how the communication occur in a microprocessor system.

Finally, the address bus must have a function in the system.

> The function or job of the address bus is to enable or select the proper path for communication in the system.

What does this mean? From Figure 1-4 we know that the microprocessor will communicate with either MEMORY or I/O. The address bus will select the memory space or I/O device with which the microprocessor will be communicating. We will show how this is actually accomplished in a later chapter. If the microprocessor is performing an internal manipulation where memory or I/O are not used, then the address bus is not used. For now you should know what the job of the address bus is during a system communication with memory or I/O.

In review, you know these facts about the system address bus;

1. The Z80, address bus is 16 bits wide.
   a. 6800 and 6502 address buses are 16 bits wide.
   b. 8080 and 8085 address buses are 16 bits wide.
   c. An 8086/8088 address bus is 20 bits wide.
   d. A 68000 address bus is 24 bits wide.
   e. A Z8000 address bus is 16 or 24 bits wide.
2. The address bus is unidirectional. (This will be altered at a later time.)
3. The address bus originates at the microprocessor.
4. The function or job of the address bus is to enable the proper path for communication with memory or I/O.

With this information in hand concerning the system address bus, let us examine the system data bus.

1-4 THE DATA BUS

In a microprocessor system information is transferred between memory and the CPU and I/O and the CPU. This information is transferred via the *data bus*. Whichever path the address bus has selected for communication, the information will be transferred on the data bus lines.

The Z80, 8080, 8085, 6800, and 6502 microprocessors have a data bus that is 8 bits wide, which accounts for the name 8-bit microprocessor. Microproces-

Microprocessor

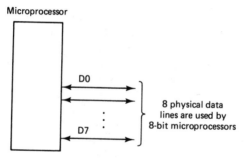

Figure 1-6 Diagram showing the 8 physical data lines of the microprocessor system. These 8 data lines give these microprocessors the name 8-bit microprocessors.

sors like the 68000 and 8086 have data buses that are 16 bits wide; they are called 16-bit microprocessors. Further, the 80386 and 68020 have data buses that are 32 bits wide and are referred to as 32-bit microprocessors. The data bus can transfer all the bits of its width, 8, 16, or 32, at the same time in a parallel fashion, as shown in Figure 1-6.

Regardless of the size of the data bus, the microprocessor can write information to memory or I/O as well as read information from memory and I/O. The information written and read by the microprocessor will occur on the same physical lines. During a write operation, the microprocessor is outputting information on the data bus to the system. During a read operation, the microprocessor is inputting information on the data bus. See Figure 1-7.

In Figure 1-7 you see that the information on the data bus will move in two directions. Therefore, the data bus is *bidirectional*. However, information will move in only one direction at any given time in the system. This may sound complicated to think of information moving in two directions on a single wire, but think of a single line of the data bus as a water hose. One end of the hose you hold and the other end is held by someone else, as shown in Figure 1-8.

When you wish to send water (information) to the other person, you pour water in the hose and it comes out the other end. When the other person wishes

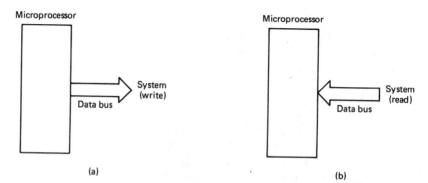

Figure 1-7 The information on the system data bus travels in two (bi) directions: (a) from the microprocessor to the system (write); (b) from the system to the microprocessor (read).

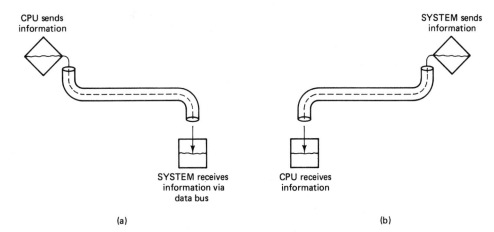

Figure 1-8 Information on the system data bus travels in two directions on the same physical lines. This concept is shown by the water traveling in two directions in a single hose. The water flows in only one direction at a time.

to send water (information) to you, that person pours water in that end and the water comes out at your end. Information (water) is traveling in two different directions but only in one direction at any given time. Further, the water travels in two directions on the same physical path.

The two directions that data (information) in a microprocessor system travels are

1. From the CPU to the system (write).
2. From the system to the CPU (read).

This is shown in Figure 1-9.

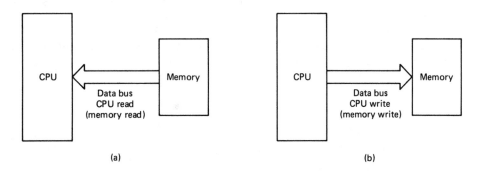

Figure 1-9 When the microprocessor is reading information from memory, data flows from the system memory to the microprocessor on the system data bus lines. When the microprocessor is writing information into memory, the data flows from the microprocessor to the memory.

In review of the system data bus, the following facts are given:

1. The Z80 data bus is 8 bits wide.
   a. A 6800 data bus is 8 bits wide.
   b. 8080 and 8085 data buses are 8 bits wide.
   c. An 8088 data bus is 8 bits wide.
   d. An 8086 data bus is 16 bits wide.
   e. A 68000 data bus is 16 bits wide.
   f. A Z8000 data bus is 16 bits wide.
   g. The 80386 data bus is 32 bits wide.
   h. The 68020 data bus is 32 bits wide.
2. Data moves in two (bi) directions on the bus:
   a. From the CPU to the system (write).
   b. From the system to the CPU (read).
3. The job or function of the data bus is to transfer the information between the CPU and memory or I/O.

We now have only one more bus to discuss in the microprocessor system, the CONTROL BUS.

1-5 THE CONTROL BUS

We know that the address bus selects the correct path for communication and the data bus transfers the data to or from that path. The *control bus* starts and stops the communication and defines the type of communication. The system control bus is unidirectional, like the system address bus. It does not come directly from the microprocessor like the data and address bus; rather, it is derived from signals from the microprocessor. This is shown in Figure 1-10.

The control bus width is not defined exactly as the address bus or the data bus width. Control bus width depends on how many events the CPU can perform in the system. The maximum number of events that will be presented in this text is seven:

1. Read memory
2. Write memory
3. Read input
4. Write output
5. Internal manipulation
6. Interrupt (not yet discussed)
7. DMA (not yet discussed)

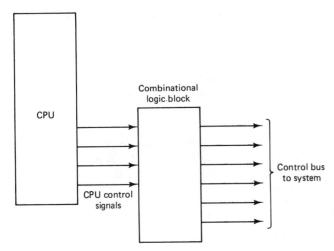

Figure 1-10 The system control bus is constructed from logically combining different signals from the microprocessor.

This list is valid for all microprocessors. However, when studying 16- or 32-bit microprocessors, there are other operations that must be considered. These involve reading and writing selected 8-bit groups from the 16- or 32-bit data bus.

From this list the only event that does not require a special control line is 5, an internal manipulation. When all the activity is occurring inside the CPU, the external circuits on the system need not be involved. However, any of the other events involve many external circuits. These circuits need to be electrically informed as to the what the CPU is doing. The CPU will inform the external circuits via the system control bus. See Figure 1-11.

In Figure 1-11 we see that each event has a unique control signal. When

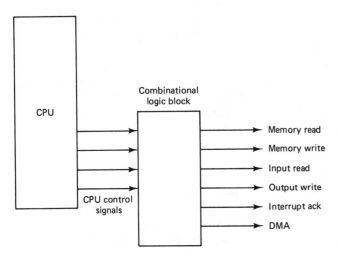

Figure 1-11 The system control lines are (1) memory read, (2) memory write, (3) input read, (4)output write, (5) interrupt acknowledge, (6) (DMA).

the CPU is performing one of the seven events, the proper control line is *asserted,* which means the signal line is set to a logical 1 or a logical 0, whichever is the active state.

> One function of the system control bus is to define electrically the type of communica-tion that is occurring on the system data bus. A second function of the system control bus is to start and stop the communication.

We note that the system control bus has two functions, whereas the other two buses had only one. Further, we see that the control bus must "time" the data transfer within the system. We discuss how this timing is accomplished in a later chapter in this text. In review of the control bus, we have the following facts:

1. The control bus is unidirectional.
2. Control bus lines are derived from CPU signals.
3. The control bus starts and stops the communication.
4. The control bus defines the type of communication.

Now that the control bus has been discussed, you have been presented with all three buses in the system. Let us now examine how the buses work together to obtain reliable communication in a microprocessor system.

1-6 COMMUNICATION WITH THE 3-BUS ARCHITECTURE

To show how the buses operate together to perform reliable system communica-tion let us consider an example when the microprocessor is writing information to memory. We have not yet discussed the system memory. Let's assume at this time that information can be written to many places in a system memory. You will learn more about the system memory at a later chapter.

When the microprocessor writes data into the system memory, the mem-ory must know where the CPU will store the information. The system address bus will electrically inform the system memory where the information will be stored. The microprocessor has this information because of the software instruc-tions that are being executed. The instruction will inform the microprocessor where to store or write the information and the microprocessor will electrically perform this operation.

The actual information to be stored must be presented to the system memory. This is accomplished via the system data bus.

Finally, the system memory must know when to store the information. The system control bus will electrically inform the system memory when to store the information. Figure 1-12 shows the important concepts of this operation.

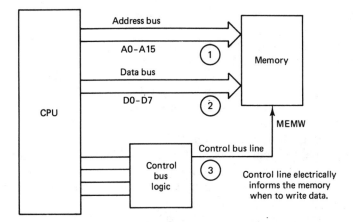

Figure 1-12   When the microprocessor is writing information to memory, (1) the address bus outputs the correct address of memory, (2) the data bus outputs the information to write at the selected address, and (3) the system control bus starts and stops the data transfer.

From the preceding discussion, you see that each bus has a different function in the overall system communication. Each bus is independent of the other buses, but each must do its job or the communication will fail. Keep in mind this 3-BUS SYSTEM architecture as you proceed through this text. As instructions are introduced you can see how they will affect the different buses.

For the person starting in microprocessors, it is important to understand the function of each bus during a system communication. This gives you a clearer picture of what is actually occurring when the microprocessor is executing instructions. We did not discuss the hardware for the buses. Clearly, each bus must be realized with hardware.

1-7 DECIMAL NUMBER SYSTEM

We begin our discussion of the number systems with the decimal system. This is the system with which we are most familiar. Once you have some basic facts in hand we will discuss two other number bases that are used in microprocessor system study.

In the decimal system each number written can be expanded as follows

$$1248 = 1 \times 10^3 \quad + \quad 2 \times 10^2 \quad + \quad 4 \times 10^1 \quad + \quad 8 \times 10^0$$
$$1000 \quad + \quad 200 \quad + \quad 40 \quad + \quad 8$$

We do not often take the time to expand the numbers like this. As we are taught to work in the decimal system we use the numbers without thinking of the expansion. Nevertheless, this is what the numbers mean. Each place in a decimal number is really a coefficient, which is multiplied by 10 (decimal) raised to an exponent.

The 1s, 10s, 100s, 1000s, 10,000s and so on, are in increasing order starting with 0. Ten (or any number) raised to the zero power is equal to 1. The number raised to an exponent in a number system is called the *base* of the system. Therefore, in the decimal system the base is 10. Further, the base tells us how many digits to expect in the number system. Base 10 has 10 digits, 0, 1, 2, 3, 4, 5, 6, 7, 8, and 9.

Not all the numbers in the base 10 system are written as whole numbers. We have numbers such as 1.23 and 457.906. That is, we have numbers that contain a decimal point. The numbers to the left of the decimal point can be expanded as we have just shown. However, what do we do with the numbers to the right of the decimal point?

These numbers also fit into the same pattern of the number base raised to an exponent. The exponents used with the numbers to the right of the decimal point starts with (-1).

Therefore, the number 16.23 may be expanded like this:

1×10^1	+	6×10^0	+	2×10^{-1}	+	3×10^{-2}
10	+	6	+	$\frac{2}{10}$	+	$\frac{3}{100}$
10	+	6	+	.2	+	.03

We see that 10 raised to a negative exponent is actually 1/10 raised to the exponent. Thus the digits in any decimal number are really coefficients multiplied by the power of 10 represented by the digit.

Thinking of the decimal number in these terms will provide a solid foundation with which to study other number bases. Regardless of the number base, the formation of the number within the base is the same. That is, forming a number in base 2 is done in exactly the same way as for base 10, with the exception that the base changes to 2. You will see this in the following section.

1-8 BINARY NUMBERS—BASE 2

We stated in Section 1-7 that the binary number system followed the same rules as the decimal system. Let us show how this is true. First, the binary system uses the base of 2. This means that only 2 digits are used to represent all numbers in the binary system. These 2 digits are 0 and 1. You will see that all numbers may be represented with only these 2 digits.

Each number in the binary system is written like the numbers in the decimal system, but only 2 different digits are used. One number in this system is

$$1100$$

This number represents the following

1×2^3	$+$	1×2^2	$+$	0×2^1	$+$	0×2^0
1×8	$+$	1×4	$+$	0×2	$+$	0×1
8	$+$	4	$+$	0	$+$	0

This binary number has the decimal equivalent of 12. The binary number system is important because the microprocessor system and computer systems operate using binary arithmetic. One voltage is used for a logical 1, while another voltage is used for a logical 0.

In examining the powers of 2 you can see that each place in the number is twice as large as the preceeding place. This is due to the fact that the number base 2 is being raised to increasing exponents starting with 0. Further, to determine the decimal equivalent of a binary number, you just add the values of the places that have a 1 in them because 0 times any number still equals 0. Adding 0 to any number does not change its value. Therefore, you may ignore the weight of any digit in a binary number that has the value of zero. The zero is used simply as a placeholder, similar to the function of zero in the decimal number system. For example, the binary number

$$1110010$$

is equal to the decimal number

64	32	16	8	4	2	1
1	1	1	0	0	1	0

$$64 + 32 + 16 + 2 = 114$$

Notice that only the weights where a 1 exists are added to obtain the decimal value. Note also that the zeroes are necessary to give the proper position to the 1 in the number. However, this is a shortcut, and the rules of number systems as described for the decimal number system still apply. The 1 or 0 associated with 2^0 is called the LSB (least significant bit). The 1 or 0 associated with the greatest power of 2 is called the MSB (most significant bit). For example, the MSB and LSB in this number are as shown:

MSB									LSB
1	1	1	0	0	0	0	1	1	0

Let us now examine how numbers with decimal points may be represented in the binary system. This is not difficult, since the rules used for decimal numbers are followed.

If you have a binary number like

$$0110.101$$

the digits to the left of the decimal point are represented in exactly the same manner as discussed previously. That is,

$$0 \times 2^3 + 1 \times 2^2 + 1 \times 2^1 + 0 \times 2^0$$

We now must represent the numbers to the right of the decimal point. In the decimal number system, the digits to the right of the decimal point were coefficients multiplied by 10 and raised to a negative power. This is exactly how the digits in a binary number system are represented.

The digit immediately to the right of the decimal point is multiplied by 2^{-1}, followed by 2^{-2}, 2^{-3} and so forth until all of the digits have been used. For the example given, the digits to the right of the decimal point are expanded to

$$1 \times 2^{-1} + 0 \times 2^{-2} + 1 \times 2^{-3}$$

The complete number is equal to

$$0 + 4 + 2 + 0 + \tfrac{1}{2} + 0 + \tfrac{1}{8}$$

This is a decimal equivalent of 6.625.

Let's do another example to further illustrate the concept. Calculate the decimal equivalent of the following binary number:

$$1001110.011001$$

The left side of the decimal point equals

$$64 + 0 + 0 + 8 + 4 + 2 + 0$$

The right side of the decimal point equals

$$0 + \tfrac{1}{4} + \tfrac{1}{8} + 0 + 0 + \tfrac{1}{64}$$

The complete number is equal to the sum of all bits

$$78 + .390625 = 78.390625$$

Try converting the following binary numbers to their decimal equivalents and compare your answers with those given.

a. $101.111 = 5.875$
b. $10001.00011 = 17.09375$
c. $1100.010101 = 12.328125$

1-9 OCTAL AND HEXADECIMAL SYSTEMS

This section describes two other number systems that are often used when working with microprocessors. First is the OCTAL number system. The OCTAL system uses the BASE of 8 and has 8 digits 0, 1, 2, 3, 4, 5, 6, and 7. Rules for numbers in the octal system are the same as we have previously described. For example, the number 756.34 in octal is equal to

7×8^2	+	5×8^1	+	6×8^0	+	3×8^{-1}	+	4×8^{-2}
7×64	+	5×8	+	6×1	+	$3 \times \tfrac{1}{8}$	+	$4 \times \tfrac{1}{64}$
448	+	40	+	6	+	.375	+	.0625

or 494.4375 in decimal.

You can see from this example that forming numbers in octal is exactly the same as forming numbers in decimal or binary. Octal is a useful system because converting from binary to octal and from octal to binary is straightforward. Here is how it is done.

Converting from Octal to Binary and Binary to Octal

1. Start with each digit of the octal number and convert it to its 3-bit binary number.
2. Then place all binary numbers in a row, forming a single large binary number.

For example, let us convert the number 3467 (octal) into its binary equivalent.

$$
\begin{array}{cccc}
3 & 4 & 6 & 7 \\
011 & 100 & 110 & 111
\end{array}
$$

011100110111

To convert from binary to octal you do the following;

1. Starting with the LSB, count off 3 bits. Repeat this until all bits of the number have been used. If a number does not have an even number of bits, add zeros as placeholders.

   For example, to convert the binary number 1010100011100011 into octal, start like this:

   001-010-100-011-100-011

2. Now convert each group of 3 bits into its octal equivalent.

   $$
   \begin{array}{cccccc}
   001 & 010 & 100 & 011 & 100 & 011 \\
   1 & 2 & 4 & 3 & 4 & 3
   \end{array}
   $$

   or 124343 in octal.

Let's do one more example of each type of converison. First convert the following binary number to its octal equivalent.

0010001110010110

1. Start with the LSB and divide the binary word into groups of 3 bits.

   0   010 001   110 010 110

2. Add zeroes to the left of the MSB to make an even 3 bits.

   000   010 001   110 010 110

3. Now convert each group of 3 bits into its equivalent octal digit.

$$000 \quad 010 \quad 001 \quad 110 \quad 010 \quad 110$$
$$0 \quad \quad 2 \quad \quad 1 \quad \quad 6 \quad \quad 2 \quad \quad 6$$

4. Put each octal digit together to form the complete octal number.

$$0010001110010110 = 21626 \text{ base } 8$$

Lets now do the opposite. Convert the following octal number into its equivalent binary number.

$$713602$$

1. Separate each octal digit.

$$7 \quad 1 \quad 3 \quad 6 \quad 0 \quad 2$$

2. Under each digit, form the equivalent 3-bit binary number.

$$7 \quad \quad 1 \quad \quad 3 \quad \quad 6 \quad \quad 0 \quad \quad 2$$
$$111 \quad 001 \quad 011 \quad 110 \quad 000 \quad 010$$

3. Put all 3 bit codes together to form the complete binary number.

$$713602 = 111001011110000010 \text{ base } 2$$

Try the following conversions and compare them with the answers given. Convert the following octal numbers to binary

1. 34671 base 8 = 011100110111001
2. 20745 base 8 = 010000111100101
3. 33710 base 8 = 011011111001000

Convert the following binary numbers to octal

4. 110010111 base 2 = 627 octal
5. 01110001 base 2 = 161 octal
6. 10110111000010101001 base 2 = 2670251 octal

Let us see how another number system, the hexadecimal system, is used in working with microprocessors.

The Hexadecimal System

The hexadecimal number is often used in the literature of microprocessor systems. This is due to the ease with which long binary numbers may be represented. For example a 32-bit binary number, which would consist of 32 logical 1s and 0s, may be represented by only 8 hexadecimal digits.

The name of this number system indicates that there are 16 different characters representing numbers (*hex* means six and *decimal* means ten). The 16 characters for the hexadecimal number system are listed with the equivalent binary and decimal values.

Hex	Binary	Decimal
0	0000	0
1	0001	1
2	0010	2
3	0011	3
4	0100	4
5	0101	5
6	0110	6
7	0111	7
8	1000	8
9	1001	9
A	1010	10
B	1011	11
C	1100	12
D	1101	13
E	1110	14
F	1111	15

The next numbers after F are

10	1 0000	16
11	1 0001	17
12	1 0010	18

Converting from Hexadecimal to Binary

Given any hexadecimal number, you can convert it to an equivalent binary number using a straightforward technique. Here is how it is done. Given the hexadecimal number 3CF2, each digit of the hexadecimal number is converted to its 4-bit binary equivalent. A table such as that given previously may be used for this purpose.

$$3 \qquad C \qquad F \qquad 2$$
$$0011 \quad 1100 \quad 1111 \quad 0010$$

After each digit has been converted, connect the 4-bit binary numbers together, forming a single binary word.

$$3CF2$$
$$0011110011110010$$

From this example it is easy to see why one would prefer to represent long binary numbers in a compact form such as hexadecimal. With practice, converting from hexadecimal to binary will become quite easy and you will be able to do it in your head or on scratch paper without much trouble.

Here is another example of converting from hexadecimal to binary.

Hexadecimal number = 7ED53

Split the number into digits and form the 4-bit binary equivalent of each digit.

7	E	D	5	3
0111	1110	1101	0101	0011

Now connect the equivalent 4-bit binary numbers together to form the complete binary word.

7ED53
01111110110101010011

Converting from Binary to Hexadecimal

Let us now examine the reverse process, converting from a binary number to a hexadecimal equivalent.

Binary number = 1100110011111111000

Start by counting 4 digits from the LSB (which is assumed to be the rightmost bit). The entire binary word is separated into 4 bit sections. After the partitioning, the number will appear as

110   0110   0111   1111   1000

Notice in the last set of digits there were only 3 bits remaining. To make 4 bits, add a 0 as placeholders in the MSBs until you have 4 bits. In this example the 110 would be replaced by 0110. By adding the extra zero(s), you do not change the value of the number. You are only making it a 4-bit word so you can use the conversion table given earlier.

Now that you have the binary word partitioned into 4-bit sections, convert each 4-bit word to an equivalent hexadecimal number. In this example the equivalent hexadecimal number is

6	6	7	F	8
0110	0110	0111	1111	1000

You now connect the equivalent hexadecimal digits together to form the complete hexadecimal number 667F8.

Here is another example of converting from binary to hexadecimal.

Binary number = 11010111000100

Divide into 4-bit sections, starting with the LSB:

11   0101   1100   0100

Put in placeholders in the last 4-bit section if necessary.

$$0011 \quad 0101 \quad 1100 \quad 0100$$

Now convert each 4-bit section into the hexadecimal number:

$$3 \quad\quad 5 \quad\quad C \quad\quad 4$$
$$0011 \quad 0101 \quad 1100 \quad 0100$$

Place the hexadecimal numbers together, forming the single number:

$$35C4$$

Let's try these.

Convert 10011000 base 2 to _____ base 16.

$$1001 \quad 1000$$
$$9 \quad\quad\; 8 \quad = \quad 98 \quad \text{base 16}$$

Convert 101110 base 2 to _____ base 16.

$$0010 \quad 1110$$
$$2 \quad\quad E \quad = \quad 2E \quad \text{base 16}$$

1-10 CHAPTER SUMMARY

This chapter introduced the concept of a microprocessor system. The introduction started with a block diagram showing the major components that comprise a complete working system. General information was given covering the topics of memory and input/output (I/O). Next, the important topic of system buses was presented.

The three buses—address, data, and control—were shown. As each bus was explained, you learned how it fit into the general communication scheme used in a microprocessor system. After the buses were discussed, the chapter focused on the study of four different number bases, binary, octal, decimal and hexadecimal.

Information in this chapter will be expanded as you progress through the text. Once you understand the general concepts given here, the details of each concept will be more easily understood. These details are presented in the remaining chapters of the book.

REVIEW PROBLEMS

1. What do the terms VLSI and MOS mean?
2. The 8080, 8085 and Z80 are packaged in a _____ pin DIP.

3. What are the three main functional blocks contained in a microprocessor system?

4. Information communicated between memory and I/O will usually pass through the _____ section shown in the system of Problem 3.

5. Which functional block connects the microprocessor system to the real world?

6. List four input and output devices that may be found in a microprocessor-controlled system.

7. List the five major events that will occur during the execution of a microprocessor program.

8. List the three buses in a microprocessor system.

9. Define a system bus.

10. What is the function of the address bus?

11. An 8085 has an _____ -bit address bus.

12. What is the function of the system data bus?

13. An 8080 has an _____ -bit data bus.

14. A microprocessor system data bus is bidirectional. What does bidirectional mean?

15. What is the function of the system control bus?

16. List the functions of each system bus during a memory write operation.

17. Expand the decimal number 5384.32 as a sum of coefficients times 10 raised to a power.

18. What decimal numbers do the following binary numbers represent?
    (a) 101100
    (b) 111111
    (c) 001101
    (d) 11100001

19. What decimal numbers do the following octal numbers represent
    (a) 3702
    (b) 2674
    (c) 7775
    (d) 3216

20. Convert the following hexadecimal numbers to binary.
    (a) 4A6F
    (b) 329C
    (c) 4A
    (d) 3E

21. Convert the following binary numbers to hexadecimal.
    (a) 11101011
    (b) 0110011111
    (c) 110010011100
    (d) 01101

2

PROGRAMMING THE Z80, 8080, AND 8085

In the previous chapter you were shown a functional block diagram of a microprocessor-based system. It is interesting to note that this block diagram may be used for almost any microprocessor-controlled system. For example, one microprocessor system may be used to control the flow of fuel to an automobile engine, while another may function to keep the temperature constant in a large office building. Still another may be used to control the actions of a toy. If the preceeding systems all use approximately the same block diagram, the question arises, What makes the system perform it's own unique function? There are a few parts to the answer, but the main answer involves *software,* or programming.

The correctness of this answer is easy to see when you examine the hundreds of software packages available for a personal computer. One package makes your computer system a word processor, whereas another package turns the same system into a home accountant or perhaps a sophisticated chess player. The hardware of the system remains constant, with the only change being in the software. It is this exciting area of microprocessor software to which you will be introduced in this and the following chapter.

This chapter begins by explaining what is meant by a microprocessor program. Next it shows the details of writing, loading, and running programs written for the Z80, 8080, and 8085 microprocessors. An interesting point to note at the start is that while you learn how to program one of these microprocessors, you are actually learning how to program all three! There are some differences, but the basics are the same. Further, you will see that a program written for an 8080-based system works—with no modifications—on the 8085 and Z80 systems. So, knowing that you are really learning how to program three microprocessors at the same time, let's begin.

2-1 WHAT IS PROGRAMMING?

Before you actually start writing programs, you need a clear definition of a program. This will allow you to understand better why certain details of writing programs are handled in a particular fashion. A statement of what a program is would be:

Program.   A *program* is a list of statements written in a specific format (programming language), which when executed by the microprocessor will perform a specified function with predectible results

Let's make a more detailed examination of this statement. First, let's look at "a list of statements." Programs for a microprocessor are written as a series of statements, one line following the other, as shown in Figure 2-1. Usual program flow starts from the statement at the top and proceeds in a sequential fashion toward the bottom of the list. However, there are program statements, or instructions, that interrupt the smooth flow from top to bottom. You will learn these instructions and how to use them. For now, as beginners in programming, let's assume that the program flow is from the statement nearest the top of the page to the statements at the bottom of the page.

The next part of the explanation says "written in a specific format (programming language)." Instructions that comprise the program must be written

```
LXI SP,87FFH

MVI A,00
OUT 34H
MOV B,A
MOV C,A
NOP
IN 25H
ANI 0FH
JNZ NO_INPUT
JMP FLAG_IN
```

Figure 2-1   Example of a microprocessor program showing a list of statements.

```
MVI A,34H        8080,8085
LD A,#34H        Z80
LDA A,#$34       6800
MOVE.B #$34,D0   68000
MOV AL,34H       8086/8088
```

Figure 2-2   Examples of different types of assembly languages for different microprocessors.

in a language the microprocessor will understand. The general language in which you will be writing your programs is called *assembly language.* There are many different types of programming languages, such as PASCAL, BASIC, FORTRAN, C, and FORTH, just to name a few. This text uses only assembly language programming. Each microprocessor available on the market today has its own unique set of assembly language instructions. It is by design that the three microprocessors Z80, 8085, and 8080 are able to use the same assembly language. Figure 2-2 gives an example of different assembly language instructions for some currently available microprocessors. Do not be concerned at this time if the instructions shown in Figures 2-1 and 2-2 have little meaning for you. By the end of this chapter you will be writing programs in assembly language. For now you should understand that each microprocessor has its own assembly language, and the language requires that the instructions be written in a specific format.

The final section of our definition to be discussed is "which when executed by the microprocessor will perform a specified function with predictible results." Each assembly language instruction has a very clearly defined operation. The results of each instruction are predictable. Therefore, the results of execution of many instructions are also predictable. This being the case, you can write your programs with an overall system function in mind. If you have arranged your instructions to be executed in the correct sequence, your program will force the system to behave in the manner you desire. You will become very familiar with the concept of having your instructions executed in the correct sequence once you actually start writing and running programs. A phrase that will surface quickly and become an integral part of your conversation is, I must debug my program, where *debugging* is the act of checking and changing your program so it will perform the way *you* want.

2-2 PROGRAM DEFINITIONS—FLOWCHARTING

Prior to writing an assembly language program for a microprocessor, it is helpful to learn some important points about program definition and flowcharting. The first step in writing any program is to define clearly the task and the sequence necessary to accomplish it. One of the main tools used in this phase of program development is the flowchart. A sample flowchart is shown in Figure 2-3.

Figure 2-3 shows that the flowchart gives a pictorial representation of a sequence of steps required to solve a particular problem. This particular flow-

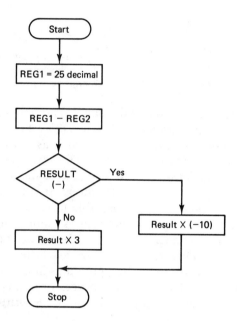

Figure 2-3 Sample flowchart showing the flow and decisions that are made during program execution.

chart is not dependent on any computer language or microprocessor. It is dependent only on the task you wish to accomplish. This type of flowchart is referred to as a TASK flowchart.

There are many symbols that can be used in writing flowcharts. However, we will concentrate on only three of these. All three are used in Figure 2-3 and are shown individually in Figure 2-4. These three symbols will get the job done for you and allow your flowcharts to be understood by others.

The symbol in Figure 2-4a will be used to start and end the flowchart. Symbol 2-4c is a decision block with one entry and two possible exits. The exits are the paths to be taken for the cases when the question asked within the block is true or false or answered by yes or no. Finally, the symbol in 2-4b is used as a process step, which is any step where program action is required, such as add, sub, shift, multiply, output, or input. Flowcharting does not have to be complicated to be useful. By limiting your symbols to these three, your flowcharts will be uncluttered, readable, and easily understood.

Most beginning programming students do not like to use flowcharts because they take too much time. While this may be true for simple programming problems, it is seldom the case for more complicated problems. Flowcharts help you organize your thoughts and solutions for a particular problem. Further,

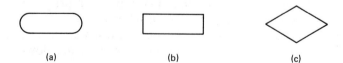

(a)   (b)   (c)

Figure 2-4 Three main symbols used in flowcharts: (a) start and stop block; (b) function block indicating an action occurs; (c) decision block.

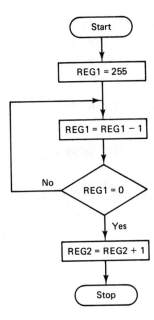

Figure 2-5 Flowchart showing a particular task to be performed by a computer.

once the flowchart is completed, you can communicate your ideas to others more easily than by showing them your program. You are encouraged to use flowcharts throughout this text for all your programs.

Once your task flowchart has been completed, you can extend the solution by tailoring it to a particular computer language or microprocessor. For example, consider the task flowchart shown in Figure 2-5. The actual computer or microprocessor program used to realize this flowchart will be dependent on the physical environment and design of the system on which the program will run. In Figure 2-6 you see the flowchart of Figure 2-5 realized using an 8080 assembly language program. The flowchart of Figure 2-5 could just as easily have been realized using a 68000 assembly language program, as shown in Figure 2-7a, or a BASIC program, as shown in Figure 2-7b.

```
            ORG 0000

            MVI B,255
BACK:       DCR B
            JNZ BACK
            INR C
            HALT
```

Figure 2-6 8080 assembly language program to realize the flowchart of Figure 2-5.

```
            ORG 3000H
            MOVE.B #255,D0
BACK:       SUBI.B #1,D0
            BNE BACK
            ADDI.B #1,D1
STOP:       JMP STOP

                (a)
```

```
10  B = 255
20  B = B-1
30  IF B=0 THEN 20
40  A = A+1
50  GOTO 50

        (b)
```

Figure 2-7 (a) 68000 assembly language program, which realizes the flowchart of Figure 2-5; (b) BASIC program, which realizes the task flowchart of Figure 2-5.

The main point is that once you have clearly defined the task to be accomplished, you can realize the entire flowchart by making each block function with a particular language. In this text we will use the 8080 assembly language and sometimes the Z80 assembly language in our solutions. You will become more familiar with flowcharts as more and more actual programs are introduced, discussed, and solved throughout the text.

2-3 PROGRAMMING MODEL FOR THE 8080, 8085, AND Z80

To begin programming any microprocessor in assembly language, you must have a good idea of what the programming model looks like. This model shows you how the internal space of the microprocessor appears to a programmer. This model is very different from an actual hardware model of the internal workings of a microprocessor. The programming model shows only the internal space to which you have access as a programmer, whereas an actual hardware model shows all the internal workings of the microprocessor.

Figure 2-8a shows the programming model for the 8080 and 8085 microprocessor, and Figure 2-8b shows the programming model for the Z80. Figure 2-8a is really a subset of Figure 2-8b. This was done by design so users of the 8080 could transfer programs over to the Z80. From a marketing standpoint, this is a very attractive feature of the Z80. In the beginning phases of learning programming in this text, the model of Figure 2-8a will be used. When this model is clearly understood, the added features contained in the Z80 will be explained, and you can appreciate the benefits of these additions.

Let's examine the model shown in Figure 2-8a and discuss exactly what information may be obtained from it. A programming model such as this is given in the literature for any microprocessor. In the first part of the programming model shown in Figure 2-8a you see the rectangles with letters A, F, B, C, D, E, and H, L. These are internal read/write locations referred to as REGISTERS. Notice further that above the registers are arrows denoting 8 bits. Each of the letters designates an 8-bit internal REGISTER.

The two registers A and F are also called the *accumulator* and *flag* registers. Remaining registers are simply referred to by their specific letters. As an assembly language programmer, you can write and read data into the registers. For example, one instruction may transfer the 8 bits of data from the B register to the C register or from the C register to the A register. Still another may add the B register to the A register. You will become familiar with all the possible manipulations of each register as you progress through this text.

The positioning of the registers in the model is also done for a particular reason. Notice that the A and F registers are shown together, as are the B and C, D and E, and H and L registers. This pairing is referred to as *register pairs*. Whenever an instruction indicates that a register pair is to be used, only the pairs shown are valid. The A and F registers are normally not used as a pair for

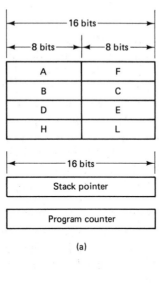

(a)

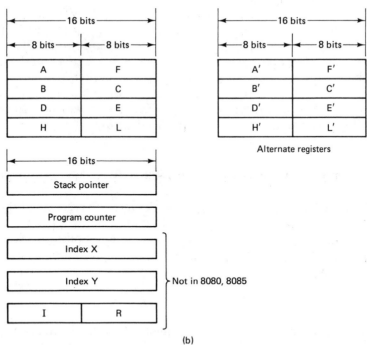

(b)

Figure 2-8   (a) Programming model for the 8080 and 8085 microprocessors; (b) programming model for the Z80 microprocessor.

reasons that will be explained later. When registers are used as pairs, the width of the new register is 16 bits instead of 8. As an example of this, consider the situation where you are using a register, say the C register, as an 8-bit counter. You could also choose the BC register pair as a 16-bit counter. Instructions that work with register pairs will be introduced to you later in the chapter. At this point in the discussion, you simply need to be aware that the internal 8-bit registers may be gouped as 16-bit register pairs.

The two remaining registers shown in Figure 2-8a are named *stack pointer* and *program counter*. We will give brief definitions of these registers now and more detailed definitions later. First notice that both of these registers are 16 bits in length. This is due to the fact that they generate special memory addresses. Recall from the discussion in Chapter 1 that the address buses for the 8080, 8085, and Z80 are 16 bits wide. During the execution of a program, these two registers may place their contents on the address bus.

The STACK POINTER keeps track of the area of memory known as the SYSTEM STACK AREA. This area is discussed in Chapter 4. The PROGRAM COUNTER keeps track of the address in memory that contains the next instruction to be executed. The importance of this function will be made clear later in the discussion.

Notice that the programming model for the Z80 shown in Figure 2-8b has an alternate set of 8-bit registers denoted by a prime and called A prime, B prime, and so on. Also shown in Figure 2-8b are two 16-bit registers labeled IX and IY. These are referred to as *index registers*. Finally, the last new registers shown in Figure 2-8b are I and R. The I register is used for special interrupt functions, and the R register is used for dynamic memory refreshing applications. All the registers shown in Figure 2-8b will be explained in detail in different sections of the text as the need arises.

2-4 STRUCTURE OF AN ASSEMBLY LANGUAGE INSTRUCTION

Now that you have an idea of what the internal registers of the microprocessor are, let's examine some actual instructions that will allow you to transfer and manipulate information in the model. To accomplish this you have to become familiar with the format of an assembly language instruction. Such an instruction is shown here:

```
MOV A,B
```

This instruction is interpreted to mean move the contents of the 8-bit register named B into the 8-bit register named A. At the conclusion of this instruction, both registers A and B will be equal.

This instruction is divided into three parts, *operation code* (op-code), *source operand*, and *destination operand*. The op-code is a three- or four-letter code that indicates the operation to be performed by the instruction. Examples of different op-codes are

```
MOV   DCX
MVI   PUSH
ORA   POP
ORI   JMP
ANI   RET
INX   XCHG
INC   SUI
ADI   OUT
```

Each of the op-codes shown will force the microprocessor to operate in a very specific manner. As stated earlier in this chapter, each microprocessor has its own unique set of op-codes. The three microprocessors discussed in this book use the same set of op-codes. Op-codes are written in a special form called a *mnemonic*. Mnemonics allow us to remember the function of the instruction. The microprocessor does not really use these mnemonics.

Associated with the op-code are the operands. There are normally two operands, source and destination. However, some instructions do not require operands, and one or none are used. The source operand is the location in the system (either inside the microprocessor or in memory) from which the data will be obtained. The destination operand for the instruction is the location in the system (either inside the microprocessor or in memory) where the data will be stored after the instruction is completed. Information in the source operand is never disturbed during an instruction. This is similar to reading a memory location. The act of reading the location does not change the information contained within the location.

The contents of the destination operand may or may not be disturbed or changed after an instruction. You can think of the destination operand as a memory location into which data will be written at the end of an instruction. Data that existed in the location before the instruction was executed is written over. It is possible that the same data may be written into the location. A good rule to remember when programming is

> During the execution of an instruction, the source operand is never changed, whereas the destination operand is always changed.

2-5 OBJECT CODE

When you write a program in assembly language you will be using the mnemonic form shown in Section 2-4. This form is only for the programmer, and the microprocessor cannot make use it. The microprocessor requires a form of instructions call *object codes*. Mnemonic forms of instructions are often referred to as a program *source* or *source code.* The object code form of the instructions are the actual 1s and 0s required for the microprocessor. The object code form of the instructions must exist because the microprocessor reads data bytes from memory and interprets these bytes as instructions.

As an example, let's consider the MOV A,B instruction given previously. We know that this instruction is to move the contents of the internal register labeled B into the internal register labeled A. This same instruction in object code form is

```
MNEMONIC FORM      OBJECT CODE FORM

   MOV A,B               78H
```

where the H on the end of the 78 indicates that the base of the number is hexadecimal. The binary form of the instruction would be

01111000

The microprocessor would read this byte from memory and perform the function of moving the contents of the B register into the A register.

You can see now why the mnemonic form is for the programmer's benefit only. The binary, or object, form of the instruction—which the microprocessor can use—has little meaning for us. The following examples further illustrate this fact.

```
MNEMONIC FORM      OBJECT CODE FORM

  LXI H,49A3H           21A349
  MVI B,25H             3625
  ORA M                 B6
```

A point to notice from these examples is that the object code form may take up more than one byte or memory location. In these examples, the object code form of the instruction takes 3, 2, and 1 bytes, respectively.

Remember that your program must be in object code form if you wish the microprocessor to execute it. So let's move on to the next section, where you will learn how this is accomplished.

2-6 HOW TO ASSEMBLE A PROGRAM

Your program must be converted into object code form before the microprocessor can run it. In this section you will learn how to accomplish this task. The function of converting your program from a SOURCE PROGRAM to OBJECT CODE is known as ASSEMBLING the program. The flowchart of Figure 2-9 shows the steps required to run a program on a microprocessor system.

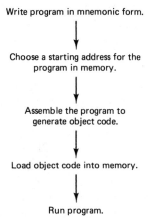

Write program in mnemonic form.

Choose a starting address for the
program in memory.

Assemble the program to
generate object code.

Load object code into memory.

Run program.

Figure 2-9 Flowchart showing the process of writing, assembling, and running an assembly language program.

One of the best ways to become familiar with the procedure necessary to assemble a source program is to do it. We will be assembling the program shown in Figure 2-10. Before doing the assembly, let's take some time to learn how the instructions given in the example operate. You will be given an extensive explanation in the later sections of this chapter.

The MVI instruction is called a *move immediate*. It moves *immediate data* from memory into the source operand. Immediate data are data that are part of the instruction and immediately follow the op-code byte. In the MVI A, 24H instruction the immediate data is hexadecimal 24 to be loaded into the A register. An H is used to indicate the data will be in hexadecimal format. If no H were used, the data would be decimal.

After execution of the first instruction, the A register would contain the hexadecimal value 24. The next instruction shown in Figure 2-10 is the MOV B,A. This instruction moves the contents of the source operand (A register) into the destination operand (B register). When this instruction has been executed, the A or B registers will both contain the hexadecimal number 24.

The third instruction in the sequence is ADD B, which adds the contents of the B register to the A register and stores the result into the A register. The source operand is specified in the instruction, in this instance B register, whereas the destination operand is implied (A register). When the microprocessor has executed this instruction, the B register will contain the value 24 hexadecimal and the A register will contain the value 2 × 24H = 48H, or 72 decimal.

```
MVI A,24H
MOV B,A
ADD B
HLT
```

Figure 2-10 Sample 8080 assembly language program to be assembled.

The final instruction given in Figure 2-10 is HLT (halt). When the microprocessor encounters the HLT instruction, it will stop execution.

Keep in mind that the explanations given here will be greatly enhanced when we actually begin study of the instruction set. These explanations are intended to place you on familiar ground for the purposes of assembling a program. With this understanding, let's begin the assembly process.

Start your assembly by taking each instruction one at a time. First the MVI A,24H instruction must be divided into its three parts. These are op-code (MVI), destination operand (A), and source operand (24H). From the op-code chart given in Appendix B, find the MVI A instruction. This instruction is under the move immediate column. Next to the MVI A, you see a hexadecimal number 3E. This is the object code equivalent of the MVI A instruction.

The assembly of this instruction is not complete because the source operand has not been encoded. Notice in the op-code chart that all the MVI instructions have a D8 following them. This indicates that 1 data byte must be included in the object code form. In our example the data byte will be the source operand of 24H. Therefore, the complete object code form of the MVI A,24H instruction contains two bytes, which are equal to 3E 24. Because this instruction requires 2 bytes to define it completely in object code, it is called a 2-byte instruction.

Instructions for the 8080 and 8085 are either 1-, 2-, or 3-byte instructions. For the Z80 there are some instructions that have more than 3 bytes. These are referred to as *extended op-codes*.

Continuing in the assembly process for the program of Figure 2-10, you encounter the MOV B,A instruction. Referring to the op-code chart under the heading of MOVE, you see that the object code form of the instruction is 47, further notice that this instruction requires only 1 byte to define it fully in object code.

The next instruction in the program is ADD B. Referring to the op-code chart under the heading of ACCUMULATOR, you find the object code equivalent for ADD B is 80.

Finally, the HLT instruction, found under the heading of CONTROL, has an object code equivalent of 76. Putting the pieces of this program together, you have a complete object code program:

```
3E
24   MVI A,24H

47   MOV B,A

80   ADD B

76   HLT
```

The hexadecimal data of the object code will be loaded into memory and the microprocessor will execute this program. Mnemonics were used only to allow you to put the program into the object code form for the microprocessor.

Converting a program from source (mnemonics) to object format can become very tedious and prone to error if the number of instructions is very large, say greater than 20. Many microprocessor application programs comprise hundreds of instructions. Therefore, the task of converting a source program into object code form can be accomplished by a computer. Here is the way it is done.

First, you write your source program using an *editor* program and store the program on a disc. Next you run a computer program called an *assembler,* which takes the source program from your disc and uses it to generate a new file, which is the object code file. Finally, you use another program that loads the object code file into the computer's memory space and runs the object code. This process is shown in Figure 2-11.

When you examine the process of converting from mnemonics to object code, it becomes clear that it is a matter of looking up equivalent values for instructions. The computer is ideally suited to perform this task. When you write your assembly language programs on the job or perhaps at home on your personal computer, you will probably be using the computer to convert to object code for you.

However, in this text (and probably in your laboratory experiments), you will be required to hand-assemble all your programs. Therefore, it is wise to become knowledgeable about the procedure. To that end, let's discuss a few

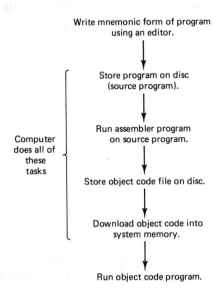

Figure 2-11 Expanded flowchart showing how the assembly language program of 2-10 could be assembled and run using a computer.

more examples of assembling a program, keeping in mind that you may not know the purpose of all the instructions.

Example 2-1

The first program to assemble is shown in Figure 2-12. We have discussed the MVI and the HLT instructions. The remaining instruction, SUB B, subtracts the contents of the B register from the contents of the A register and stores the result into the A register. The overall effect of this program will be to subtract 35H from 36H and store the answer in the A register. With this brief introduction, let's assemble the program.

```
MVI A,36H
MVI B,35H
SUB B
HLT
```

Figure 2-12 8080 assembly language program to assemble in Example 2-1.

First, locate the MVI A instruction from the op-code chart. You find that it has an equivalent of 3E. Further, this instruction requires an additional byte, which will be the data to load into the A register. The total assembly of the MVI A,36H instruction is 3E 36. This same process may be used for assembling the MVI B,35H instruction. After performing this operation, you find that the object code equivalent for the MVI B,35H is 06 35.

The SUB B instruction is located under the heading of ACCUMU-LATOR and has a single-byte object form of 90. Finally, the HLT instruction is located on the op-code chart under the heading of CONTROL and has a single-byte equivalent of 76. This completes the entire assembly process, and you will have a program that looks like this:

```
OBJECT CODE     MNEMONICS

   3E
   36           MVI A,36H

   06
   35           MVI B,35H

   90           SUB B
   76           HLT
```

Example 2-2

In this example, shown in Figure 2-13, there are two new instructions, DCR D, NOP. When the microprocessor executes the DCR instruction, it will decrement by one the register specified. In this case the microprocessor will decrement the D register. The second new instruction is the NOP,

```
MVI D,01H
DCR D
MOV B,D
NOP
HLT
```

Figure 2-13 8080 assembly language program to assemble in Example 2-2.

which stands for *no operation*. This is a 1-byte instruction, which does nothing. The microprocessor simply reads the instruction and gets set to read the next instruction. You will see later why NOPs can be important instructions. With this introduction try to assemble the program yourself using the op-code chart. Compare your answer to the one given here:

OBJECT CODE	MNEMONICS
16	
01	MVI D,01H
15	DCR D
42	MOV B,D
00	NOP
76	HLT

You have now had some practice in assembling programs. The steps are not difficult, and with a little patience you can assemble even large programs in an error-free manner. There are still details that need to be covered concerning three byte instructions; these will be given as the need arises.

2-7 STORING THE INSTRUCTIONS IN MEMORY

You have been shown that the microprocessor will execute instructions in object code format. These instructions are read from system memory, interpreted, and then executed. In this section you will be shown the last piece of the puzzle, where the object code instructions are stored in memory, and how the microprocessor knows where to get them in the memory.

When the microprocessor first powers up or has the RESET pin asserted, certain internal operations occur. One of these is that the microprocessor will output a specific address on the address lines and expect the first op-code to be located at this address. For the 8080, 8085, and Z80 this address is 0000H.

Recall from Chapter 1 that the entire system address space may be modeled as shown in Figure 2-14. You see in Figure 2-14 that the available memory

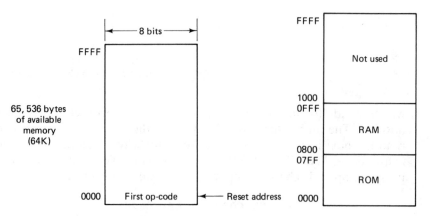

Figure 2-14 The microprocessor system address space may be modeled as a list of addresses starting with 0000 and ending with FFFF.

Figure 2-15 Diagram showing how the memory space of a microprocessor system may be divided up. This is called a memory map.

addresses are from 0000 to FFFFH. Further, the address space for a microprocessor system is usually divided up into some ROM (Read only Memory) and RAM (Random access memory) area, an example of which is shown in Figure 2-15. Notice that the ROM is located at address 0000. This is to accomodate the fact that the first instruction must be at address 0000 when the microprocessor is first powered up.

The microprocessor systems used in your laboratory experiments will indicate to you where you can begin loading your own programs. Programs are loaded into the system RAM space. The system is executing instructions from the instant it is powered up. Some of these instructions allow you to enter your own program from the keyboard. You must inform the microprocessor to start running the program you have loaded into RAM. All the details about loading and running your program on a particular microprocessor system may be obtained from the system documentation.

Let's assume that on a microprocessor system the documentation indicates that user programs may be loaded into address spaces 8200H to 86FFH. This gives 4FF, or 1279 decimal, bytes of memory available for you to enter your program. Where in all of these available bytes do you load the program? In most systems it will not matter where you start because there is usually some type of system command that lets you start running a program from any address you specify. As an arbitrary choice, we will choose address 8200H as the beginning address at which to load our programs.

By choosing this address as the start, all memory locations for every byte of the program are now defined. Take, for example, the program you assembled in the previous section. That program is listed here for your convenience.

```
16   MVI  D,01H

01

15   DCR  D

42   MOV  B,D

00   NOP

76   HLT
```

If address 8200H is chosen for the first op-code of the program, then all the following bytes are loaded as shown:

```
ADDRESS   BYTE   INSTRUCTION
8200      16
8201      01     MVI  D,01H

8202      15     DCR  D

8203      42     MOV  B,D

8204      00     NOP

8205      76     HLT
```

As you can see from this example, the only information you load into system memory is hexadecimal data. Some of this information involves instruction op-codes, whereas other information is data to be used by the instruction. For example, the 16 at address 8200 is an op-code, but the 01 at address 8201 is data used by the instruction and not an op-code. The only way to distinguish between these two types of data is to know the starting address of the first op-code in the program.

Once you choose a starting address for your program, you must load all remaining bytes of the program at the correct locations. If you do not, the microprocessor will fail to run correctly. This is because the internal hardware of the microprocessor will be expecting you to follow certain conventions when loading your object code. As an example of how you might cause the program to malfunction, consider the following.

Suppose you choose 8200 as the starting location for your program. However, you neglected to load the 01 for the MVI D, 01H instruction. Instead you loaded the following.

ADDRESS	DATA	INSTRUCTION
8200	16	
8201	15	MVI D,15H
8202	42	MOV B,D
8203	00	NOP
8204	76	HLT

Notice how the program was changed by deleting the 01 that should have been loaded at address 8201. This change occurs because inside the microprocessor there is a register called the PROGRAM COUNTER, which keeps track of the memory address from which to read the next instruction. When the microprocessor reads the byte 16 and interprets it, the internal hardware sets the program counter (PC) to 8202 because the instruction 16 requires data located at the address 8201. That is, the internal hardware of the microprocessor knows that 16 (MVI D) is a 2-byte instruction.

In this instance, the program will still run, but it will not give the correct results. In some cases where you make an error such as this, the program will not run at all. A good practice is always to verify at which address you are loading data on your system and compare this address with the one written on paper next to your assembled program.

All the preceeding information has been leading you to the point of writing, assembling, running, and perhaps debugging your first microprocessor program. The only point left is for you to learn some of the instructions available on the 8080, 8085, and Z80. This information is given in the next section.

2-8 8080, 8085, AND Z80 INSTRUCTION SET (INTRODUCTION)

In the preceding sections of this chapter, you have been informally introduced to some of the available mnemonic instructions for the 8080, 8085, and Z80 microprocessors. This was done to allow you to see how a program was arranged in memory and how to translate from a source program to object code. These two facts are very important in the task of writing and running a program on a real system. Let us assume that you understand how a program is loaded into memory and how to translate it to a form compatible with the microprocessor. From this starting point, the instructions for the microprocessor will be formally introduced and explained here and in Chapter 3.

All the available instruction mnemonics for the microprocessor are given in the op-code chart in Appendix B. When you are assembling a program, this chart is a very handy reference. However, when learning the instructions, a different description is required to explain exactly what the instruction does.

This type of description is given in Appendix B. There each instruction is listed in its mnemonic form, and a detailed description of the instruction is given. Appendix B is mentioned because you may wish to refer to it when you are writing a program and are unsure of what instruction is needed to do the job you desire. You will be referred to this appendix at various places in this chapter and the next.

The microprocessor instructions are divided into five major groups:

1. Data transfer group
2. Arithmetic group
3. Logical group
4. Branch group
5. Stack, I/O, and machine-control group

The data transfer group and the I/O instructions will be discussed first.

The *data transfer group* moves data betwween internal registers or between system memory and internal registers. You have already seen some of the instructions in this group, namely, the MOV and MVI. In the following discussion, each instruction is shown in 8080/8085 and Z80 assembly language format. The first instruction is:

```
          MOVE REGISTER

   8080/8085        Z80
   MOV r1,r2       LD r1,r2
```

This instruction will move the contents of the source operand (r2) into the destination operand (r1). The contents of r2 are not changed. In examining the programming model for the microprocessor, you see that there are many different combinations of MOV that can occur. Any 8-bit register, A, B, C, D, E, H, L, may be moved to any other 8-bit register, including itself. That is, the instruction MOV B,B is a valid instruction. Its usefulness is limited, but the microprocessor will execute it.

Figure 2-16 shows the general object code form for the MOV r1,r2 instruction. You see in this figure that DDD and SSS are 3-bit codes, which stand for the destination register and the source register. The register codes are

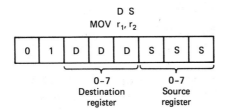

Figure 2-16   Object code form of the MOV r1,r2 instruction.

111	=	A	register
000	=	B	register
001	=	C	register
010	=	D	register
011	=	E	register
100	=	H	register
101	=	L	register

In making use of these codes you can form the object code for any possible MOV instruction by inserting the correct 3-bit code for the source and destination registers.

For example, suppose you wish to form the object code for MOV A,E. In this example the A register is the destination and the E register is the source. The correct object code form is

$$01 \quad 111 \quad 011 = 01111011 = 7BH$$

You can check this answer by referring to the op-code chart in Appendix B and looking up the code for MOV A,E.

MOVE FROM MEMORY INTO REGISTER

```
8080/8085      Z80
 MOV r,M      LD r,(HL)
```

This instruction will move a byte from memory into the destination register specified. Any internal register, A, B, C, D, E, H, L, may be specified. In order to read the data in memory, a 16-bit memory address must be supplied by this instruction. The address for memory is implied and exists in the H and L registers, with H = high-order bits (A15–A8) and L = low-order bits (A7–A0).

When this instruction is executed, the contents of the H and L registers are automatically placed onto the microprocessor address lines, and the byte of data at that address is read and stored into the register specified.

For example consider this: L register = 24H and H register = 5AH. When the MOV A,M instruction is executed, address 5A24 will be placed onto the microprocessor address bus and data at that address will be read into the microprocessor. Let us say that at address 5A24, data = 8CH. At the conclusion of this instruction the A register will have the value 8CH stored in it.

MOVE REGISTER TO MEMORY

```
8080/8085      Z80
 MOV M,r      LD (HL),r
```

This instruction does exactly the opposite of the MOV r,M in that the information in an internal register is written into memory at the address specified by the H and L registers. Any internal register, A, B, C, D, E, H, L, may be specified. Examples of this instruction are:

```
8080/85        Z80

MOV M,A        LD (HL),A
MOV M,B        LD (HL),B
MOV M,C        LD (HL),C
MOV M,H        LD (HL),H
```

MOVE IMMEDIATE DATA TO REGISTER

```
8080/85           Z80
MVI r,data     LD r,data
```

The move immediate instruction consists of 2 bytes, as shown in Figure 2-17. The first byte is the op-code containing the 3-bit code of the internal 8-bit destination register. Immediately following the first byte is the actual 8-bit data, which is to be loaded into the destination register. This instruction is useful for initializing internal registers to specific values or loading constants into the registers.

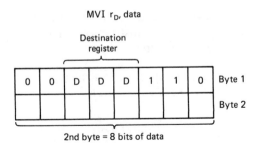

MVI r_D, data

Destination register

| 0 | 0 | D | D | D | 1 | 1 | 0 | Byte 1 |

2nd byte = 8 bits of data

Figure 2-17 Two bytes for the MVI instruction. The first byte is the op-code, whereas the second byte is the immediate data to move.

Examples of this instruction are:

```
8080/85         Z80

MVI A,34H      LD A,34H
MVI H,2AH      LD H,2AH
```

Note: The H is used to inform an assembler program of the type of data (immediate) and its number base (H). Since you will be hand-assembling your programs, you need not put it in. However, it is a good habit to get into because if you ever write assembly programs that will use an assembler program, the H may be required. This symbol will be used throughout this book, as it is good programming practice.

```
MOVE IMMEDIATE DATA TO MEMORY
     8080/85              Z80
   MVI M,data     LD (HL),data
```

This instruction is very similiar to the move immediate data to an internal register, except it will move the data to an address in memory. The address to store the data resides in the H, L register pair. (As a reminder, the H register contains memory address bits A15–A8 and the L register contains memory address lines A7–A0).

An example of this instruction is:

```
       8080/85              Z80

     MVI H,83H       LD H,83H
     MVI L,20H       LD L,20H
     MVI M,55H       LD (HL),55H
```

At the conclusion of this three-instruction sequence, the data 55H will be stored into the memory location 8320H.

Three-Byte Instructions

Thus far in our discussions we have concentrated on 1- and 2-byte instructions. We now turn our attention to some 3-byte instructions. These instructions are contained in the *data transfer group*. The first of these is:

```
     LOAD IMMEDIATE 16 BITS OF DATA
          INTO THE REGISTER PAIR
       8080/85              Z80
   LXI rp,data 16     LD rp,data 16
```

Figure 2-18 shows how the three bytes of object code are formed for this instruction. The first byte comprises the op-code with the 2-bit register pair indicator inserted. The register pairs are coded as follows:

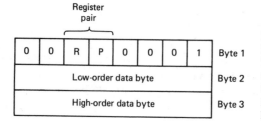

Figure 2-18 Three bytes of object code for the single instruction LXI rp,data. The first byte is the op-code, whereas the second and third bytes are the 16 bits of data to be moved. Byte 2 is the lower-order byte and byte 3 is the higher-order byte.

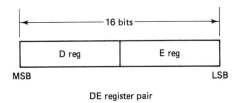

Figure 2-19 The D and E registers may be used together to form a 16-bit register. This is also true of the BC and HL registers.

$$00 = B,C$$
$$01 = D,E$$
$$10 = H,L$$
$$11 = SP$$

For each of the register pairs, the register listed first is the high-order byte. For example, if the D, E register pair were treated as a 16-bit register, it would appear as shown in Figure 2-19. The SP register was shown in the programming model and is a 16-bit register used to generate the stack pointer address when the stack is used. We have not discussed the stack area of memory and the stack pointer register is mentioned here only for completeness of the LXI instruction. A complete discussion of the system stack is given in Chapter 4.

When the LXI instruction object code is formed, the data that will be loaded into the low-byte register of the pair is placed into the second byte of the instruction. Data to be loaded into the high-order register of the pair is placed into the third byte of the instruction. For example, consider the object code for the instruction LXI H,5A3CH.

The first byte would equal 00   10   0001 = 00100001 = 21H. This is the op-code for LXI H. The second byte will be the data to be loaded into the L register, which is 3CH. This leaves the third byte equal to data loaded into the H register, which is 5AH. The total object code for the instruction is:

                     21H
                     3CH
                     5AH

It appears that the last two bytes should be reversed according to the way the instruction is written in mnemonic form, LXI H,5A3CH. This is a point to be remembered when forming the object code for 3-byte instructions.

Examples of the LXI instruction are:

8080/85	Z80
LXI H,2356H	LD HL,2356H
LXI SP,59A3H	LD SP,59A3H
LXI B,0000	LD BC,0000H
LXI D,1245H	LD DE,1234H

```
STORE ACCUMULATOR DIRECT
8080/85          Z80
STA addr     LD (addr),A
```

This is a 3-byte instruction which will store the contents of the A (accumulator) into the 16-bit memory address specified by the instruction. For example, suppose you wish to store the value of the A register into memory location 53A8H. The mnemonic form for this instruction is:

```
8080/85          Z80

STA 53A8H    LD (53A8H),A
```

In this instruction the 53A8 is not immediate data but rather is a memory address. The assembler will logically determine this because of the type of instruction.

Forming the object code for this instruction follows the rules of object code for all 3-byte instructions. That is, the op-code is contained in the first byte of data. The lower-address byte or the lower-register data is given in the second byte and the high-address byte or high-register data is given in the third object code byte. For this instruction the object code is:

```
32H      op-code LDA
A8H      lower-address byte
53H      high-address byte
```

Again, it appears that the second and third bytes of data are reversed from the instruction as presented in mnemonic form. Some microprocessor object codes do put them in the order of high byte and then low byte—for example, the 6800 microprocessor. But the 8080, 8085, and Z80 all use the object code form given here, and you will need to remember this when writing, running, and loading your own assembly language programs.

```
LOAD ACCUMULATOR DIRECT
8080/85          Z80
LDA addr     LD A,(addr)
```

This 3-byte instruction will load the data at the memory location specified to the A register (accumulator). For example, if you wish to load the data at memory location 34D9H into the accumlator, you would write the mnemonic instruction

```
8080/85          Z80

LDA 34D9H    LD A,(34D9H)
```

The object code form is:

```
3AH     op-code
D9H     lower address byte
34H     upper address byte
```

```
        LOAD HL DIRECT

8080/85              Z80
LHLD addr       LD HL,(addr)
```

This 3-byte instruction will load the H and L registers with data stored in memory at the address specified. Since there are two registers to be loaded, it will require two memory locations. The instruction specifies only one memory location. The second memory location is the address specified plus 1.

Data at the memory address specified in the instruction will be loaded into the L register, whereas data at memory address specified plus 1 will be loaded into the H register. As an example, consider this:

```
MEMORY ADDRESS     DATA

    237EH          39H
    237FH          2DH
```

The instruction LHLD 237EH is executed. At the conclusion of this instruction the L register is 39H and the H register is 2DH.

Formation of the object code for this instruction is accomplished like the STA and LDA instructions. The first byte is the op-code, and the second byte equals the lower 8 bits of the 16-bit address; the third byte is set to the upper 8 bits of the 16-bit address. For the instruction given, the three bytes of the op-code are

```
2AH     op-code
7EH     lower address byte
23H     upper address byte
```

```
        STORE HL DIRECT

8080/85              Z80
SHLD addr       LD (addr),HL
```

This 3-byte instruction will perform the opposite function of the LHLD instruction just shown. The contents of the H and L registers will be stored in memory at the address and address + 1 designated. The L register is stored in the address specified and the H register is stored in the address + 1.

Let's consider the same memory locations used for the previous instruction. Suppose you wanted to save the values of the H and L registers in memory, starting at memory address 237EH. The mnemonic instruction is

```
        8080/85            Z80

     SHLD 237EH      LD (237EH),HL
```

Object code for the instruction is

```
     22H      op-code
     7EH      lower 8 bits of address
     23H      upper 8 bits of address
```

Further, suppose these instructions were executed;

```
        8080/85            Z80

     MVI H,58H       LD H,58H
     MVI L,41H       LD L,41H
     SHLD 237EH      LD (237EH),HL
```

The resulting operation would place the following bytes in memory at these addresses:

```
     ADDRESS    DATA

      237EH     41H
      237FH     58H
```

More 1-Byte Instructions in the Data Transfer Group

```
     LOAD THE ACCUMULATOR INDIRECT

        8080/85            Z79
        LDAX rp     LD A,(rp)
```

This load instruction will load the A register (accumulator) with the data from memory addressed by the register pair. This is similar to the MOV A,M instruction, where the memory is addressed by the H,L register. In LDAX instruction, however, the register pair may be BC or DE.

Mnemonic form for this instruction is

```
        8080/85            Z80

        LDAX B      LD A,(BC)
        LDAX D      LD A,(DE)
```

Only the BC and DE register pairs may be used for this instruction.

The object code form of this instruction is

```
            00RP 1010
```

where RP is the 2-bit code for the register pair. For this instruction RP may be equal to 00 (BC) or 01 (DE). The two object code forms of the LDAX instruction are:

```
00001010=LDAX B

00011010=LDAX D
```

As an example of using this instruction consider the following instruction sequence:

```
8080/85        Z80

MVI B,43H     LD B,43H
MVI C,28H     LD C,28H
LDAX B        LD A,(BC)
```

Suppose the data in memory was as follows:

```
MEMORY ADDRESS     DATA
    4328           F4H
```

At the conclusion of this three-instruction sequence, the data in the A register (accumulator) would be F4H. The program first loads the B and C registers with the data; then the LDAX instruction uses the data in these registers as a 16-bit memory address from which to read information and load it into the internal A register.

```
STORE THE ACCUMULATOR INDIRECT
8080/85          Z80
  STAX        LD (rp),A
```

This 1-byte store instruction will do just the opposite of the LDAX instruction discussed earlier. STAX will store the contents of the A register into the memory address specified by the BC or DE register pair. Mnemonic form for the STAX instruction is:

```
8080/85          Z80

STAX B        LD (BC),A
STAX D        LD (DE),A
```

Object code form for these two instructions is:

```
02H  STAX B
12H  STAX D
```

To see the effect of this instruction consider the following program:

```
8080/85          Z80

MVI D,43H        LD D,43H
MVI E,28H        LD E,28H
MVI A,32H        LD A,32H
STAX D           LD (DE),A
```

At the conclusion of this program the data in memory would appear as:

```
MEMORY ADDRESS    DATA

     4328H          32
```

The data in the accumlator was written into memory using the D, E register pair as the 16-bit address. The D register was the high-order byte A15–A8 and the E register was the low-order byte A7–A0.

The final instruction in the data transfer group that we will discuss is the following.

```
EXCHANGE HL WITH DE

8080/85          Z80
XCHG             EX DE,HL
```

This 1-byte exchange instruction exchanges the contents of the HL register pair with the DE register pair. To show how this instruction operates, consider this program.

```
8080/8085        Z80

LXI H,239AH      LD HL,239AH
LXI D,4567H      LD DE,4567H
XCHG             EX DE,HL
```

Just prior to the XCHG instruction H = 23H, L = 9AH, D = 45H, and E = 67H. After execution of the XCHG instruction, H = 45H, L = 67H, D = 23H, and E = 9AH. Notice that no information was lost; it was just exchanged between the register pairs.

This concludes the discussion of the data transfer group of instructions. The use of these instructions may not become clear until you actually start writing some of your own application programs. However, knowing these instructions exist will help you better decide how to make the microprocessor accomplish the task you wish.

2-9 I/O INSTRUCTIONS

In this section you will be shown how the input and output instructions operate for the 8080, 8085, and Z80 microprocessors. These instructions are shown at this point in the text because your laboratory experiments will probably require them. Detailed discussion of how the hardware uses the input and output instructions is reserved for Chapter 5. This presentation concentrates on how the software of the input and output operations works.

In Chapter 1 you saw a block diagram of a typical microprocessor-based system. One of the major blocks of this system was labeled I/O. The I/O block of the system was physically different from system memory. Whenever you wish to move data between the microprocessor and the I/O block, you use the input and output instructions.

Input and output devices in a microprocessor-based system are devices like these:

Floppy disk
Keyboard
LED display
CRT display
Voice processing
Relay control
Temperature control and monitoring

In general, any destination for data during an output operation and any source of data during an input operation is defined as I/O. The display and keypad in the systems you are using for your laboratory experiments are I/O devices.

Output

Every output device in a microprocessor-based system is assigned a unique address. This address is similar to a memory address, which is assigned to each storage location in memory. The 8080, 8085, and Z80 use only the lower 8 bits of the address bus to define the system I/O space. This means there are 256 (00–FFH) unique output addresses in the system.

When you use the output instruction, the address of the output device must be specified. The address of the device is sometimes referred to as a *port address*. These two words *port address* and *I/O address* are used in the literature to mean the same thing. Figure 2-20 shows a block diagram of several output devices with their addresses given.

Information to be written to the output device comes from the A register in the microprocessor. This is always true for the 8080 and 8085, but the Z80 is

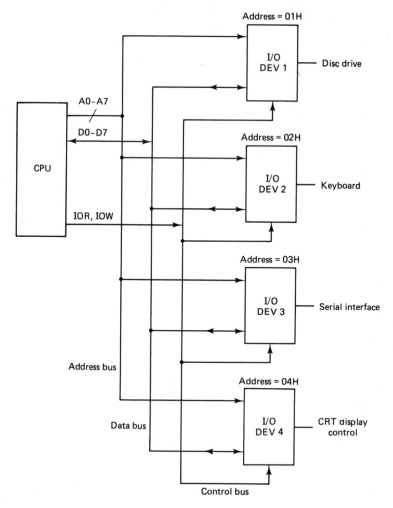

Figure 2-20 Block diagram showing four different I/O devices and their respective I/O addresses.

capable of writing information from other registers. You will be shown this in Chapter 4, where extended op-codes are discussed. For now it is safe to assume that all information written to an output device comes from the A register.

All 8 bits of the A register are written to the output port in a parallel fashion. This is exactly the same type of operation that occurs when a byte of data is written onto system memory. The hardware of the output port electrically captures the data, as you will see later. For now you should remember that when you wish to write information to an output address (port), the data must first be placed into the A register.

The mnemonic form of the output instruction is

```
8080/85          Z80

OUT addr     OUT (addr),A
```

Where the address is a single byte of data specifying the logical value of the address lines A7–A0 to be used as the port address. As an example, suppose you wished to write data to port 53H. The mnemonic form of the instruction would be:

```
8080/85          Z80

OUT 53H     OUT (53H),A
```

The object code format for this instruction requires 2 bytes, the first for the op-code and the second for the port address. Object code form for the instruction just given is:

```
D3   op-code
53   port address
```

To show how the output instruction may be used in a program, suppose you wished to write the data 34H to port 67H. A program to do this is:

```
8080/85          Z80

MVI A,34H    LD A,34H
OUT 67H      OUT (67H),A
```

Notice that the 8080/85 output instruction does not specify from where the data is coming that will be written into port 67H. The source operand is implied and is equal to the A register.

In most laboratory training systems, there are individual light-emitting diodes, LEDs (8 of them), which can be turned on and off under program control. Further, each LED is usually assigned to an individual bit of the byte that is written to the port. This is shown in Figure 2-21. With this type of display you can check what data you are actually writing to an individual port. Or if some hardware is connected to this port, you can monitor the data flow from the microprocessor to the port.

Input

The opposite of the output instruction is the input instruction. This instruction allows you to read a byte of data from a specified port address. Like the output instruction, the input instruction will communicate with a possible 256 different addresses (00–FFH).

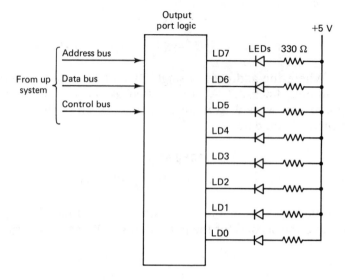

Figure 2-21   Diagram showing an output port that has a LED connected to each bit of the port. When the bit written to the output port if a logical 1, the LED will light. By examining which LEDs have been lighted, you can see exactly what data was written to the port.

Whenever data is read from an input port, it is stored in the A register. The mnemonic form for the input instruction is;

```
      8080/85          Z80

      IN addr     IN A,(addr)
```

or

```
      IN 24H      IN A,(24H)
```

where the address is 00–FF in hexadecimal. The object code form for the input instruction consists of 2 bytes, as follows:

```
      DB        op-code
      24H       port address
```

Most laboratory systems have a switch for input data, which resides at some predetermined port in the system. Such a system is shown in the block diagram of Figure 2-22. Each switch may be set to a 1 or 0 position. This allows the user to set individual bits of the input byte to specific values.

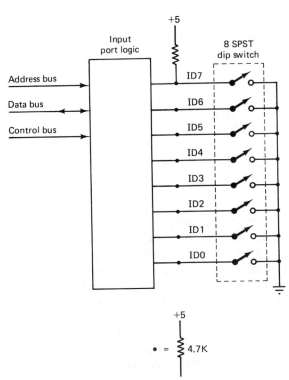

Figure 2-22 Diagram showing an input port with a switch connected to each bit of the port. With the switch, you can set any data desired to be read from the port.

2-10 JUMP INSTRUCTION

We present the JMP, or jump, instruction at this point in the text because in some early programs you write, you may wish to use this instruction. The IN and OUT instructions were introduced for much the same reason.

In a microprocessor program the flow of instruction execution is from the instructions residing in low memory addresses to the instructions residing in higher memory addresses. This flow is shown in Figure 2-23.

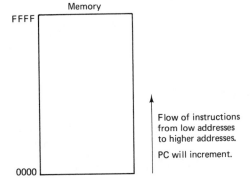

Figure 2-23 The flow of program instructions is from the lowest memory address to the highest, as shown in this memory map.

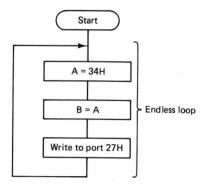

Figure 2-24   To alter the normal program flow from lowest memory address to highest, you can use the jump instruction. This program requires the program to loop back to a previous instruction.

Using the JMP instruction you can alter the flow of the program from this normal low to high flow. For example, suppose you wish to realize the flowchart of Figure 2-24. You see in this figure that the program will execute normally until the last instruction is reached. From that point you wish the microprocessor to execute the first instruction over again, thus creating a loop. This loop will continue until you reset the processor or turn the power off. A loop that will continue in this manner is called an *endless* loop.

To show how the loop of Figure 2-24 is formed using the JMP instruction, let's actually do the steps necessary to assemble and run the program.

The first step is to write the mnemonics for each step of the flowchart, as shown in Figure 2-25. The last mnemonic is JMP, which is the jump. However, the jump instruction requires a 16-bit address to which to jump. This address will be included in the instruction. Therefore, the object form of the JMP instruction appears as;

```
JMP        C3H
byte 1     lower address bits A7-A0
byte 2     upper address bytes A15-A8
```

The complete mnemonic form is:

```
      8080/85                   Z80

JMP 16 bit address      JP 16 bit address
```

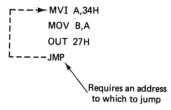

Figure 2-25   The jump instruction will force the microprocessor to go to a specified memory location to fetch the next op-code.

An example is

$$\text{JMP 2679H} \qquad \text{JP 2679H}$$

Object form of this instruction is:

```
C3H    op-code
79H    lower address byte
26H    upper address byte
```

Returning to our original problem, what address do we use for the jump instruction? You must specify the address of the op-code for the instruction to which you wish to jump. In this case you use the address for the op-code of MVI A,34H.

However, we have not yet assigned a memory location to this op-code. Assume the program will start at address 8200H. With this address as the start, the program will assemble in the following manner.

ADDRESS	DATA	INSTRUCTION
8200H	3EH	MVI A
8201H	34H	immediate data 34
8202H	78H	MOV B,A
8203H	D3H	OUT
8204H	27H	output port address
8205H	C3H	JMP
8206H	00H	low-address byte
8207H	82H	high-address byte

You see from this example that the JMP instruction will jump to memory address 8200H. You must specify the memory address of an op-code for the instruction to jump to. The address of this op-code is sometimes referred to as a *target address*.

Suppose you wanted the program to loop on the OUT instruction instead of the MVI instruction. The address for the OUT op-code is 8203H. Therefore, the object form of the JMP instruction would be:

```
C3H    op-code JMP
03H    lower address byte
82H    upper address byte
```

Labels

When you are writing your programs, the memory addresses may not have been determined. Therefore, you can use *labels* to define the jump locations and fill in the memory addresses at a later time. For example, instead of using memory address 8200H as the jump address, you could write your program like this:

```
            8080/85                    Z80

START    MVI A,34H      START    LD A,34H
         MOV B,A                 LD B,A
         OUT 27H                 OUT (27H),A
         JMP START               JP START
```

START is first shown as a label associated with the MVI op-code. When START is used as a target address, the value of the memory address with which START is associated will be used. In this case you define a starting address for the MVI instruction and START becomes that address.

In the example used previously, START was associated with, or equal to, memory address 8200H. Therefore, whenever START is used in the program as a target address the value 8200H is inserted. Using labels in this manner helps keep the program uncluttered with memory addresses. The only time you need to assign actual memory addresses is when you assemble and load the program.

If you are fortunate enough to have a computer program that will assemble the program for you, labels can be kept in your source code, and the assembler will automatically insert the correct memory address. You will be shown later that labels may also be used for other purposes, not only addresses but constants as well. Figure 2-26 shows a printout of an actual computer assembled source program for the example used.

Summation for Instruction Introduction

In the preceeding sections of this chapter you have been shown many details concerning programming of a microprocessor. Along with these details you were also shown examples of how to generate object code from a source program. In the next section you will practice with this information by examining more examples of source programs and the associated object code.

```
 1                          ;;;;;;;;;;;;;;;;;;;;;;;;;;;;;;;;
 2                          ;
 3                          ; COMPUTER ASSEMBLY OF FIGURE 2-25
 4                          ;
 5                          ;;;;;;;;;;;;;;;;;;;;;;;;;;;;;;;;;
 6                          ;
 7  0100                            ORG 0100H
 8  0100   3E 34            START   MVI A,34H      ;LABEL START
 9  0102   47                       MOV B,A
10  0103   D3 27                    OUT 27H
11  0105   C3 00 01                 JMP START      ;JUMP TO TARGET LABEL
12  0108                            END            ;END FOR ASSEMBLER
```

Figure 2-26 Printout of an actual program which uses labels as targets for the jump instruction.

2-11 ASSEMBLER DIRECTIVES AND PROGRAMMING EXAMPLES

The following are programming examples of different 8080 and Z80 instructions. These examples were written with an editor, and then an ASSEMBLER program generated the object code. As you go through these examples you will note the use of instructions that are in the OP-CODE field but have not been discussed as 8080/85 or Z80 instructions. An example of this is the ORG. This type of instruction is called an *assembler directive.*

Assembler directives are instructions meant for the assembler program only, they will not be assembled as instructions. There are several directives for an assembler. You will be introduced to them as the need arises in this text. At the present time we will discuss only the ORG directive.

Recall from an earlier discussion that when you begin to assemble a program, you must choose a starting address in memory. Once this address is chosen all the instructions will fall into place in the correct memory locations. When using an assembler program to generate the object code, you must inform the assembler what address you wish to use as the starting address.

Once this address has been selected, the assembler knows exactly where to place the object code bytes in memory. The starting address is chosen by the ORG, or *origin* directive. This directive is placed in the same field as the op-code.

Lets assume that you wish your starting address to be location 0400H in memory. To inform the assembler of this starting address you would write the ORG directive as

```
ORG 400H
```

If no ORG directive is encountered prior to a legal instruction, the assembler will assume the starting address is 0000.

With this introduction to the ORG directive, study the programming examples that follow. The examples are shown in their 8080/85 and Z80 formats.

```
                                        ; 8085 EXAMPLE 1
 1                                      ; INITIALIZE ALL REGISTERS TO 00
 2                                      ;
 3    0100                                      ORG  0100H   ;ORIGIN = 100H
 4    0100    3E 00                             MVI  A,00H   ;A REG = 00
 5    0102    47                                MOV  B,A     ;B=0
 6    0103    4F                                MOV  C,A     ;C=0
 7    0104    57                                MOV  D,A     ;D=0
 8    0105    5F                                MOV  E,A     ;E=0
 9    0106    67                                MOV  H,A     ;H=0
10    0107    6F                                MOV  L,A     ;L=0
11    0108    C3 08 01            BACK          JMP  BACK    ;ENDLESS LOOP
```

```
 1                              ; Z80 EXAMPLE 1
 2                              ; INITIALIZE ALL REGS TO 00
 3                              ;
 4     0100                            ORG 100H    ;SET ORIGIN
 5     0100      3E 00                 LD A,00H    ;A = 0
 6     0102      47                    LD B,A      ;B=0
 7     0103      4F                    LD C,A      ;C=0
 8     0104      57                    LD D,A      ;D=0
 9     0105      5F                    LD E,A      ;E=0
10     0106      67                    LD H,A      ;H=0
11     0107      6F                    LD L,A      ;L=0
12     0108      C3 08 01       BACK JP BACK       ;ENDLESS LOOP
```

```
 1                              ;8085 EXAMPLE 2
 2                              ;INITIALIZE ALL REGS = 00
 3                              ;
 4     0100                            ORG 0100H
 5     0100      01 00 00              LXI B,0000H ;BC = 0
 6     0103      11 00 00              LXI D,0000H ;DE = 0
 7     0106      21 00 00              LXI H,0000H ;HL = 0
 8     0109      C3 09 01       BACK   JMP BACK    ;ENDLESS LOOP
```

```
 1                              ;Z80 EXAMPLE 2
 2                              ;INITIALIZE ALL REGS = 00
 3                              ;
 4     0100                            ORG 100H     ;SET ORIGIN
 5     0100      01 00 00              LD BC,0000H  ;BC = 0
 6     0103      11 00 00              LD DE,0000H  ;DE = 0
 7     0106      21 00 00              LD HL,0000H  ;HL = 0
 8     0109      C3 09 01       BACK JP BACK        ;ENDLESS LOOP
```

```
 1                              ;8085 EXAMPLE 3
 2                              ;CLEAR MEMORY LOCATIONS
 3                              ;8345H, 823AH, 823BH
 4                              ;
 5     0100                            ORG 0100H
 6     0100      21 45 83              LXI H,8345H ;HL=8345H
 7     0103      06 00                 MVI B,00    ;B = 00
 8     0105      70                    MOV M,B     ;CLEAR MEM
 9     0106      21 00 00              LXI H,0000H ;HL = 0000
10     0109      22 3A 82              SHLD 823AH  ;CLEAR MEM
11     010C      C3 0C 01       BACK   JMP BACK
```

```
 1                              ;Z80 EXAMPLE 3
 2                              ;CLEAR MEMORY LOCATIONS
 3                              ;8345H, 823AH, 823BH
 4                              ;
 5     0100                            ORG 0100H
 6     0100      21 45 83              LD HL,8345H   ;HL=8345H
 7     0103      06 00                 LD B,00       ;B = 00
 8     0105      70                    LD (HL),B     ;CLEAR MEM
 9     0106      21 00 00              LD HL,0000H   ;HL = 0
10     0109      22 3A 82              LD (823AH),HL ;CLEAR MEM
11     010C      C3 0C 01       BACK JP BACK
12     010F                            END
```

```
1                                      ;8085 EXAMPLE 4
2                                      ;MOVE DATA FROM ONE MEM
3                                      ;LOCATION TO ANOTHER
4                                      ;
5   0100                               ORG 0100H
6   0100   3A 45 87                    LDA 8745H    ;DATA TO A
7   0103   32 9E 87                    STA 879EH    ;DATA TO MEM
8   0106   C3 06 01        BACK    JMP BACK
```

```
1                                      ;Z80 EXAMPLE 4
2                                      ;MOVE DATA FROM ONE MEM
3                                      ;LOCATION TO ANOTHER
4                                      ;
5   0100                               ORG 0100H
6   0100   3A 45 87                    LD A,(8745H)    ;DATA TO A
7   0103   32 9E 87                    LD (879EH),A    ;DATA TO M
8   0106   C3 06 01        BACK    JP BACK
```

```
1                               ;;;;;;;;;;;;;;;;;;;;;;;;;;;;;
2                               ;
3                               ; 8085 EXAMPLE 5
4                               ;
5                               ; MOVE 16 BITS OF DATA IN MEMORY
6                               ; TO DIFFERENT LOCATIONS
7                               ;
8                               ;;;;;;;;;;;;;;;;;;;;;;;;;;;;;
9                               ;
10  0100                               ORG 100H    ;SET ORIGIN
11  0100   2A F0 86                    LHLD 86F0H  ;GET DATA FROM MEM
12  0103   22 00 82                    SHLD 8200H  ;STORE DATA TO MEM
13  0106   C3 06 01        BACK    JMP BACK    ;ENDLESS LOOP
14  0109                               END
```

```
1                               ;;;;;;;;;;;;;;;;;;;;;;;;;;;;;
2                               ;
3                               ; Z80 EXAMPLE 5
4                               ;
5                               ; MOVE 16 BITS OF DATA IN MEMORY
6                               ; TO DIFFERENT LOCATIONS
7                               ;
8                               ;;;;;;;;;;;;;;;;;;;;;;;;;;;;;
9                               ;
10  0100                               ORG 100H        ;SET ORIGIN
11  0100   2A F0 86                    LD HL,(86F0H)   ;GET DATA FROM MEM
12  0103   22 00 82                    LD (8200H),HL   ;STORE DATA TO MEM
13  0106   C3 06 01        BACK    JP BACK         ;ENDLESS LOOP
14  0109                               END
```

```
 1                              ;;;;;;;;;;;;;;;;;;;;;;;;;;;;;
 2                              ;
 3                              ; 8085 EXAMPLE 6
 4                              ;
 5                              ; SAVE ALL REGISTERS TO MEMORY
 6                              ;
 7                              ;;;;;;;;;;;;;;;;;;;;;;;;;;;;;
 8                              ;
 9                              ; REGISTER LOCATIONS IN MEMORY
10                              ;
11                              ; 8400H = A
12                              ; 8401H = B
13                              ; 8402H = C
14                              ; 8403H = D
15                              ; 8404H = E
16                              ; 8405H = H
17                              ; 8406H = L
18                              ;
19  0100                               ORG 100H
20  0100    32 00 84                   STA 8400H   ;SAVE A REG
21  0103    7D                         MOV A,L
22  0104    32 06 84                   STA 8406H   ;SAVE L REG
23  0107    7C                         MOV A,H
24  0108    32 05 84                   STA 8405H   ;SAVE H REG
25  010B    EB                         XCHG        ;DE TO HL
26  010C    22 03 84                   SHLD 8403H  ;SAVE DE REGS
27  010F    60                         MOV H,B
28  0110    69                         MOV L,C     ;BC TO HL
29  0111    22 01 84                   SHLD 8401H  ;SAVE BC REGS
30  0114    C3 14 01           BACK    JMP BACK
31  0117                               END
```

```
 1                              ;;;;;;;;;;;;;;;;;;;;;;;;;;;;;
 2                              ;
 3                              ; Z80 EXAMPLE 6
 4                              ;
 5                              ; SAVE ALL REGISTERS TO MEMORY
 6                              ;
 7                              ;;;;;;;;;;;;;;;;;;;;;;;;;;;;;
 8                              ;
 9                              ; REGISTER LOCATIONS IN MEMORY
10                              ;
11                              ; 8400H = A
12                              ; 8401H = B
13                              ; 8402H = C
14                              ; 8403H = D
15                              ; 8404H = E
16                              ; 8405H = H
17                              ; 8406H = L
18                              ;
19  0100                               ORG 100H
20  0100    32 00 84                   LD (8400H),A   ;SAVE A REG
21  0103    7D                         LD A,L
22  0104    32 06 84                   LD (8406H),A   ;SAVE L REG
23  0107    7C                         LD A,H
24  0108    32 05 84                   LD (8405H),A   ;SAVE H REG
25  010B    EB                         EX DE,HL       ;DE TO HL
26  010C    22 03 84                   LD (8403H),HL  ;SAVE DE REGS
27  010F    60                         LD H,B
28  0110    69                         LD L,C         ;BC TO HL
29  0111    22 01 84                   LD (8401H),HL  ;SAVE BC REGS
30  0114    C3 14 01           BACK    JP BACK
31  0117                               END
```

2-12 CHAPTER SUMMARY

This chapter was the first step in learning how to program the 8080, 8085, and Z80 microprocessors. You were first given a definition of programming, and then flowcharting was introduced. Proper flowcharting will allow others easily to understand the intent of your program. After flowcharting you were shown a programming model for the microprocessor. This model is used as a memory aid to determine which internal registers you can access via software instructions.

The topics of source and object codes were covered next. You write your programs in source code and then convert to object code. The process of conversion from source to object code is called assembly. Several examples were given that allowed you to practice converting from source to object code.

After learning how to assemble a program you studied several microprocessor instructions. These were 1-, 2-, and 3-byte instructions. The last instruction shown was the jump. With the instructions given in this chapter you can write many elementary microprocessor programs. Several examples of such programs were given at the end of the chapter.

Chapter 3 will build on the information given and continue the discussion on progrmming.

REVIEW PROBLEMS

1. Write a definition of the term programming.
2. T/F Programs written for the 8080 can run with the 8085 and Z80 microprocessors with no modifications.
3. Draw a flowchart for the following operations:
   Read the flowrate setting on switches; then compute and set the flowrate valve. Next read the actual flowrate, compute the delta flowrate = flowrate on switches-actual flowrate, and finally adjust the actual flowrate such that the absolute value of the delta flowrate is less than or equal to .01. Continually loop on this program.
4. Draw the programming model for the 8080 and 8085.
5. Draw the programming model for the Z80.
6. List the 8-bit registers in the 8080 and 8085.
7. List the 8-bit register in the Z80.
8. What is the F register for the 8080?
9. List the 16-bit registers in the 8080 and 8085.
10. Show the op-code, source operand, and destination operand in the MOV C,D instruction.
11. What does mnemonic refer to in a microprocessor assembly language instruction?
12. The _____ operand is never changed when executing an instruction, whereas the _____ operand may be changed.

13. Generate the object code for the following program.

```
MVI     A,32H
MOV     B,A
MOV     D,C
```

14. Assume a starting address of 8200H. Show the object code and memory addresses for Problem 13.

15. Assume a starting address of 5301H and assemble the following program.

```
MVI     A,53H
ADD     B
SUB     C
HLT
```

16. Assume a starting address of 3506H and assemble the following program.

```
MVI     A,22H
ADD     B
SUB     C
LXI     H,0055H
MOV     M,A
HLT
```

17. Write two different programs that write 00 to memory location 3576H, 3577H, 3578H.

18. What is the difference between the LDA and STA instruction?

19. Explain the difference between the LDAX and MOV M,A instructions.

20. Show how the XCHG instruction works.

21. There are _____ different input and output ports in the 8080, 8085, and Z80 systems.

22. Generate the object code for the instruction OUT 35H.

23. Generate the object code for the instruction IN 3AH.

24. Generate the object code for the following program assuming a starting address of 5000H.

```
START       MVI A,22H
            ADD A
            SUB C
            IN 29H
            OUT 43H
            JMP START
```

3

PROGRAMMING FOR THE Z80, 8080, AND 8085 (PART 2)

This chapter continues with additional programming examples, instruction types, and special instructions for the three microprocessors. The previous chapter concentrated on the basics of programming. Now that you are familiar with these basics, many new concepts can be added that will enhance the application programs you can write for the microprocessor. Let us start with a presentation of the different addressing modes available for the instructions.

3-1 ADDRESSING MODES

The term *addressing mode* refers to the manner in which data may be read or written from one operand to the other. These operands may be internal registers or a memory address. Addressing modes covered in this section will apply to almost any microprocessor used in industry today, including 16- and 32-bit microprocessors.

To explain how the different addressing modes operate, the MOV instruction will be used. We use this instruction because you are already familiar with the function of the MOV from Chapter 2. With your understanding of the MOV instruction, you can concentrate on the new information given.

There are five different addressing modes for the Z80, and four of these are available for the 8085 and 8080. The five modes are

1. Register
2. Direct
3. Immediate
4. Register indirect
5. Indexed (Z80 only)

Let us discuss each mode one at a time. You will see that you have already used these modes of addressing without even knowing it.

Register

The *register* mode of addressing is used when you wish to move data from a source operand that is an internal register to a destination operand, which is another internal register. Examples of this type of instruction are:

```
        8080,8085       Z80

        MOV A,B     LD  A,B
        MOV C,D     LD  C,D
```

Both operands are internal registers. Notice that the microprocessor does not have to access memory for either of the operands.

Direct

Direct addressing mode is when you specify the address of memory directly as part of the instruction. These are 3-byte instructions. Examples of this type of instruction are:

```
        8080,8085       Z80

        LDA 2345H   LD  A,(2345H)
        STA 4567H   LD  (4567H),A
```

The LDA instruction will read the memory address specified in the instruction (2345H) and then store this data in the A register. The STA instruction will write the data from the A register to the memory address specified in the instruction (4567H).

In both of these instructions, the address of the operand was explicitly given as part of the instruction.

Immediate

Immediate addressing is where the operand is specified immediately following the op-code. These are 2- or 3-byte instructions. Examples of immediate addressing are:

```
      8080,8085           Z80

    MVI B,23H          LD B,23H
    MVI H,0FH          LD H,0FH
    LXI D,49A3H        LD DE,49A3H
```

If you form the object code for each of these instructions, the data to be moved into the register follows the op-code immediately. For the MVI B, there is only 1 byte of immediate data, whereas the LXI D requires two immediate bytes for the E and D registers.

Register Indirect

Register indirect addressing is a mode that uses an internal register (16 bit) or register pair as the source of the memory address. An example of this type of instruction is:

```
      8080,8085           Z80

    MOV M,A           LD (HL),A
```

This is a 1-byte instruction that uses the contents of the HL register pair as the memory address for storing the data in the A register. In forming the object code for the instruction, take note of the fact that it is only 1 byte. The address is *not* directly written as part of the instruction. See Figure 3-1.

Another example of the indirect addressing mode is:

```
      8080,8085           Z80

    LDAX D            LD A,(DE)
    STAX B            LD (BC),A
```

In the first instruction (LDAX) the A register is loaded with the data from the memory address contained in the DE register pair. The STAX instruction will store the data in the A register into the memory location addressed by the data in the BC register pair.

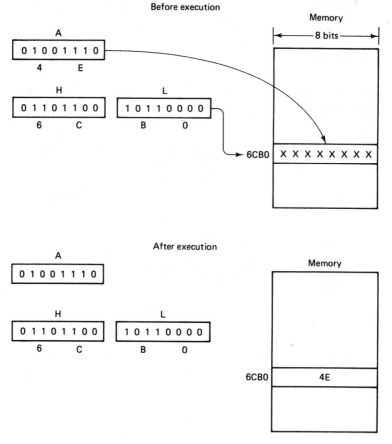

Figure 3-1   With the correct values in the HL registers, the MOV M, A instruction
will write indirectly to memory. After the instruction is executed, the location 6CB0
is loaded with the byte of data in the A register.

Indexed

The preceding addressing modes are common for the 8080, 8085, and Z80
microprocessors. A fifth addressing mode will be presented, which is available
only in the Z80 microprocessor. This is due to the different internal architecture
of the Z80.

There are two 16-bit registers in the Z80 that are not in the 8080 or 8085;
these are called the index (IX and IY) registers. We have not discussed how
these registers are used or manipulated. You are asked to accept, for now, that
the IX and IY registers may be loaded to any value you choose. With the
registers properly loaded, they can be used to address memory indirectly. An
example is:

```
LD      A,(IX+1)
LD      (IY+3),B
```

These instructions will use the value of the IX or IY registers and add a constant to this value. This new value will be used as the memory address for reading or writing data. For example, assume IX = 25CDH and you use the instruction

```
LD  (IX+2),B
```

The Z80 microprocessor would write the contents of the B register into the memory location 25CD+2 = 25CF. See Figure 3-2. You will be formally introduced to this instruction later in the chapter. It is given here to complete the types of addressing modes available for the microprocessors.

You will be making use of these five different types of addressing modes whenever you write any assembly language program. Each instruction will access data using one of the modes presented here.

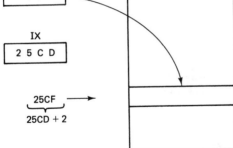

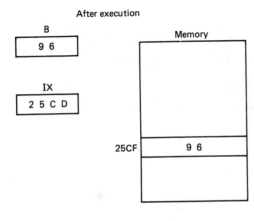

Figure 3-2 The index register in the Z80 is used for indirectly writing data to memory.

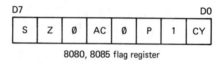

Figure 3-3   Programming model of the 8080 and 8085 flag register.

3-2 FLAGS

In the programming model for the 8080 8085, and Z80 microprocessors given in Chapter 2, an internal register called the flag register was shown. In this section we present this register in more detail so you will be able to use it in your assembly language programs.

The flag register for the 8080 and 8085 is shown in Figure 3-3. All the flags shown in this figure exist for the Z80 as well. However, the Z80 has some additional flags that will be explained in a later example. Referring to Figure 3-3, the flags are:

```
CY     carry flag
P      parity flag
AC     auxiliary carry flag
Z      zero flag
S      sign flag
```

You should be aware that these flags are actually bits in a register that are set to a logical 1 = T or a logical 0 = F. The updating, setting, or resetting of the flag occurs on any instruction which affects the flags. This implies that some instructions do not alter the flags. Data books on a particular microprocessor will indicate if an instruction affects the flags and which flags it affects. With this introduction, let's examine exactly what each flag indicates.

CY: Carry Flag

The carry flag is set true whenever the microprocessor generates an internal carry into the carry bit. The carry bit may be thought of as an extra ninth bit which is used for many arithmetic and logical instructions. The carry bit is shown in pictorial form in Figure 3-4.

As an example of how the carry bit may be set, consider the problem of adding two 8-bit numbers. If the sum of the numbers is greater than 255 decimal, then there is a carry generated.

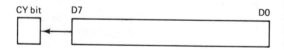

Figure 3-4   The carry bit may be thought of as a ninth bit of the accumulator that reflects results of different instructions.

$$
\begin{array}{r}
234 \\
+ \quad 22 \\
\hline
256
\end{array}
\quad = \quad 1 \quad 00000001
$$

carry

Another example of generating a carry is when you are subtracting two 8-bit numbers and the difference is less than 0. In this case a borrow will be generated, which sets the carry bit true.

P: Parity Flag

This bit is set true whenever even parity exists in the result. Even parity is when the total number of 1s in the result is even. Odd parity is when the total number of 1s in the result is odd. For example 00011010 is odd parity and 00101101 is even parity. In the following addition example, the parity flag would be set:

$$
\begin{array}{r}
01010000 \\
+ \quad 00010000 \\
\hline
01100000
\end{array}
\quad = \quad \text{even parity}
$$

AC: Auxiliary Carry

The zero flag is set true whenever there is a carry from bit 3 to bit 4 in the result. The main use of this flag is for the DAA, or decimal adjust accumulator, instruction. The decimal adjust deals with 4-bit quantities, and a carry from bit 3 to bit 4 is important in these types of numbers.

Z: Zero Flag

The zero flag is set true whenever the result of an operation is zero. Whenever the result of an operation is not zero, the zero flag is set false. As an example of setting the zero flag true, consider this subtraction problem.

$$
\begin{array}{r}
01110000 \quad = \quad 112 \text{ decimal} \\
- \quad 01110000 \quad = \quad 112 \text{ decimal} \\
\hline
00000000 \quad = \quad 0
\end{array}
$$

After the subtraction occurs, the zero flag will be set true. If the problem were $112 - 111$, the result would be 1 and the zero flag would be set false = 0.

S: Sign Flag

The sign flag is set true if the MSB of the result is a 1. It is called the sign flag because in signed arithmetic, the MSB is used to determine the sign of a number.

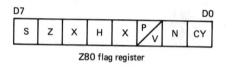

Figure 3-5 Programming model for the Z80 flag register.

Z80 Flags

The flag register for the Z80 is shown in Figure 3-5. You see in this figure all the previous flags mentioned for the 8080 and 8085 plus two additional flags. These new flags are labeled P/V and N. The H flag (half-carry) is the same as the AC (auxiliary carry) flag in the 8080 and 8085 flag register.

P/V Flag

The P/V (parity/overflow) flag in the Z80 has a variety of functions associated with it. One function is parity. This type of function was described for the 8080 and 8085. For instructions that affect the parity flag, this bit of the Z80 flag register is set or cleared in exactly the same way as described for the 8080 and 8085.

A second function of the P/V flag is to indicate an overflow operation. An overflow is an inadvertant sign change, which may occur as the result of adding or subracting signed numbers. Let's examine these two situations in which an overflow occurs. With 8-bit signed numbers the largest negative and positive numbers are −128 (10000000) and +127 (01111111), respectively. The following two examples illustrate how the overflow bit may be set.

Example 3-1 (Adding two positive numbers)

$$
\begin{array}{rll}
  01111110 & +126 & \\
+\ 00000010 & +2 & \\
\hline
  10000000 & -128 & \text{the sign bit changed}
\end{array}
$$

The answer is invalid because there was an overflow.

Example 3-2 (Adding two negative numbers)

$$
\begin{array}{rl}
  10000010 & (-126) \\
+10001000 & (-120) \\
\hline
  00001010 &
\end{array}
$$

1
carry

The result should be negative, but because of the overflow the sign of the result changed.

In general, the overflow bit will be set in the following cases:

  a. If there is a carry into bit 7 from bit 6 *and* there is no carry generated from bit 7 to the carry bit.
  b. If there is a carry generated into the carry bit from bit 7 *and* no carry from bit 6 into bit 7.

Whenever an overflow condition exists, the P/V bit is set to a logical 1 = true. It is up to the programmer to test the result of this bit after performing an addition or subtraction problem.

Two More Functions of the P/V Bit

The P/V bit has two more uses in the Z80 besides parity and overflow. These uses are in conjunction with some special instructions. We will introduce you to this information now, and a more detailed explanation will be given in the section of the chapter that deals with the use of these instructions.

For block transfer instructions (LDD, LDDR, LDI, LDIR) and search instructions (CPD, CPDR, CPI, CPIR), the P/V flag will indicate when the counter register is 0. With block instructions that decrement, P/V = 0 if BC register = 0. With block instructions that increment, P/V flag will = 0 if BC register $-1 = 0$ prior to execution of the instruction. In this case the register will equal 0 after the instruction is executed. Knowing the state of this flag will indicate when the block loop is complete. More information is given on this topic in Chapter 4.

The final use of the P/V flag is by the two instructions LD A,I and LD A,R. In one case the interrupt register I is loaded into the A register, and in the other the refresh register R is loaded into the A register. In both of these instructions the P/V flag is set equal to the contents of the interrupt enable flip-flop (IFF2). This flip-flop is discussed further in the chapter that covers interrupts.

N bit for the Z80.   This flag is normally used by the Z80 processor during BCD (binary coded decimal, 4-bit) operations. The DAA instruction will adjust the result of a decimal operation. However, the adjustment is different for a subtract operation as compared to an add operation. This means the DAA instruction will operate differently depending on the state of the N bit. The N bit is set to a logical 1 after a subtraction and cleared to a logical 0 after an addition operation.

3-3 ARITHMETIC INSTRUCTIONS

In the previous section, the microprocessor flags were presented. We next discuss many new instructions, which, when executed, will affect the flags. After examining these instructions and understanding how the flags may be set or cleared, you will learn how to make decisions based on the present state of the system flags using *conditional* instructions. The group of instructions presented at this time are called the *arithmetic group of instructions* and are used to perform different arithmetic operations with the microprocessor. During this discussion keep in mind that all the addressing modes discussed previously apply.

It is interesting to note that the most complicated arithmetic operation that can be performed with the microprocessors is addition or subtraction. (Subtraction is really 2s complement addition.) All other operations, such as multiplication and division, square root, and logarithms, must be accomplished by programs that add or subtract.

With such a limited arithmetic set of instructions, a greater burden is placed on the programmer who must make the device perform higher math functions. However, the postive side to this is that learning and understanding how to use the arithmetic group of instructions for the microprocessor is a fairly easy task. Let's start with the instructions that perform addition.

In execution of the addition subgroup of instructions, *all* the following flags are affected: CARRY, ZERO, SIGN, PARITY, and AUXILIARY CARRY. We make this assumption, and the instructions that vary from this will be noted.

Add Register to Accumlator

```
8080/8085      Z80

ADD r        ADD A,r
```

The add register instruction adds the contents of the register specified to the contents of the accumulator. The result is stored in the accumulator. Examples of this mnemonic are:

```
8080/8085      Z80

   ADD B      ADD A,B
   ADD D      ADD A,D
```

Add Memory to ACC

```
ADD M      ADD A,(HL)
```

ADD M adds the contents of the memory location addressed by the HL register pair to the contents of the ACC. The result is stored in the ACC. The addressing for the source operand is exactly like the addressing that was presented in Chapter 2 for the MOV M instruction.

ADI Data (Add the immediate data)

ADI data is a 2-byte instruction that adds the immediate data specified in the instruction to the contents of the ACC. The result is stored in the ACC.

```
        8080/8085          Z80

        ADI 23H       ADD A,23H
```

The three instructions

```
        8080/8085                    Z80

    ADC r                    ADC A,r
    ADC M                    ADC A,(HL)
    ACI immediate data       ADC A,immediate data
```

operate exactly the same way as the ADD instructions, with one important exception. The contents of the carry flag are added to the result of the addition. As an example of how the ADC instruction may be used, consider the following:

Example 3-3

You wish to add two 16-bit numbers. The microprocessors are 8-bit machines, so one way to accomplish this task is using two consecutive adds, one without carry and the following with carry. Assume that the two 16-bit numbers are stored in memory at these locations:

```
        8400H       LSByte of #1
        8401H       MSByte of #1
        8402H       LSByte of #2
        8403H       MSByte of #2
        8404H       LSByte of result
        8405H       MSByte of result
```

To put some numbers with this example, suppose you are adding 01F3H and 03A6H = 499 + 934 = 1433. With these numbers the data in the memory locations specified would look like this:

```
        8400H       F3
        8401H       01
        8402H       A6
        8403H       03
        8404H       XX
        8405H       XX    where XX = don't care
```

The following program would add the two 16-bit numbers.

COMMENTS

```
LXI H,8400H        address of first number LSByte
LXI D,8402H        address of second number LSByte
MOV A,M            get data of first number
XCHG               swap addresses
ADD M              add two numbers: result in A
                   reg,carry bit
STA 8404H          store result of LSByte
INX H              address next byte
MOV A,M            get MSByte at address 8403H
XCHG               swap addresses
INX H              address = 8401H
ADC M              add numbers with carry bit
STA 8405H          store MSByte of result
```

Notice that two different add instructions were used. The first, ADD M, simply added two bytes and generated a carry if needed. Following the ADD M instruction, the STA, INX, MOV, XCHG instructions did *not* affect the flags. Therefore, the carry bit was still equal to the result set by the ADD M instruction. It is important to know which instructions affect the flags and which do not.

The ADC M instruction added the most significant bytes plus carry from the least significant byte. If one of the preceeding instructions affects the carry flag, then the ADC instruction may or may not have had the correct result.

After this program is executed, the memory locations would appear as:

```
8400H        F3
8401H        01
8402H        A6
8403H        03
8404H        99
8405H        05
```

Now that you have seen how the add instructions operate, it is an easy matter to understand how the subtract instructions work. These instructions are formatted in exactly the same way as the add. The six subtract instructions are:

8080/8085	Z80
SUB r	SUB A,r
SUB M	SUB A,(HL)
SUI immediate data	SUB A,immediate data
SBB r subtract with borrow	SBC A,r
SBB M	SBC A,(HL)
SBI immediate data	SBC A,immediate data

The subtract instruction will subtract (8 bits) the source operand specified from the contents of the A register. The result is stored in the A register. Subtract with borrow performs the subtraction as indicated and then subtracts the contents of the carry bit from the result. The final result is stored in the A register. Subtract with borrow can be used when subtracting numbers that are greater than 8 bits.

Example 3-4

As an example, let's consider the opposite of the 16-bit addition problem given earlier, subtracting 16-bit numbers. One difference in the problem is that you must be careful to keep the memory addresses consistent—that is, you must keep track of which number is being subtracted from the other. It did not matter when performing addition because addition is commutative, that is, $A + B = B + A$. The memory locations for the subtraction problem are:

```
8400H      LSByte of #1
8401H      MSByte of #1            V
8402H      LSByte of #2
8403H      MSByte of #2            T
8404H      LSByte of result
8405H      MSByte of result
```

The subtraction problem will be $V - T$, where $V = 0100H$, and $T = 00FFH$. This gives a subtraction of $0100H - 00FFH = 0001$. The program for subtracting these values is:

```
                        COMMENT
LXI H,8400H     address of V LSByte
LXI D,8402H     address of T LSByte
MOV A,M         get data of V
XCHG            swap addresses
SUB M           subtract V-T LSByte
STA 8404H       store result of LSByte
XCHG            swap address
INX H           address next byte
MOV A,M         get MSByte at address 8401H
XCHG            swap address
INX H           address = 8403H
SBB M           sub numbers with borrow bit
STA 8405H       store MSByte of result
```

After this program is executed, the memory locations are:

```
8400H      00
8401H      01      V
8402H      FF
8403H      00      T
8404H      01      V-T LSByte
8405H      00      V-T MSByte
```

Now that you have seen how the 8-bit add and subtract instructions operate, here is a 16-bit add instruction, DAD.

Double Add a Register Pair

```
     8080/8085          Z80

     DAD rp       ADD HL,rp
```

The double add instruction performs a 16-bit add of the register pair specified (BC, DE, HL, SP) to the HL register pair. The result is stored in the HL register pair. Mnemonics for this instruction are:

```
     8080/8085          Z80

     DAD B        ADD HL,BC
     DAD D        ADD HL,DE
     DAD H        ADD HL,HL
     DAD SP       ADD HL,SP
```

Example 3-5

Using this new instruction, let's see how the 16-bit addition problem given earlier may be solved. The memory addresses and data remain the same.

```
     8400H    F3
     8401H    01
     8402H    A6
     8403H    03
     8404H    XX
     8405H    XX      where XX = don't care
```

One possible solution is

```
     LHLD  8400H    HL = 01F3
     XCHG           DE = 01F3
     LHLD  8402H    HL = 03A6
     DAD   D        HL = HL+DE
     SHLD  8404H    write result to mem
```

As can be seen from this example, the program for adding two 16-bit numbers is much simpler in concept when the DAD instruction is used. Compare this example with the 16-bit add example given previously.

Increment and Decrement Instructions

The last instructions that fall into the arithmetic group are the ones used for *incrementing* and *decrementing* operands. Incrementing and decrementing are the functions of adding one to or subtracting one from an operand. There are increment and decrement instructions for 8-bit and 16-bit operands.

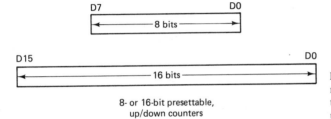

Figure 3-6 For the increment and decrement instructions you may think of the registers as 8- or 16-bit presettable, up/down counters.

8- or 16-bit presettable, up/down counters

To illustrate how the increment and decrement instructions operate, a simple model for the operand is shown in Figure 3-6. For 8-bit operands, use an 8-bit counter as the model, and for 16-bit operands, use a 16-bit counter. These counters are up/down counters. The only special cases for incrementing and decrementing are when the counter is at its maximum—all 1s—or at its minimum—all 0s.

Let's examine the case of the counter being at its maximum count. Figure 3-7a shows the counter, 8 or 16 bits, equal to all 1s. On the next increment instruction, the counter will reset to all 0s, as shown in Figure 3-7b.

The opposite case occurs when the counter is at its minimum count and another decrement instruction is executed. This is shown in Figure 3-8. Figure 3-8a shows the counter prior to the decrement and Figure 3-8b shows the resulting value of the register. Note the carry bit is *not* affected with this instruction. You see in Figure 3-8b that the counter goes to all 1s. This is exactly the type of operation one would expect from a typical hardware up/down counter.

The following mnemonics are used for the increment instructions. Addressing modes are exactly the same as discussed for the previous instructions.

	8080/8085		Z80
INR r	increment 8 bit register		INC r
INR A	example		INC A
INR M	increment memory (address=HL)		INC (HL)

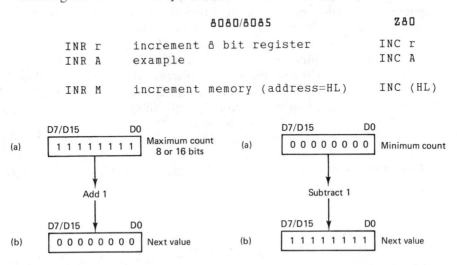

Figure 3-7 If the counter is at maximum count 11111111, the next increment will set the counter to 00000000.

Figure 3-8 If the counter is at its minimum count 00000000, the next decrement will set the count to 11111111.

The 16-bit increments are:

```
INX  B        increment register pair        INC BC
INX  D                                        INC DE
INX  H                                        INC HL
INX  SP                                       INC SP
```

The decrement instructions are as follows:

```
DCR  r     decrement 8-bit register          DEC r
DCR  M     decrement memory (address=HL)      DEC (HL)
```

The 16-bit increments are:

```
DCX  B     decrement register pair           DEC BC
DCX  D                                        DEC DE
DCX  H                                        DEC HL
DCX  SP                                       DEC SP
```

Note: The 16-bit increment and decrement instructions do not affect the flags.

3-4 CONDITIONAL BRANCHING

In Sections 3-2 and 3-3 we showed what the system flags are and how they are affected. This section shows how the flags can be used to make decisions in a program. Recall from previous discussions that instructions are executed from those op-codes residing at the low address toward the op-codes residing at a higher memory address. The only way to make the microprocessor execute in a different order is to use the JMP, or jump, instruction.

The JMP instruction requires 3 bytes, 2 of which are the memory address or target address to jump to when the instruction is executed. One drawback to the JMP instruction is that the jump is always taken. That is, whenever the instruction is executed, the next instruction is the one located at the target address. This is called an *unconditional jump*. No matter what the condition of the microprocessor flags, the jump is always taken.

Another type of jump instruction is the *conditional jump*. This type of jump first tests the condition of the system flags and, based on these logical values, the jump may or may not be taken. See Figure 3-9. To further

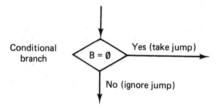

Figure 3-9 When using the jump on condition instructions, the jump may or may not be taken. This is shown by the decision block of the flowchart.

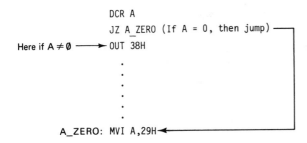

Figure 3-10 Program showing how the jump on zero instruction will operate.

illustrate this, let's examine one of the conditional jump instructions, *jump if zero* (JZ).

The mnemonic of the jump if zero is JZ. The object code format for the JZ instruction is formed exactly like the JMP instruction. There are 3 bytes, with the first being the op-code (CA) and the remaining 2 bytes being the target address for the jump.

When the JZ instruction is executed, the microprocessor examines the logical state of the zero flag. If the flag is true, the jump is taken. If the flag is not true, the jump is *not* taken, and the next instruction after the JZ is executed as shown in Figure 3-10. The following is a sample program where the JZ instruction is used.

Example 3-6

```
         MVI A, OFFH    load ACC = FF
BACK     OUT 02         output acc to port 02
         DCR A          a=a-1
         JZ NEXT        a=0! yes then jump to next
         JMP BACK       jump to label back
NEXT     MVI A, OFFH    load acc=FF
         OUT 03         output acc to port 03
         HLT            halt the program
```

This will output the value of the ACC to port 02 until the A register is equal to zero. The DCR A instruction will affect the zero flag each time it is executed. The JZ instruction will test the value of the zero flag. When it is true (A = 0), the jump is taken to the target label NEXT. If the zero flag is false, the JMP instruction is executed, and the program is directed back to the OUT 02 instruction.

Note that the conditional jumps make a decision on the flags at the time the instruction is executed. Therefore, you must ensure that the test on the flags is done for the conditions you want. For example, in the program just given, the JZ will test the results of the DCR instruction. However, if another instruction were placed before the JZ instruction, say a INR C, the JZ would make a decision based on the value of the C register. It is the programmer's responsibility to ensure the conditions are correct for the test wanted.

The following assembly of the program completes the example given. This shows you exactly how the conditional jump instructions are assembled into object code.

```
add    data

8200   3E          MVI A, 0FFH      load ACC = FF
8201   FF
8202   D3     BACK OUT 02          output acc to port 02
8203   02
8204   3D          DCR A            a=a-1
8205   CA          JZ NEXT          a=0! yes then jump to NEXT
8206   0B
8207   82
8208   C3     JMP BACK             jump to label back
8209   02
820A   82
820B   3E     NEXT MVI A, 0FFH     load acc=FF
820C   FF
820D   D3          OUT 03           output acc to port 03
820E   03
820F   76          HLT              halt the program
```

Notice that the JZ and the JMP instruction are assembled in exactly the same way.

Example 3-6 showed only the testing of the zero flag for true. However, the zero flag may be tested for false using the JNZ (*jump if not zero*) instruction. The following list shows all the conditional jump instruction available.

```
JZ     jump if zero
JNZ    jump if not zero

JC     jump if carry true
JNC    jump if no carry (false)

JPE    jump if parity even (true)
JPO    jump if parity odd (false)

JM     jump if minus (sign bit true)
JP     jump if plus (sign bit false)
```

As you can see from this list, both conditions for the particular bit may be tested by the microprocessor. Which sense of the test you use will depend on the application. Remember that all these conditional jumps are 3-byte instructions with the second and third byte being the address for the jump. The first byte is the op-code, which can be obtained from the op-code chart under the *jump* heading.

Z80 Representations for Conditional Jumps

The mnemonics for the conditional jump instructions for the Z80 are slightly different from the 8080 and 8085, but the object code form is exactly the same. The following is a list of the mnemonics for the Z80.

```
JP  Z, address         jump if zero true
JP  NZ, address        jump if zero false

JP  C, address         jump if carry true
JP  NC, address        jump if carry false

JP  PE, address        jump if parity even
JP  PO, address        jump if parity odd

JP  M, address         jump if minus
JP  P, address         jump if plus
```

3-5 LOGICAL GROUP OF INSTRUCTIONS

In this section we present the group of instructions that perform logical operations on the operands. The first subgroup is the instructions that perform the following functions;

    AND
    OR
    Exclusive OR

All these types of instructions employ the three addressing modes *register, register indirect* (HL), and *immediate*. Also, when these instructions are executed you may think of the operation as occurring on a bit-by-bit basis. For example, if the AND instruction were executed between the A and B registers, these two registers would appear as shown in Figure 3-11. When the instruction is executed, each bit of the A register is ANDed with its corresponding bit of the B register. The result of the ANDing is stored in the destination registers corresponding bit. This is shown in Figure 3-12.

The example given for the AND instruction illustrates exactly what occurs for OR and XOR operations on a bit-by-bit basis. With this introduction, let's examine the instructions for ANDing. Mnemonics for the AND instruction are:

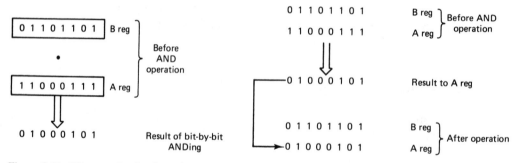

Figure 3-11 Diagram showing how the AND instruction operates.

Figure 3-12 Diagram showing the A and B registers before and after the AND instruction is executed.

```
        8080/8085                                      Z80

ANA  r (AND A reg with reg r)                      AND  r
ANA  M (AND A reg with Memory (HL))                AND  (HL)
ANI  data (AND A reg with immediate date)          AND  data
```

The destination register for the AND instructions is always the A register. Further, all flags are affected by the AND instruction.

Since the ANDing operation is performed on a bit-by-bit basis, it is easy to understand what the result will be by examining a simple truth table. Such a truth table is shown in Figure 3-13. You see in Figure 3-13 that the result will be true (1) only if both the inputs A and B are true (1).

The AND operation is used often in microprocessor application programs where you wish to set specific bits of an operand to the logical 0 state. As an example, suppose you had a system that input data from a port. When data is input from a port, 8 bits are input to the A register in a parallel fashion. In this example system, only the first 2 bits of the port are actually used. See Figure 3-14.

Therefore, the other input bits, D2–D7, may be 0s or 1s. Of course, you could hardwire these bits to 0s or 1s, but this requires an extra hardware step. You could force these unused bits to be logical 0s by simply ANDing them with logical 0s. The following is a small program that would do this:

```
IN 45        input data from port 45
ANI 03H      AND with 0000 0011
```

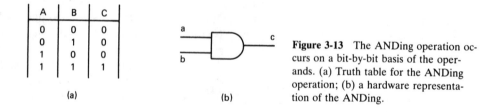

Figure 3-13 The ANDing operation occurs on a bit-by-bit basis of the operands. (a) Truth table for the ANDing operation; (b) a hardware representation of the ANDing.

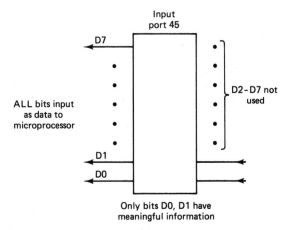

Figure 3-14 The AND instruction may be used to MASK off unused bits of the data bus.

The result of this ANDing would be that bits D2–D7 will equal 0, and bits D0–D1 will equal the logical value read from the input port. This technique is called *masking off bits*.

ORing

The three instructions for ORing data are:

8080/8085	Z80
ORA r (OR A register with register r)	OR r
ORA M (OR A register with memory (HL))	OR (HL)
ORI data (OR A register with immediate data)	OR data

The destination register for the OR instruction is always the A register. All flags are affected by this instruction.

Oring is accomplished on a bit-by-bit basis like the AND instruction, as shown in Figure 3-15. The OR instruction is often used in microprocessor programs to set desired bits equal to a logical 1.

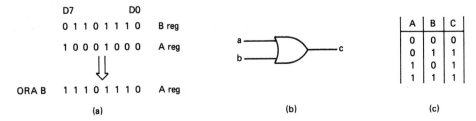

Figure 3-15   (a) Diagram showing how the OR instruction operates. (b) schematic symbol for OR gate; (c) truth table for an OR function.

Example 3-7

A copy of the byte that is output to port 34 is saved in memory. Let's say that the microprocessor wishes to set only the bit that turns on the heater (bit 5); all other bits will remain constant. Here is a small program to accomplish that. It is assumed in this program that a copy of the output byte has been saved in memory location labeled C_control.

```
LDA  C_control      get data presently at port
ORI  20H            set bit 5 0010 0000=1
OUT  34H            output to control port
STA  C_control      update copy in memory
```

XORing (Exclusive ORing)

Exclusive ORing of data occurs on a bit-by-bit basis between the data in the A register and the source operand specified by the instruction. Figure 3-16 shows the truth table for exclusive ORing. The mnemonics for this instruction are:

8080/8085		Z80
XRA r (XOR A reg with register r)		XOR r
XRA M (XOR A reg with memory (HL))		XOR (HL)
XRI data (XOR A reg with immediate data)		XOR data

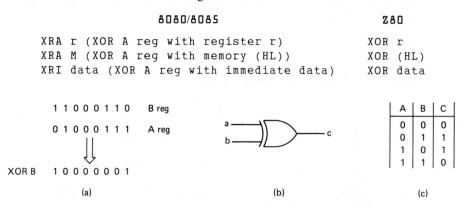

Figure 3-16   (a) Diagram showing how the XOR instruction operates; (b) schematic symbol for an exclusive OR gate; (c) truth table for an XOR function.

Example 3-8

One use of the XOR instruction is during a test program to indicate failed bits. To illustrate this, the following program is given. This program will write a 1010 1010 pattern to memory location and then read the data back. The program will then test the data and write the failed bits to output port 41H. A logical 1 represents a failed bit.

```
MVI  A, AAH         data 1010 1010 to A reg
MOV  B,A            save the data for later
LDA  TEST_LOC       read the test data
XRA  B              test data, failed bits = 1
OUT  41H
HLT
```

The Z80 mnemonics are:

```
LD A,AAH          data 1010 1010 to A reg
LD B,A            save the data for later
LD A,(TEST_LOC)   read the test data
XOR B             test data
OUT (41H),A
HALT
```

When the program executes the XRA (XOR) instruction, the exclusive OR operation occurs between the A and B registers. The B register has the known good data, and the A register has the unknown data from memory. Figure 3-16 shows the exclusive OR operation. Notice in Figure 3-16 that if any bits are different between the A and B registers, the resulting bit will be a logical 1.

If you wish to check automatically whether the test data are good, then a JNZ or JZ command is used to check the result of the XRA. If the result is zero, then the test is good. If it is not zero, the test fails.

A very useful way to clear the A register and set the flags to a known state is to execute an XRA A instruction. This sets the A register to 0000 0000, sets the zero flag, clears the carry, indicates parity is true, and shows the sign bit as 0.

Compare Instructions

The next subgroup of the logical instructions is the *compare* instructions. These compare two operands and set the condition flags with the results of the compare. All compare operations are performed in the following manner:

1. The source operand is compared against the data in the A register.
2. To perform the compare, the source operand is subtracted from the A register. Condition flags are affected to indicate the result of the subtraction.
3. Neither the source operand or the A register data is changed after the execution of the compare instruction.

As an example, suppose you wished to compare the present 8-bit value in the B register against the value in the A register. You would specify the instruction CMP B. Further, suppose the B register had the value 45H and the A register had the value 44H. The following would be the result after the CMP B instruction was executed:

1. B is the source operand and it is subtracted from the A register. This gives the result of $44H - 45H = -1 = FFH$.

2. The result is not zero: zero flag = false = 0.
3. The result generated a borrow: carry flag = true = 1.

Using the zero flag and the carry flag the programmer can automatically decide if

Source operand = A reg; zero flag = true
Source operand < A reg; carry flag = false
                        zero flag = false
Source operand > A reg; carry flag = true

Using this information it is an easy matter to test the flags after the compare instruction and determine the relationship between the source operand and the A register.

There are three main addressing modes used for the compare instruction:

```
        8080/8085                                    Z80

CMP r  compare A with register r              CP r
CMP M  compare A with memory (HL)             CP (HL)
CPI data compare A with immediate data        CP data
```

Example 3-9

To illustrate the use of the CMP instruction, the following program inputs two numbers from different memory locations and then writes information to output port 34H, which indicates the relationship between the numbers. These numbers will be stored in memory locations NUM_1 and NUM_2. The output information will determine the relationship between NUM_1 compared to NUM_2.

```
        IF NUM_1 = NUM_2     PORT 01 = 1
        IF NUM_1 < NUM_2     PORT 01 = 3
        IF NUM_1 > NUM_2     PORT 01 = 2

                LDA NUM_2       get number 2 from memory
                LXI H, NUM1     address of number 1 in HL
                CMP M           subtract NUM_1 from NUM_2
                JC LESS_1       if C then NUM_1<NUM_2
                JNZ GREAT_1     if not zero,NUM_1>NUM_2
                MVI A, 01       equal, output 01
                JMP END         stop
LESS_1          MVI A, 03       less than, output 03
                JMP END
GREAT_1         MVI A, 02       greater than: output 02
END             OUT 34H         output data to port
                HLT             stop program
```

The Z80 form of program is:

```
          LD  A,(NUM_2)     get number 2 from memory
          LD  HL,NUM_1      address of number 1 in HL
          CP  (HL)          subtract NUM_1 from NUM_2
          JP  C,LESS_1      if C then NUM_1<NUM_2
          JP  NZ,GREAT_1    if not zero,NUM_1>NUM_2
          LD  A,01          equal, output 01
          JP  END           stop
LESS_1    LD  A,03          less than, output 03
          JP  END
GREAT_1   LD  A,02          greater than: output 02
END       OUT (34H),A       output data to port
          HALT              stop program
```

Rotate Instructions

The next subgroup of logical instructions to be presented is called *rotate* instructions. This subgroup operates on data contained in the accumulator (A register). There are two main types of rotate instructions:

1. Rotate without the carry bit.
2. Rotate using the carry bit.

Figure 3-17 shows the effect of a rotate instruction on the data in the A register. Let's first discuss the details of the rotate instruction that does not use the carry bit as part of the data. This instruction has the mnemonic form of;

8080 Mnemonic	Z80 Mnemonic	
RLC	RLCA	Rotate left
RRC	RRCA	Rotate right

Figure 3-18 shows the effect that the RRC and RLC instruction has on the data in the A register. Notice in Figure 3-18 that the carry bit is *not* part of the rotating process. Rather the carry bit is set or reset, depending on the data that

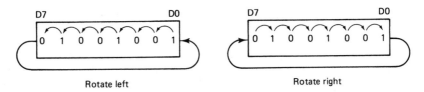

Rotate left   Rotate right

Figure 3-17   This figure shows how the rotate instructions operate. You can specify instuctions to rotate left or right.

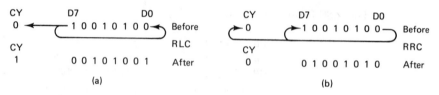

Figure 3-18 The RLC and RRC rotate the date into the carry bit.

is rotated through the A register. Further, information in the carry bit will not affect any bits of the A register during the execution of the RRC or RLC instruction.

The second form of the rotate instruction makes use of the carry bit as part of the rotate, as shown in Figure 3-19. Mnemonic form of this instruction is:

```
8080                Z80 Mnemonics

RAL     RLA     Rotate left through carry
RAR     RRA     Rotate right through carry
```

One may think of this instruction as actually rotating the data in a 9-bit register, the carry bit being the ninth bit.

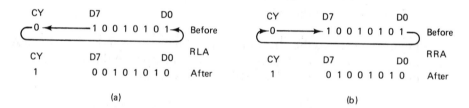

Figure 3-19 RLA and RRA use the carry bit as part of the roate register. You may think of these instructions as 9-bit rotates.

Complement Accumulator, Carry, and Set Carry

The final three instructions we discuss in this chapter affect the A register and the carry flag. First, the instruction that complements the A register is, in mnemonic form:

```
8080 Mnemonics          Z80 Mnemonics

CMA          CPL   1s complement of A register
```

When this instruction is executed, the 8 bits of data contained in the A register are complemented. That is, the 1s become 0s and the 0s become 1s. This operation is shown in Figure 3-20. No flags are affected by this instruction.

The two instructions that affect the carry flag are complement carry and set carry true. In mnemonic form these appear as:

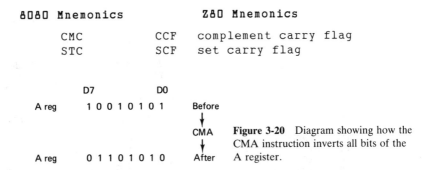

8080 Mnemonics	Z80 Mnemonics	
CMC	CCF	complement carry flag
STC	SCF	set carry flag

Figure 3-20 Diagram showing how the CMA instruction inverts all bits of the A register.

To set the carry flag true prior to an operation, you simply use the STC instruction. However, to clear the carry flag you must use both instructions, STC and then CMC. These two instructions first set the carry flag to a logical 1 and then complement it to a logical 0.

You have now been shown a majority of instructions for the three microprocessors. Of course, the Z80 has many special instructions to which you have not yet been introduced. In the next section we take the instructions given in this and the last chapter and present some application programs. These programs will allow you to see different examples of how the instructions could be used. After examining these applications you will be prepared to write your own programs to make the microprocessor perform as you wish.

3-7 PROGRAMMING EXAMPLES

To illustrate how the instructions given in this chapter may be used in an actual program application, the following examples are given.

```
 1                         ;;;;;;;;;;;;;;;;;;;;;;;;;;;;;;;
 2                         ;
 3                         ; 8085 PROGRAM EXAMPLE
 4                         ;
 5                         ; INCREMENT 8 BIT REGISTER
 6                         ; AND OUTPUT VALUE TO I/O PORT
 7                         ; 25H.  DELAY TO LET LEDS BE
 8                         ; VISIBLE.
 9                         ;
10                         ;;;;;;;;;;;;;;;;;;;;;;;;;;;;;;;
11                         ;
12   0100                         ORG 100H
13   0100   3E 00                 MVI A,00H   ;A=0
14   0102   67          AGAIN     MOV H,A     ;SAVE A
15   0103   D3 25                 OUT 25H     ;OUTPUT A REG
16   0105   01 FF 0F              LXI B,0FFFH ;GET DELAY COUNT
17   0108   0B          BAC1      DCX B       ;DEC BC NO FLAGS AFFECTED
18   0109   79                    MOV A,C     ;MOVE C TO A
19   010A   B0                    ORA B       ;OR B AND C
20   010B   C2 08 01              JNZ BAC1    ;IF NOT ZERO KEEP GOING
21   010E   7C                    MOV A,H     ;RECALL A
22   010F   3C                    INR A       ;INC TO NEXT VALUE
23   0110   C3 02 01              JMP AGAIN   ;DO IT AGAIN
24   0113                         END
```

```
 1                        ;;;;;;;;;;;;;;;;;;;;;;;;;;;;;;;
 2                        ;
 3                        ; Z80 PROGRAM EXAMPLE
 4                        ;
 5                        ; INCREMENT 8 BIT REGISTER
 6                        ; AND OUTPUT VALUE TO I/O PORT
 7                        ; 25H.   DELAY TO LET LEDS BE
 8                        ; VISIBLE.
 9                        ;
10                        ;;;;;;;;;;;;;;;;;;;;;;;;;;;;;;;
11                        ;
12   0100                        ORG 100H
13   0100   3E 00                LD A,00H   ;A=0
14   0102   67          AGAIN    LD H,A     ;SAVE A
15   0103   D3 25                OUT (25H),A ;OUTPUT A REG
16   0105   01 FF 0F             LD BC,0FFFH ;GET DELAY COUNT
17   0108   0B          BAC1     DEC BC     ;DEC BC NO FLAGS AFFECTED
18   0109   79                   LD A,C     ;MOVE C TO A
19   010A   B0                   OR B       ;OR B AND C
20   010B   C2 08 01             JP NZ,BAC1 ;IF NOT ZERO KEEP GOING
21   010E   7C                   LD A,H     ;RECALL A
22   010F   3C                   INC A      ;INC TO NEXT VALUE
23   0110   C3 02 01             JP AGAIN   ;DO IT AGAIN
24   0113                        END
```

```
 1                        ;;;;;;;;;;;;;;;;;;;;;;;;;;;;;;;
 2                        ;
 3                        ; 8085 PROGRAM EXAMPLE
 4                        ;
 5                        ; TRAVELING LIGHT PROGRAM
 6                        ; THIS PROGRAM WILL MAKE THE
 7                        ; LEDS AT OUTPUT PORT 25
 8                        ; APPEAR TO BE TRAVELING FROM
 9                        ; THE LOW BIT TO THE HIGH BIT
10                        ;
11                        ;;;;;;;;;;;;;;;;;;;;;;;;;;;;;;;
12                        ;
13   0100                        ORG 100H
14   0100   3E 01                MVI A,01H  ;A=1
15   0102   67          AGAIN    MOV H,A    ;SAVE A
16   0103   D3 25                OUT 25H    ;OUTPUT A REG
17   0105   01 FF 0F             LXI B,0FFFH ;GET DELAY COUNT
18   0108   0B          BAC1     DCX B      ;DEC BC NO FLAGS AFFECTED
19   0109   79                   MOV A,C    ;MOVE C TO A
20   010A   B0                   ORA B      ;OR B AND C
21   010B   C2 08 01             JNZ BAC1   ;IF NOT ZERO KEEP GOING
22   010E   7C                   MOV A,H    ;RECALL A
23   010F   07                   RLC        ;NEXT VALUE
24   0110   C3 02 01             JMP AGAIN  ;DO IT AGAIN
25   0113                        END
```

```
 1                          ;;;;;;;;;;;;;;;;;;;;;;;;;;;;;;;
 2                          ;
 3                          ; Z80 PROGRAM EXAMPLE
 4                          ;
 5                          ; TRAVELING LIGHT PROGRAM
 6                          ; THIS PROGRAM WILL MAKE THE
 7                          ; LEDS AT OUTPUT PORT 25 APPEAR
 8                          ; TO BE TRAVELING FROM THE LOW
 9                          ; BIT TO THE HIGH BIT.
10                          ;
11                          ;;;;;;;;;;;;;;;;;;;;;;;;;;;;;;;
12                          ;
13     0100                        ORG 100H
14     0100   3E 01                LD A,01H  ;A=1
15     0102   67          AGAIN    LD H,A    ;SAVE A
16     0103   D3 25                OUT (25H),A ;OUTPUT A REG
17     0105   01 FF 0F             LD BC,0FFFH ;GET DELAY COUNT
18     0108   0B          BAC1     DEC BC  ;DEC BC NO FLAGS AFFECTED
19     0109   79                   LD A,C  ;MOVE C TO A
20     010A   B0                   OR B    ;OR B AND C
21     010B   C2 08 01             JP NZ,BAC1 ;IF NOT ZERO KEEP GOING
22     010E   7C                   LD A,H  ;RECALL A
23     010F   07                   RLCA    ;NEXT VALUE
24     0110   C3 02 01             JP AGAIN ;DO IT AGAIN
25     0113                        END
```

```
 1                          ;;;;;;;;;;;;;;;;;;;;;;;;;;;;;;;
 2                          ;
 3                          ; 8085 PROGRAM EXAMPLE
 4                          ;
 5                          ; BOUNCING LIGHT PROGRAM
 6                          ; THIS PROGRAM WILL MAKE THE
 7                          ; LEDS AT OUTPUT PORT 25
 8                          ; APPEAR TO BE TRAVELING FROM
 9                          ; THE LOW BIT TO THE HIGH BIT
10                          ; AND BACK TO THE LOW BIT AGAIN
11                          ;
12                          ;;;;;;;;;;;;;;;;;;;;;;;;;;;;;;;
13                          ;
14     0100                        ORG 100H
15     0100   3E 01                MVI A,01H  ;A=1
16     0102   67          AGAIN    MOV H,A    ;SAVE A
17     0103   D3 25                OUT 25H    ;OUTPUT A REG
18     0105   01 FF 0F             LXI B,0FFFH ;GET DELAY COUNT
19     0108   0B          BAC1     DCX B   ;DEC BC NO FLAGS AFFECTED
20     0109   79                   MOV A,C ;MOVE C TO A
21     010A   B0                   ORA B   ;OR B AND C
22     010B   C2 08 01             JNZ BAC1 ;IF NOT ZERO KEEP GOING
23     010E   7C                   MOV A,H ;RECALL A
24     010F   FE 80                CPI 80H ;CHECK FOR LAST BIT
25     0111   CA 18 01             JZ GORIGHT ;YES,TRAVEL DOWN
26     0114   07          GOLEFT   RLC     ;NEXT VALUE
27     0115   C3 02 01             JMP AGAIN ;DO IT AGAIN
28                          ;
29                          ;
30     0118   0F          GORIGHT  RRC      ;ROTATE RIGHT
31     0119   67                   MOV H,A  ;SAVE A
32     011A   D3 25                OUT 25H  ;OUTPUT A REG
33     011C   01 FF 0F             LXI B,0FFFH ;GET DELAY COUNT
34     011F   0B          RIGHT1   DCX B   ;DEC BC NO FLAGS AFFECTED
35     0120   79                   MOV A,C ;MOVE C TO A
36     0121   B0                   ORA B   ;OR B AND C
37     0122   C2 1F 01             JNZ RIGHT1 ;IF NOT ZERO KEEP GOING
38     0125   7C                   MOV A,H ;RECALL A
39     0126   FE 01                CPI 01H ;CHECK FOR LAST BIT
40     0128   CA 14 01             JZ GOLEFT ;YES,TRAVEL UP
41     012B   C3 18 01             JMP GORIGHT ;GO RIGHT AGAIN
42     012E                        END
```

```
 1                              ;;;;;;;;;;;;;;;;;;;;;;;;;;;;;;;;;
 2                              ;
 3                              ; 8085 PROGRAM EXAMPLE
 4                              ;
 5                              ; MEMORY TEST -
 6                              ; THIS PROGRAM WILL TEST
 7                              ; RAM LOCATIONS FROM
 8                              ;    1000H TO 17FFH
 9                              ; IF TEST PASSES THEN OUTPUT
10                              ; TO PORT 10H.
11                              ; IF TEST FAILS THEN OUTPUT
12                              ; TO PORT 11H.
13                              ; MUST MONITOR OUTPUT PORT
14                              ; WITH HARDWARE TO SEE IF
15                              ; TEST PASSED OR FAILED.  THIS
16                              ; PROGRAM IS GOOD FOR SYSTEMS
17                              ; WHICH DO NOT HAVE A CRT SCREEN
18                              ; TO REPORT RESULTS.
19                              ;
20                              ;;;;;;;;;;;;;;;;;;;;;;;;;;;;;;;;;;;;
21                              ;
22    0100                              ORG 100H     ;SET ORIGIN
23    0100    21 00 10                  LXI H,1000H  ;SET START ADDRESS
24    0103    01 00 08                  LXI B,2048   ;SET AMOUNT OF MEM
25                              ;
26                              ; FIRST WRITE BACKGROUND DATA
27                              ;
28    0106    1E 00                     MVI E,00     ;DATA FOR MEM
29    0108    73              WRITE1 MOV M,E         ;DATA TO MEM
30    0109    0B                       DCX B
31    010A    79                       MOV A,C
32    010B    B0                       ORA B         ;CHECK BC FOR 0
33    010C    C2 12 01                  JNZ WRITE2   ;DO NEXT BYTE
34    010F    C3 16 01                  JMP NEXT     ;DO NEXT PART
35    0112    23              WRITE2 INX H           ;BUMP TO NEXT ADD
36    0113    C3 08 01                  JMP WRITE1   ;LOOP AGAIN
37                              ;
38                              ;NEXT PART OF PROGRAM
39                              ;
40    0116    21 00 10        NEXT   LXI H,1000H     ;START ADDRESS
41    0119    01 00 08               LXI B,2048      ;NUMBER OF LOCATIONS
42    011C    7E              READ1  MOV A,M         ;GET DATA FROM MEM
43    011D    B7                     ORA A           ;TEST FOR ALL 0,S
44    011E    C2 4F 01               JNZ FAIL        ;JUMP TO FAIL
45    0121    2F                     CMA             ;COMPLEMENT DATA
46    0122    77                     MOV M,A         ;WRITE DATA TO MEM
47    0123    79                     MOV A,C
48    0124    B0                     ORA B           ;CHECK BC FOR 0
49    0125    C2 2B 01               JNZ READ2       ;NEXT READ ADD
50    0128    C3 30 01               JMP LAST        ;LAST PART OF TEST
51    012B    23              READ2  INX H           ;NEXT ADD
52    012C    0B                     DCX B
53    012D    C3 1C 01               JMP READ1       ;LOOP IT
54                              ;
55                              ; NOW READ AND WRITE LAST TIME
56                              ;
57    0130    21 00 10        LAST   LXI H,1000H     ;START ADDRESS
58    0133    01 00 08               LXI B,2048      ;NUMBER OF LOCATIONS
59    0136    7E              READ3  MOV A,M         ;GET DATA FROM MEM
60    0137    2F                     CMA
```

```
61    0138   B7                         ORA A          ;TEST FOR ALL 0,S
62    0139   C2 4F 01                   JNZ FAIL       ;JUMP TO FAIL
63    013C   77                         MOV M,A        ;WRITE DATA TO MEM
64    013D   79                         MOV A,C
65    013E   B0                         ORA B          ;CHECK BC FOR 0
66    013F   C2 45 01                   JNZ READ4      ;NEXT READ ADD
67    0142   C3 4A 01                   JMP PASS       ;LAST PART OF TEST
68    0145   23             READ4       INX H          ;NEXT ADD
69    0146   0B                         DCX B
70    0147   C3 36 01                   JMP READ3      ;LOOP IT
71                          ;
72                          ;
73    014A   D3 10          PASS        OUT 10H
74    014C   C3 4A 01                   JMP PASS       ;KEEP LOOPING
75    014F   D3 11          FAIL        OUT 11H
76    0151   C3 4F 01                   JMP FAIL       ;KEEP LOOPING
77    0154                              END
```

3-8 CHAPTER SUMMARY

This chapter was a continuation of the programming information given in Chapter 2. It began with the topic of addressing modes for the microprocessor. These modes are the different ways an address may be formed to access information in a system. After studying address modes, you were introduced to the concept of a flag. The flag register was shown and discussed.

After the flag register was given, you concentrated on the arithmetic set of instructions. These instructions allowed the microprocessor to add and subtract in different ways. Several examples of using these instructions were given.

The next topic, conditional jumps, made use of the flag register. These types of instructions allow the microprocessor to make decisions under program control. Another group of instructions covered in this chapter was the logical group. These instructions performed boolean functions such as AND, OR, and XOR.

Following this group of instructions you were shown how the rotate left and rotate right instructions affect the bits of the accumulator. The final group of instructions complemented the accumulator and Carry or set the Carry bit.

This chapter concluded with several programming examples, which were designed to show how the instructions given in this chapter can be applied to a real problem.

REVIEW PROBLEMS

1. Define the term addressing mode.
2. Name the five addressing modes presented in this chapter.
3. Write an instruction, in mnemonic form, for each of the five addressing modes listed in Problem 2.

4. Write the addressing mode represented by each of the following.
   (a) LXI D, 49A3H
   (b) LD B,A
   (c) LDA 4567H
   (d) MOV M,A

5. Indexed addressing is used on which microprocessors, the 8080, 8085, Z80, or on all three microprocessors?

6. Define the term flags, or flag register.

7. Draw a block diagram of the flag register for the 8080.

8. Draw a block diagram of the flag register for the Z80.

9. True or false: The flags are affected on every instruction executed by the microprocessor.

10. In the following 8080 program, which flags are affected?

```
MVI     A,23H
MOV     B,A
MOV     C,A
LXI     H,2453H
HLT
```

11. Write a small program that affects the carry flag. Your program may affect other flags.

12. List two instructions that affect the zero flag.

13. True or false: A flag bit may be set to a 1 or a 0 by all instructions.

14. Write a program that sets the sign bit to a 1.

15. What are the two conditions that set the overflow flag true on the Z80?

16. Insert a parity bit that makes the following binary word even parity.

101110001P

17. Write a program that adds the two 16-bit numbers 2356H and 5321H and stores the result in the HL register pair.

18. Write a program that subtracts 0121H from 0523H and stores the results in memory locations 8305 (MSB) and 8306 (LSB).

19. Draw a flowchart that shows how a conditional branch instruction is executed by the CPU.

20. Write a program that delays for a count of 1F.

21. Write a program that delays for a count of FFFF.

22. Write a program that inputs data from port 51H and masks all bits except D0. Test this result for zero. If the result is zero, then halt. If the result is not zero, then jump back to once again input data from port 51H.

4

STACK AREA, SUBROUTINES, AND Z80 SPECIALS

This chapter continues with the microprocessor software discussion, focusing on the topics of stacks, subroutines, and special instructions for the Z80. We start by introducing the system stack area and then explain what subroutines are and how they are used. The stack is an integral part of subroutine usage, and the programmer must first know how to use the stack.

After these two topics have been covered, a presentation of instructions common only to the Z80 is given. These instructions involve the alternate register set as well as some extended op-code instructions. Programming examples are given to show how the instructions can be used.

At the completion of this chapter you will have been introduced to and used the major instruction types available on the Z80, 8080, and 8085 microprocessors.

4-1 STACK AREA

Many times when you are writing microprocessor application programs, you want to save the contents of an internal register for use at a later time. For example, a section of your program may be multiplying two numbers using a certain technique. This technique requires saving of temporary or intermediate results that will be used later in the solution. The question is, How can you save the contents of an internal register? You have already learned two different methods for doing this. We illustrate by showing how the B register may be saved.

```
MOV  A,B
STA  TEMP_VALUE        save B

LXI  H,TEMP_VALUE      get address in HL
MOV  M,B               save B
```

In both of these examples, two operations occurred that could complicate the problem:

1. Other internal registers were changed.
2. The address for the save has to be given. That is, you must keep track of exactly where in memory the B register was saved. This may not seem like a big problem, it would be if you wanted to save all the internal registers.

A solution to this problem and others, is the formation of a special area in memory called the *stack area*. Within this area or address range of memory, internal registers may be saved and recalled with a single instruction.

The stack area is shown pictorially in Figure 4-1. You see in Figure 4-1 that

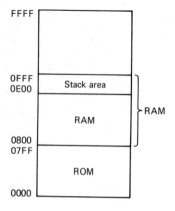

Figure 4-1 Memory map showing the address location of the system stack area. The stack area may be of any length and located anywhere within the addressing space of the microprocessor.

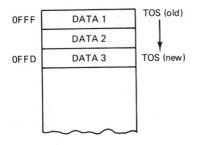

Figure 4-2 As data is written to the stack, the SP register will always point to the last entry, which will be on the top of stack.

the stack area may be any address range from 0000H to FFFFH. A special internal 16-bit register keeps track of the address within the stack area. This register is called the SP, or *stack pointer*. The SP register always points to the current TOS, or top of stack. As information gets written into the stack, the TOS decrements in address. This is shown in Figure 4-2.

The TOS, or *stack pointer,* must be initialized in your program. When the microprocessor first powers up, the SP register, like most of the internal registers, comes up with a random address. The instruction to initialize the SP register is:

```
LXI SP,16 bit value      8080code
LD  SP,16 bit value      Z80  code
```

There are two main instructions that are used to store and retrieve information in the stack area:

```
PUSH      saves data on stack
POP       gets data off stack
```

Valid forms of the PUSH instruction are;

8080	Z80	FUNCTION
PUSH B	PUSH BC	save BC pair
PUSH D	PUSH DE	save DE pair
PUSH H	PUSH HL	save HL pair
PUSH PSW	PUSH AF	save A reg and Flags

Where PSW = processor status word. To illustrate exactly what occurs during the execution of a PUSH instruction, let's take an example. The program that will be executed sets the SP register to 3FFFH.

```
LXI SP,3FFFH      initialize stack pointer (TOS)
MVI B,45H         45 to B reg
MVI C,2AH         2A to C reg
PUSH B            save B and C reg on stack
```

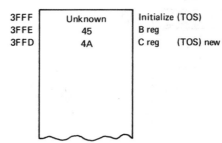

Figure 4-3 Before the PUSH B instruction, the SP register pointed to address 3FFF. After the PUSH B instruction, the SP register points to the new TOS, which is address 3FFD.

Using this partial program, Figure 4-3 shows what the stack area looks like prior to and after the PUSH B instruction. Referring to Figure 4-3, notice that the first operation performed by the microprocessor is to decrement the SP register. This is to keep from writing information into a variable that was previously placed onto the stack. In this case there was no variable.

After the SP register is decremented, the data for the higher-order register (B) is written into memory. The SP register is decremented again, and the data for the lower-order register (C) is written into memory. Notice that the current TOS is pointing to the last data written into memory, 3FFD. The PUSH instruction always writes 16 bits to the STACK. You cannot save only one register using PUSH.

We continue in our example to save another register;

```
LXI SP ,3FFFH     initialize stack pointer (TOS)
MVI B,45H         45to B reg
MVI C,2AH         2A to C reg
PUSH B            save B and C reg on stack
MVI D,89H         89H to D
MVI E,1EH         1EH to E
PUSH D
```

Figure 4-4 shows how the stack would appear after saving the DE register pair. Notice that the first operation was to decrement the stack pointer. In the first section, when saving the BC register, this was an unnecessary operation.

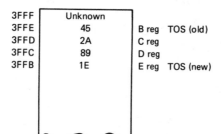

Figure 4-4 System stack area and stack pointed value after the PUSH DE register instruction.

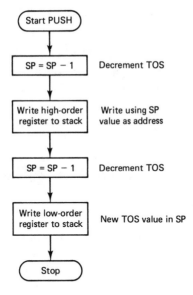

Figure 4-5 Flowchart showing the sequence of events that occur each time the PUSH instruction is executed. Notice that 2 bytes are written onto the stack each time it is accessed.

However, now you can clearly see why this step is important. Next, the microprocessor writes the data for the higher-order register (D) into memory. The SP register is again decremented and the data for the lower-order register (E) is written into memory. The SP points to the new TOS, which is 3FFB.

The flowchart of Figure 4-5 shows the sequence of events that occur each time the PUSH instruction is executed.

It is interesting to note that the stack actually grows downward in memory addresses. See Figure 4-6. When writing a program, the code grows upward to higher memory addresses. When using the stack, it is important to keep the two areas separated. If your stack grows down into the program area or the program area grows upward into the stack, then unpredictable results will occur. A good practice when writing programs is to initialize your stack pointer early in the program—almost the first thing you do.

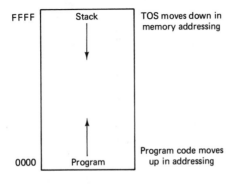

Figure 4-6 As you write your program code, the memory addresses increase in value. When data is written to the stack area the memory addresses decrease. The stack area and program area must *not* overlap or your program could malfunction.

Getting Data Off the Stack (POP)

Placing data on the stack is one very useful operation that can be performed by the microprocessor. However, we must be able to retrieve the data. This is accomplished by the reverse of the PUSH, the POP instruction. Valid forms of the POP instructions are:

```
8080        Z80

POP B       POP BC     restore BC register pair
POP D       POP DE     restore DE register pair
POP H       POP HL     restore HL register pair
POP PSW     POP AF     restore A reg and flags
```

To illustrate how the POP instructions operate in a system environment, let's continue with the example started previously. In that example the BC, DE registers were saved on the stack and the stack area appeared as shown in Figure 4-7. Using the POP instructions we can restore the register values. The following example shows how that is accomplished (continued from the previous example).

```
LXI B,0000     zero out the BC registers
LXI D,0000     zero out the DE registers
POP D          restore DE regs
POP B          restore BC regs
```

Notice in this example that the data must be popped from the stack in the reverse order in which it was pushed. This is consistent because the stack pointer will be pointing to the data that was last pushed onto the stack. This type of memory operation is sometimes referred to as LIFO (last in, first out).

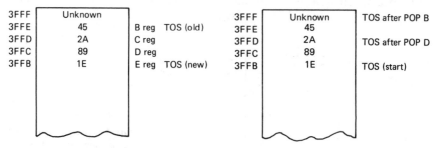

Figure 4-7 System stack area and stack pointer value after the PUSH DE register instruction.

Figure 4-8 The date on the system stack remains even after the POP instruction is executed. Data will only change when new data is pushed onto the same stack address.

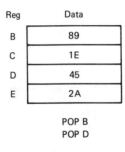

Reg	Data
B	89
C	1E
D	45
E	2A

POP B
POP D

Figure 4-9   When popping data off the stack you normally pop off in the reverse order you pushed. However, you can change the order of popping data to modify the value of the microprocessor internal registers.

After the pop D instruction, the stack pointer points to the new top of stack 3FFDH. This is the stack address for the BC registers. Further notice in Figure 4-8 that the data in the stack is still there; only the stack pointer was incremented.

When the POP instruction is executed, the microprocessor does not know which register was placed on the stack by a PUSH. It only knows to get the data off the top of the stack and place it in the registers specified by the POP instruction. In this example, if the POP B and POP D instructions were reversed, then data from the stack would appear in the internal registers of the microprocessor, as shown in Figure 4-9. Popping data off the stack in reverse order is sometimes used in a program to exchange the data between registers.

4-2 SUBROUTINES

In typical microprocessor programs, software instructions are executed from the instruction residing at the lower memory location to the ones residing at higher memory locations. You have seen how the JMP instruction can change this order of execution. This section presents another type of instruction that will also alter the flow of instruction execution, resulting in fewer program instructions. These types of instructions are the CALL subroutine.

To understand what a subroutine is, consider the case of a program that executes the same block of instructions in many different places. This is shown in Figure 4-10. You see in Figure 4-10 that the block of instructions for getting a character from a keyboard is executed over and over in different places in the program. One solution to this problem is to simply rewrite the instructions "in line" as many times as necessary. This is the solution shown in the figure.

Another solution to this problem is to write the instructions necessary to perform the specific task one time. Then, each time the program requires that task to be performed, that same section of instructions is executed. When the task is completed, the program resumes execution from the place the block of instructions was needed. This is shown in block diagram form in Figure 4-11.

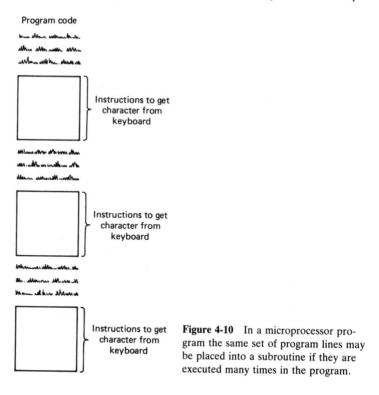

Figure 4-10 In a microprocessor program the same set of program lines may be placed into a subroutine if they are executed many times in the program.

You see in Figure 4-11 that this solution results in fewer instructions being required by the program (not fewer instructions executed). A further benefit of this technique, though not obvious, is that the program becomes more readable and structured.

The block of instructions is referred to as a *subroutine*. Microprocessors support the use of subroutines by making use of two instructions, CALL and RET. These are the mnemonics for *CALL subroutine* and *Return from subroutine*. The CALL instruction is similar in operation to a JMP. That is, the CALL is a 3-byte instruction with the first byte an op-code and the last 2 bytes the 16-bit address of the first op-code in the subroutine. However, when the CALL instruction is executed, one additional operation is performed by the microprocessor. In this additional operation, the contents of the PC are pushed onto the system stack. The reason for this is the following: At the time the microprocessor fetches the CALL instruction and decodes it, the internal PC is updated to fetch the next op-code from memory. That is, the PC value is a 16-bit address of the next sequential op-code to execute. This address is exactly the address to which the program will return after executing the subroutine instructions. See Figure 4-12.

Program code

CALL GET_KEY

CALL GET_KEY

CALL GET_KEY

GET_KEY

Section of
code to
get key from
keyboard

Subroutine

Figure 4-11 The program section GET KEY is placed into a subroutine. The main program can access this subroutine whenever necessary. Having a subroutine results in less programming lines and a "cleaner" program.

The address pushed onto the stack during the execution of the CALL instruction is referred to as the *return address*. After the return address is pushed onto the system stack, the microprocessor performs a jump to the CALL address. The instruction residing at the CALL address is the next one to be executed. This is shown in Figure 4-13.

After all the instructions in the subroutine have been executed, the microprocessor must get back to the return address and start executing op-codes from

Mnemonics

CALL GET_KEY
MVI   A, 25H

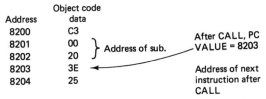

Address	Object code data		
8200	C3		
8201	00	} Address of sub.	After CALL, PC VALUE = 8203
8202	20		
8203	3E	◄	Address of next instruction after CALL
8204	25		

Figure 4-12 When the CALL instruction is executed, the PC register is pointing to the memory location of the op-code that follows the CALL. This is the memory location to which you wish the program to return after the subroutine has been executed.

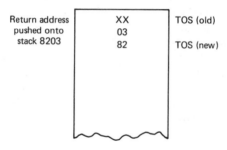

Return address pushed onto stack 8203

XX   TOS (old)
03
82   TOS (new)

Figure 4-13 When the CALL instruction is executed, the return address is automatically written to the system stack.

there. This action is accomplished by executing the RET instruction. When this instruction is executed, the microprocessor POPs the two bytes off the TOS and places these into the PC register. The microprocessor then begins execution of instructions from the new PC register value. Figure 4-14 shows a flowchart of the events that occur during a CALL and RET operation.

The mnemonics for the CALL and RET instructions follow. While you are reviewing this list, remember that both the CALL and RET can be conditional on the setting of the system flags.

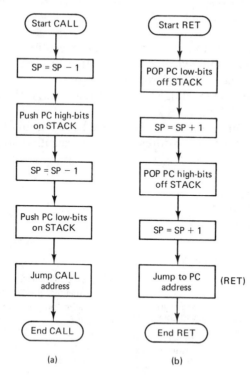

Figure 4-14 (a) Sequence of events that occur during the execution of the CALL instruction. Notice that the return address is written onto the system stack; (b) sequence of events that occur during the execution of the RET instruction. Notice that the return address is read from the system stack.

MNEMONICS for CALL and RET

8080 MNEMONICS	Z80 MNEMONICS	FUNCTION
CALL address	CALL address	unconditional call
CZ address	CALL Z,address	call if zero
CNZ address	CALL NZ,address	call if not zero
CC address	CALL C,address	call if carry true
CNC address	CALL NC,address	call if no carry
CPE address	CALL PE,address	call if parity even
CPO address	CALL PO,address	call if parity odd
CP address	CALL P,address	call if positive
CM address	CALL M,address	call if minus

RETURN instructions

8080 MNEMONICS	Z80 MNEMONICS	FUNCTION
RET	RET	unconditional return
RZ	RET Z	return if zero
RNZ	RET NZ	return if not zero
RC	RET C	return if carry
RNC	RET NC	return if no carry
RPE	RET PE	return if parity even
RPO	RET PO	return if parity odd
RP	RET P	return if plus
RM	RET M	return if minus

To illustrate further the use of subroutines, Example 4-1 is given. This example uses a delay subroutine. The routine is entered with the BC register set to the delay value.

Example 4-1

The following program has in-line code placed into a subroutine.

```
         MVI A,00H  ;A=0
AGAIN    MOV H,A    ;SAVE A
         OUT 25H    ;OUTPUT A REG
         LXI B,0FFFH ;GET DELAY COUNT
BAC1     DCX B      ;DEC BC NO FLAGS AFFECTED
         MOV A,C    ;MOVE C TO A
         ORA B      ;OR B AND C
         JNZ BAC1   ;IF NOT ZERO KEEP GOING
         MOV A,H    ;RECALL A
         INR A      ;INC TO NEXT VALUE
         JMP AGAIN  ;DO IT AGAIN
```

```
            MVI  A,00H  ;A=0
  AGAIN     MOV  H,A    ;SAVE A
            OUT  25H    ;OUTPUT A REG
            LXI  B,0FFFH ;LOAD DELAY COUNT
            CALL DELAY  ;GOTO TO SUBROUTINE
            MOV  A,H    ;RECALL A
            INR  A      ;INC TO NEXT VALUE
            JMP  AGAIN  ;DO IT AGAIN
  ;
  DLEAY     DCX  B      ;DEC BC NO FLAGS AFFECTED
            MOV  A,C    ;MOVE C TO A
            ORA  B      ;OR B AND C
            JNZ  DELAY  ;IF NOT ZERO KEEP GOING
            RET         ;RETURN FROM SUBROUTINE
```

4-3 AVOIDING PITFALLS WITH SUBROUTINES

Subroutines have many advantages over in-line code, which make their use valuable for the programmer. However, there are a few facts to remember when using subroutines, or your program could malfunction and finding the problem may be difficult.

```
 8  0100                        ORG 0100H
 9
10                     ;
11                     ; THE MAIN BODY OF THE PROGRAM WOULD GO HERE
12                     ;
13                     ; NOW TO CALL A SUBROUTINE
14                     ;
15                     ; NEED TO SAVE ALL REGS PRIOR TO CALLING SUB
16                     ;
17  0100  F5                    PUSH PSW   ;SAVE A,F
18  0101  C5                    PUSH B     ;SAVE B,C
19  0102  D5                    PUSH D     ;SAVE D,E
20  0103  E5                    PUSH H     ;SAVE H,L
21  0104  CD 0B 01              CALL SUB1  ;CALL THE SUBROUTINE
22  0107  E1                    POP H      ;RESTORE H,L
23  0108  D1                    POP D      ;RESTORE D,E
24  0109  C1                    POP B      ;RESTORE B,C
25  010A  F1                    POP PSW    ;RESTORE A,FLAGS
26                     ;
27                     ; NOTE THAT REGISTERS WERE POPPED IN REVERSE
28                     ; ORDER THAT THEY WERE PUSHED
29                     ;
30  010B  00           SUB1     NOP
31  010C  C9                    RET        ;DUMMY SUB FOR PROGRAM
32  010D                        END
```

Figure 4-15 Partial program showing how the microprocessor registers may be saved prior to calling a subroutine and restored after the return from the subroutine. This is sometimes necessary because you wish to use the register within the instructions of the subroutine.

1. Save any registers used in the subroutine prior to CALLing the routine. This is necessary because your subroutine may destroy the contents of internal registers. When you return to your main code, you may want the status of your program to be the same as when you left. If you don't care if the subroutine alters the contents of internal registers, then do not worry. Figure 4-15 shows a sample of how registers may be saved and restored before and after a subroutine call.

   Also, document (in your subroutine) which registers, if any, are altered in the subroutine itself. This type of documentation is shown in Figure 4-16.

2. Ensure that the stack has been cleaned off before RETurning from your subroutine. If you push any data during the execution of your subroutine, be sure you pop this data off before the RET instruction because of the operations that occur when the RET instruction is executed.

   The top 2 bytes on the system stack are popped off and used as the return address for the subroutine. Figure 4-17 shows the results of not properly cleaning the stack prior to returning from a subroutine.

   This problem occurs often in cases where you are returning conditionally from a subroutine. When using these types of returns, take precautions to ensure your stack area is clean.

3. You may nest subroutines as deeply as you desire. The microprocessor will keep track of which address on the top of the stack is from which nesting level. Figure 4-18 shows subroutines nested three deep and the resulting stack values.

```
 1                        ;;;;;;;;;;;;;;;;;;;;;;;;;;;;;;;;;
 2                        ;
 3                        ; SUBROUTINE FOR RETURNING THE
 4                        ; ASCII VALUE OF THE HEXADECIMAL
 5                        ; NUMBER.
 6                        ;
 7                        ; ENTRY - A REG HAS HEXADECIMAL NUMBER
 8                        ;       - LOWER NIBBLE ONLY
 9                        ;
10                        ; EXIT  - B REG HAS ASCII NUMBER RETURNED
11                        ;
12                        ; REGISTERS DESTROYED IN SUBROUTINE
13                        ;
14                        ; B,C,D,E
15                        ;;;;;;;;;;;;;;;;;;;;;;;;;;;;;;;;;;
16                        ;
17                        ; SUBROUTINE BODY GOES HERE
18                        ;
19   0000    00     ASCII    NOP
20   0001    C9              RET
21   0002                    END
```

Figure 4-16 Example of how you could document which registers are destroyed and used during the execution of the subroutine. This will help you when writing your main program and using the subroutine.

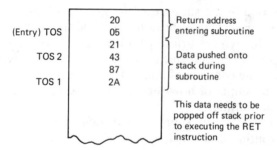

RET address could be 872A or
2143 if STACK
is not cleared
prior to RET

Figure 4-17   If you do not properly clean the stack during the execution of a subroutine, when the RET instruction is executed, the top 2 bytes of the stack will be used as the return address.

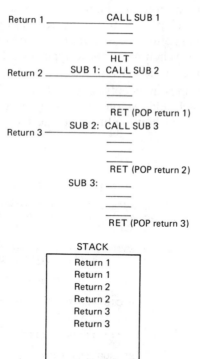

Figure 4-18   It is valid to call a subroutine from another subroutine. The system stack will operate correctly. It is up to you as the programmer to ensure that each subroutine cleans the stack prior to executing the RET instruction.

Remember, all housekeeping chores are the programmer's responsibility. The microprocessor will adhere to a rigid set of rules, which cannot be altered. As a programmer, you must ensure that you are not violating any of the rules for proper operation.

4-4 Z80 ALTERNATE REGISTER SET

In Chapter 2 you were shown a block diagram of the internal registers of the Z80 microprocessor. Contained in this block diagram were a group of registers called the alternate registers. The alternate register set is a group identical to the regular register set. However, the alternate registers may be accessed only in specific ways. Figure 4-19 shows the relationship between the alternate register set and the regular register set of the Z80.

In order to access data in the alternate register set, you must first exchange their contents with the regular register set. Once this is accomplished, you may use the standard instructions for manipulating internal data. The idea of the alternate register set is to have a convenient place to store the contents of the internal registers should the need arise. You have seen in previous examples when the need to store the contents of the internal registers can arise.

The only method for saving the contents of the internal registers is to write them to memory by placing them on the stack or writing them to a specific memory location. With the Z80 you have a third alternative: Exchange them with the alternate registers. There are two instructions designed specifically for the Z80 for use with the alternate registers:

```
EX AF ,AF'     exchange AF with AF'
EXX            exchange BCDEHL with BCDEHL'
```

Notice how fast this exchange occurs. In only two instructions the entire status of the Z80 microprocessor may be saved. The topic of interrupts has not been introduced as yet, but you will see that these instructions allow the Z80 to respond quickly to interrupt requests with no loss of data.

To show how these alternate registers operate and how the instructions work, let's take an example. In this example we save the contents of all the internal registers two different ways and compare each.

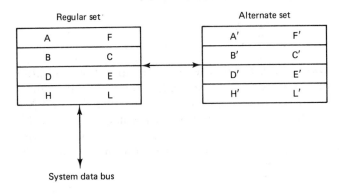

Figure 4-19   The Z80 microprocessor has an alternate register set which may be used to save the regular register set.

Example 4-2 (Saving internal registers to memory)

We assume that all internal registers, AF, BC, DE, and HL, must be saved.

```
PUSH AF
PUSH BC
PUSH DE
PUSH HL
```

Although this saving of internal registers required only four instructions, the time required was quite long. This is due to the fact that for each of these instructions, the Z80 must write to memory twice. You will see in Chapter 5 exactly how the hardware accomplishes this and the time it takes. As an easy analysis, let's divide each instruction into functions. Each PUSH instruction does the following functions:

1. Read op-code from memory (instruction fetch).
2. Decode instruction.
3. Write first byte to memory.
4. Write second byte to memory.

Let's further assume that each of these functions takes the same amount of time. You will learn that this is not the case, but we can make that assumption for this analysis. It is also valid to assume that no matter how much time it takes for each separate function, that time will be fixed regardless of the instruction. That is, it takes a specific amount of time to read the op-code from memory, to write to memory, and so on.

Using these assumptions, we see that it takes 4 time units to execute each push instruction. To save all the internal registers, it will require 4 × 4, or 16, time units. Let's now see how this time can be shortened by use of the alternate register set.

Example 4-3 (Saving internal registers to alternate registers)

```
EX AF ,AF
EXX
```

Figure 4-20 shows how the internal register sets appear before and after the execution of the EX and EXX instruction. Notice that this program requires only two instructions. Even if the time required to execute the EX or EXX instruction were the same as for a push (which it is not), you can see that this technique would be twice as fast. However, let us use the same assumptions that were given for the previous example. Under these conditions the time units required for each instruction would be

57	68	AF
41	A2	BC
B6	C9	DE
25	32	HL

05	49	A'F'
3B	4D	B'C'
8C	21	D'E'
63	A7	H'L'

EX AF, AF'
EXX

05	49	AF
3B	4D	BC
8C	21	DE
63	A7	HL

57	68	A'F'
41	A2	B'C'
B6	C9	D'E'
25	32	H'L'

Figure 4-20   Using the EX and EXX instructions you can exchange the contents of the alternate registers with the regular registers. This saves programming steps and instructions.

1. Read op-code from memory.
2. Decode instruction.

Using the assumptions given earlier, each instruction would require 2 time units. This gives a total of 4 time units for saving of the internal registers. Using the EX and EXX instructions is four times as fast as using the push instructions. This is one of the main reasons for having an alternate register set. You can completely save the internal state of the microprocessor quite quickly.

4-5 Z80 INDEX REGISTERS IX, IY

There are two internal registers on the Z80 that do not appear in the 8080 or 8085. These are called *index registers*. They are each 16 bits wide and are given the labels IX and IY. Most of the normal instruction mnemonics for the Z80, such as the LD instruction, also make provisions for using the IX and IY registers.

Example 4-4 shows how the IX or IY may be used for accessing data in memory.

Example 4-4

Assume you have a block of data in memory. This data is arranged in a specific order, as shown in Figure 4-21. Within the large block of data in the figure there are several smaller subblocks. Each subblock has data arranged in exactly the same way. One illustration of such data is files on

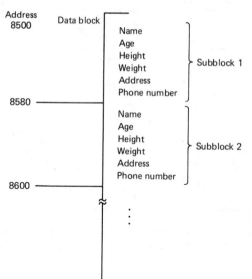

Figure 4-21   Data in your program may be arranged in specific order. This allows you to use the index instructions of the Z80 to access the data.

people. The large block of data would be all the data on all the people. Each smaller subblock would be the data on each person.

The data for each person would be arranged in the same order. One possible order is name, age, height, weight, address, and phone number. If you wished to access the data shown in Figure 4-21, you would set the IX or IY register to the address of the start of each subblock. Then you would specify an offset to be added to this address to get you to the data within the subblock. This is shown in Figure 4-22.

Mnemonics for accessing this data are:

```
LD IX,8500H
LD A,(IX+2)
```

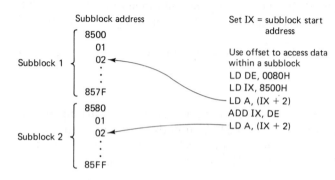

Figure 4-22   Using the index value +2 you can access the second byte of each block quite easily under program control.

All the previous rules for Z80 mnemonics are still in force. The data from memory addressed by the IX register plus the offset will be loaded into memory. In this example, the data from memory address 8502 will be loaded into the A register. The IX register may be replaced by the IY register.

4-6 EXTENDED OP-CODES

The term *extended op-codes* refers to the Z80 object code form of the mnemonics. When the 8080 and 8085 mnemonics were assembled into object code format, there was a possibility of 256 different op-codes. However, not all the 256 possibilities were used for the complete instruction set of the 8080 and 8085. The byte DDH was one of the unused op-code bytes in the 8080 and 8085 instruction set.

In examining the Z80 instruction set, you find there are many more instructions, including every instruction for the 8080 and 8085. Designers of the Z80 took advantage of the fact that not all the bytes for op-codes in the 8080 instruction set were used. This allowed the Z80 to use these "illegal" 8080 and 8085 op-codes for valid Z80 op-codes. However, in accomplishing this, the Z80 object code form of an instruction was not optimized. That is, a 3-byte instruction may take 4 bytes with the Z80.

As an example, let's examine the object code for the instruction LD IX, 835A = DDH 21H 5AH 83H. Notice that this takes 4 bytes, although all the other 16-bit immediate instructions we have discussed required only 3 bytes. This instruction takes 4 bytes because the first byte informs the Z80 that this is an extended op-code instruction. By using the extended op-code form, two instructions may even have the same first byte.

For example, the instruction LD IX ,(835A) has the object code form DDH 2AH 5AH 83H. The first byte gets the microprocessor into the level of decoding the remaining bytes as extended op-code instructions. As you use the Z80 mnemonics, you will find that there are other extended op-code instructions, which are due to the expanded instruction set and complete compatibility with the 8080 and 8085.

4-7 BLOCK INSTRUCTIONS FOR THE Z80

The Z80 has unique instructions designed for performing operations on blocks of data in memory. For example, suppose you have a block of data in memory and you wish to search for a specific byte. An instruction for the Z80 will allow you to specify the address bounds for the search, and complete *the* search, and stop when there is a match or logically inform you there was no match. This concept is shown in Figure 4-23.

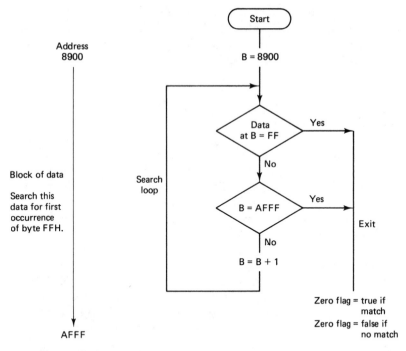

Figure 4-23 Flowchart showing the sequence of events that occurs during the execution of the block search instruction.

Mnemonics for the block instructions available for the Z80 are as follows:

```
LDI      load and increment
LDD      load and decrement
LDIR     load increment and repeat
LDDR     load decrement and repeat

CPI      compare and increment
CPD      compare and decrement
CPIR     compare increment and repeat
CPDR     compare decrement and repeat
```

Let's examine each of these instructions and show exactly how it is implemented with the Z80 microprocessor. All block transfer instructions use the BC, DE, HL register pairs. The BC register is used as a 16-bit counter, HL is used as a memory pointer for the source operand, and DE is used as a memory pointer for the destination operand.

LDI

The LDI instruction moves only 1 byte each time it is executed. The data pointed to by the HL (source) is moved to the memory location pointed to by the DE (destination). After this instruction is executed, HL = HL + 1, DE = DE + 1, and BC = BC − 1. Once you understand exactly how this instruction operates, it will be an easy matter to see how similar block instructions work. To that end, let's get into some detail about using the LDI instruction.

Example 4-5

In this example we wish to move the 1024 bytes of data starting at memory address 8100H to the memory address starting at 3000H.

There are several methods for accomplishing this function, and the technique shown here makes use of the LDI instruction. Prior to using the LDI instruction the approriate registers must be set up, as follows:

```
LD HL,8100H     set up source pointer
LD DE,3000H     set up destination pointer
LD BC,1024      set up count, or number of
                bytes to move
```

With these registers set up, the following instruction loop will perform the block move.

```
B_loop     LDI
           JP PE,B_loop     check the parity flag
```

During the execution of the LDI instruction, the P/V flag will show the results of the BC register after the instruction is executed. The flag will be set to a 0 (odd parity) if the BC register will equal zero after execution of the instruction. That is, the flag actually shows BC−1.

Notice that once the source and destination registers are set up, they are automatically incremented. Therefore, you must keep track only of how many bytes to move; the instruction will increment the address pointers for you. Figure 4-24 shows a flowchart of how the preceeding program operates. Note the value that is placed into the BC register in order to move a specified number of bytes.

LDD

The load and decrement instruction operates in the same manner as the LDI with this exception. The source and destination memory pointers are *decremented* rather than incremented after each execution of the instruction. This

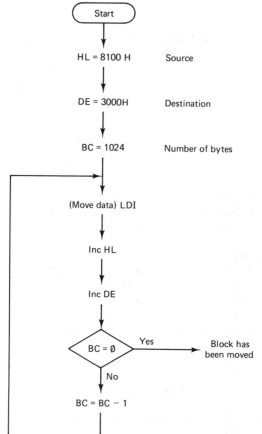

Figure 4-24   Flowchart showing the sequence of events that occur during the execution of the block move instruction.

allows you to move data starting at the highest memory location. The following example shows how this instruction may be used.

Example 4-6

We will move a 1024 block of data from address space 8000–83FF to 7400–77FF. The following is a program to accomplish this.

```
LD BC,1024              set up how many bytes to move
LD HL,83FFH             set up maximum address of
                        source
LD DE,77FFH             set up maximum address of des-
                        tination
;
B_loop LDD              load and decrement first byte
       JP PE,B_loop     keep going until P/V=0
```

This program is very similar to the one presented in Example 4-7, with the exception that the source and destination registers are decremented instead of incremented after each execution.

LDIR, LDDR

We present these two instructions at the same time because their use is very similar. This single instruction will move an entire block of data. The BC, DE, and HL registers have the same function as presented for the LDI and LDD instructions. Example 4-7 shows how these instructions are used.

Example 4-7

In this example we will move a block 512 bytes of data from address 9000H–91FF to address B000H–B1FFH.

```
            LD BC,512        load byte counter
            LD HL,9000H      load source pointer
            LD DE,B000H      load destination pointer
B_loop      LDIR             move the entire block
```

Example 4-8

In this example we will move the same block of data using the LDDR instruction instead of the LDIR instruction.

```
            LD BC,512        load byte counter
            LD HL,91FFH      load source pointer
            LD DE,B1FFH      load destination pointer
B_loop      LDDR             move the entire block
```

As the block of data is being transferred, the source and destination registers are decremented toward 9000 and B000.

In examining these instructions, there seems to be little programming difference between the LDI, LDIR, LDD, and LDDR. The loop required to move the block of data requires only one extra instruction with the LDI as compared to the LDIR. While this is true of the programming, the hardware operations are very different. There are two main points to be considered:

1. Response time for an interrupt
2. Refresh address output

We have not discussed either of these two main points in this text. However, let's briefly mention it now and reconsider it at the point in the text where these topics are discussed in detail. The point about interrupt response time is this: When the microprocessor is issued an external interrupt, it will respond after execution of the pending instruction. For the LDIR and LDDR, this could be quite a long time because of the enormous amount of memory reads and writes that could occur.

If the block move loop were accomplished with the LDI and LDD instructions, the interrupt would be sampled after each instruction. Therefore, the time for interrupt acknowledge would be shorter with the LDI loop than with the LDIR loop.

The second point about the refresh address has to do with the situation where the Z80 is using and controlling dynamic RAMs. We mention this briefly here and give a more in-depth explanation in the chapter of the text that discusses using dynamic RAMs.

Dynamic RAMs require addressing at minimum time intervals to retain the information written into them. The Z80 is designed to output a refresh address in the R register during each instruction cycle. However, during execution of the LDDR and LDIR instructions, there will only be one refresh address output when the Z80 first reads the instruction from memory. The next address will be output after the LDIR and LDDR instruction is complete and the next instruction is executed. This could be in several hundred milliseconds, depending on the number of bytes transferred.

The dynamic RAM devices normally require a refresh address to be output in intervals of tens of microseconds. This usually occurs during the execution of a program. It is only during the execution of block instructions that the Z80 may have hardware problems with dynamic RAMS. If you are unsure of exactly what a refresh address does, do not be concerned at this time. It is mentioned only to complete the discussion about these instructions. Be assured that this topic will be discussed in the chapter on dynamic RAMs. Also keep in mind that the interrupt topic will be covered in detail in the chapter on interrupts. For now, just acknowledge these facts about the use of the LDIR and LDDR block instructions.

Block Compare Instructions

Like the block move instructions just presented, there are four forms of the block compare instructions. The mnemonics for these are:

```
CPI     compare with increment
CPD     compare with decrement
CPIR    compare, increment, and repeat
CPDR    compare, decrement, and repeat
```

There are only two main register pairs used with the block compare instructions, HL and BC. The HL register is used as a memory pointer into the compare block. The BC register is used as a byte counter to indicate how many memory addresses to check.

The contents of the data at the memory location addressed by HL are compared against the data in the A register. After execution, the flags show the result of the compare: HL = HL $\pm$1, BC = BC $-$ 1.

To show how the CPI and CPD instructions operate, consider Example 4-9.

Example 4-9

In this example we will use the CPI and CPD instructions to search a block of 2048 memory locations for a specific byte of data. When this byte of data is reached, the loop will stop, with the HL registers pointing to the address of the compare byte minus 1. We will assume that the block of data to search in memory is contained between memory locations A000–A7FF. Further, let's assume that the correct byte of data resides at memory location A489H. In a real example, you would not know where the byte existed. It is given here simply to show how the registers will appear as you exit the loop.

```
       LD BC,2048          set byte count
       LD HL,A000H         starting address for block
       LD A,55H            byte to compare against in A
                           reg

       ;

B_loop CPI                 check for compare
       JP Z,MATCH          was a good compare
       JP PE,B_loop        no match, not finished with
                           loop
```

At the section of program labeled MATCH, the HL register would be equal to A48A, which is one higher than the actual match data. Notice that the zero flag must be checked to see if there was a match. The P/V flag will determine if the loop is finished, as was the case for the LDI and LDD instructions. If the program does not exist the loop with the JP Z instruction, then there was no match with the data in the A register over the entire block of memory.

Example 4-10

The following program is identical to that of Example 4-9, except that the search will be from the highest memory location to the lowest. When the loop is exited, the HL register will point to the matched memory location minus 1.

```
       LD BC,2048          set byte count
       LD HL,A7FFH         starting address for block
       LD A,55H            byte to compare against in A
                           reg

       ;
```

```
B_loop CPI              check for compare
       JP Z,MATCH       was a good compare
       JP PE,B_loop     no match, not finished with
                        loop
```

At the match location, HL would equal A488H.

CPIR, CPDR

The two instructions CPIR and CPDR are very similar in their operations, with the exception that one increments (CPIR) the HL pair, whereas the other (CPDR) decrements the HL pair.

Each of these instructions will continue to loop until one of the following conditions is met:

1. The BC register pair is 0.
2. The A register matches the (HL) data in memory.

When you exit the loop, the program must check the condition of the flags to determine how the exit was accomplished. That is, if the zero flag is set, then the exit occurred because of a match. If the parity flag is false, then the exit occurred because BC = 0. The following example shows how the CPIR and CPDR instructions may be used to check a block of memory.

Example 4-11 (CPIR)

```
LD BC,2048      set byte count
LD HL,A000      starting address for block
LD A,55H        byte to compare against in A
                reg
;
CPIR            check for compare
JZ MATCH        if true the exit was match
```

The value of the HL register will be the match address plus 1 due to the extra increment that occurs after each instruction.

Example 4-12 (CPDR)

```
LD BC,2048      set byte count
LD HL,A7FFH     starting address for block
LD A,55H        byte to compare against in A
                reg
;
CPDR            check for compare
JZ MATCH        if true the exit was match
```

The HL register will equal the match address minus 1 due to the extra decrement that occurs with this instruction.

4-8 A PROGRAMMING EXAMPLE

The following example shows the use of a subroutine that returns the hexadecimal ASCII (American Standard Code for Information Interchange) values of the two binary nibbles in the accumulator.

```
 1                          ;;;;;;;;;;;;;;;;;;;;;;;;;;;;;;;
 2                          ;
 3                          ; SUBROUTINE FOR CONVERTING AN
 4                          ; 8 BIT BINARY NUMBER TO TWO
 5                          ; ASCII CHARACTERS FOR TRANSMISSION
 6                          ; ON A SERIAL BUS.
 7                          ;
 8                          ;;;;;;;;;;;;;;;;;;;;;;;;;;;;;;;;;;
 9                          ;
10                          ; ENTER SUBROUTINE WITH
11                          ; A = 8 BIT DATA TO CONVERT
12                          ;
13                          ; RETURN WITH ASCII DATA IN
14                          ; B = MSB
15                          ; C = LSB
16                          ;
17                          ; DESTROYS REGS A,B,C
18                          ;
19                          ;;;;;;;;;;;;;;;;;;;;;;;;;;;;;;;;;;
20                          ;
21   0000   F5         ASCII   PUSH PSW    ;SAVE A REG
22   0001   E6 0F              ANI 0FH     ;MASK OFF HI NIBBLE
23   0003   FE 09              CPI 09H     ;CHECK FOR NUMBER
24   0005   DA 0D 00           JC ALPHA1   ;IF JUMP ALPHA A-F
25   0008   F6 30              ORI 30H     ;CHANGE TO 30H-39H
26   000A   C3 0F 00           JMP HI      ;DO HI NIBBLE
27   000D   C6 37      ALPHA1  ADI 37H     ;CHANGE TO 41H-46H
28   000F   4F         HI      MOV C,A     ;LOWER BYTE IN C
29   0010   F1                 POP PSW     ;RESTORE A REG
30   0011   07                 RLC
31   0012   07                 RLC
32   0013   07                 RLC
33   0014   07                 RLC         ;HI NIBBLE TO LO NIBBLE
34   0015   E6 0F              ANI 0FH     ;MASK OFF HI NIBBLE
35   0017   FE 09              CPI 09H     ;CHECK FOR NUMBER
36   0019   DA 21 00           JC ALPHA2   ;IF JUMP ALPHA A-F
37   001C   F6 30              ORI 30H     ;CHANGE TO 30H-39H
38   001E   C3 23 00           JMP LAST    ;FINISHED WITH CONVERT
39   0021   C6 37      ALPHA2  ADI 37H     ;CHANGE TO 41H-46H
40   0023   47         LAST    MOV B,A     ;SAVE INTO B REG
41   0024   C9                 RET         ;DONE WITH CONVERSION
42   0025                      END
```

For example, suppose the data in the accumulator were equal to 01101100, which is equivalent to the hexadecimal characters 6C. The ASCII equivalent of these two characters is 6 = 00110110 and C = 01000011. The program will return the two 8-bit ASCII bytes.

4-9 CHAPTER SUMMARY

This chapter started with the general topic of a stack area for a microprocessor system. You learned that this is an area of memory reserved for special CPU operations. Topics covered included how to push and pop information from the stack. Once you understood how the stack worked, subroutines were introduced. You were shown how the microprocessor calls and returns from a subroutine.

After introducing these general topics for all microprocessors, the chapter focused on special topics that apply only to the Z80 microprocessor. These topics included the alternate register set, special Z80 registers, and extended op-codes. From here, you learned about block transfer instructions common only to the Z80. These instructions allow the microprocessor to move, search, and replace large blocks of memory data.

The chapter finished with a programming example showing how to convert an 8-bit number into two ASCII characters for transmission on a serial bus.

This chapter concludes the formal study of programming a microprocessor system. In the remaining chapters of this text, you will use the information given in Chapters 2, 3, and 4 to program and control different hardware of the system. The hardware and software of the microprocessor system must operate together in a complete system application.

REVIEW PROBLEMS

1. Define the term stack area.
2. Show a memory map with a stack area of 1024 bytes and the highest address of the stack of 7FFF.
3. Show the 8080 and Z80 mnemonic instructions to set the stack pointer register to 7FFF.
4. What 8080 instruction writes the BC registers to the stack?
5. What registers are written onto the stack when the PSW is pushed?
6. What 8080 instruction removes data from the system stack?
7. Is a stack described as a FIFO (first in, first out) or LIFO? Why?
8. Write a set of 8080 mnemonics that exchange the contents of the BC and DE registers using stack operations.
9. True or False: To properly use subroutines the microprocessor must first have the stack pointer initialized.
10. What instruction is used to jump to a subroutine? How many bytes is the instruction?
11. What information, if any, is written to the system stack when the microprocessor jumps to a subroutine?
12. True or false: Subroutines require the microprocessor to execute fewer instructions.
13. List four 8080 instructions that force the microprocessor to execute a subroutine.

14. List the sequence of events that occur when the microprocessor returns from a subroutine.
15. What is a problem that may occur when the stack area is accessed during a subroutine?
16. List the Z80 alternate registers.
17. Show the Z80 instructions that access the alternate registers.
18. What is an advantage of having an alternate register set?
19. What is an extended op-code? Give an example.
20. List all the Z80 block instructions.

5

HARDWARE
FUNDAMENTALS

In Chapter 1 you were introduced to the 3-bus system architecture for a typical microprocessor system. This architecture included the address, data, and control buses. At that time you were asked to accept that these buses would indeed operate and allow the system software to execute properly. No details concerning actual hardware for construction of the buses was given.

This chapter reviews and describes the 3-bus system architecture. Knowledge of this architecture will allow you to understand basic hardware orginization, interfacing, and troubleshooting of microprocessor systems. Many examples of interfacing to microprocessors are given in this text and all rely on the 3-bus architecture presented here. Further, almost all microprocessor-based systems make use of these three buses. This is true for 8-, 16-, and 32-bit microprocessors. The three busses for each miroprocessor, 8080, 8085, and Z80, will be presented and discussed. At the end of these discussions you should be able to relate the details to any other microprocessor based system.

Let us begin with a definition of a system bus.

System bus.   A collection of electronic signals and signal lines or paths that are grouped according to function. Each signal line of the bus has the same point of origin and destination. The width of a bus is the number of signal lines contained in the group.

5-1 REVIEW OF THE 3-BUS ARCHITECTURE

The three major buses used to describe the digital action in a microprocessor-controlled system are the

1. Address bus
2. Data bus
3. Control bus

Every hardware action that takes place in a microprocessor-controlled system can be performed using the 3-bus approach. Notice that 3-bus system architecture is not a simplified structure used to describe a complex action; rather, it is an accurate model that presents the complex action of a microprocessor system in a different, easier-to-understand way.

As stated earlier, the 3-bus model can accurately describe the following seven hardware actions that occur in microprocessor controlled systems:

1. Write data to system memory from the CPU.
2. Read data from system memory to the CPU.
3. Write data to system output devices from the CPU.
4. Read data from system input devices to the CPU.
5. Handle interrupt activity by the CPU.
6. Control DMA (direct memory access) activity.
7. Manipulate internal registers contained in the CPU.

These hardware activities occur in a system as a result of the software that is being executed by the CPU. This is true with the exception of events 5 and 6, which are usually initiated by external hardware.

Each hardware activity in a microprocessor system falls into one of the seven catagories shown. However, microprocessor systems do not have to use all seven hardware operations to be of value; many very useful systems are designed to employ only three or four of the seven possible actions. But no matter how complex the operation or how long a controlling software program is, every system is executing only the seven hardware operations listed.

Figure 5-1 is a block diagram for a typical microprocessor system. Notice

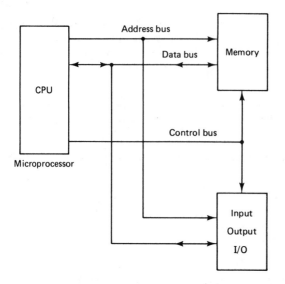

Figure 5-1 Block diagram of a typical microprocessor system.

the three major system buses that are used. In the discussions that follow, you will be shown how these three buses can be realized with hardware. Once you understand how these three buses are designed with hardware, each type of system communication (hardware activity) listed previously will be presented. These presentations will show the use of the three buses and how they are electrically connected to perform the needed hardware operation.

For the present you are asked to accept that these three buses will perform all of the hardware functions that were listed. This will be made quite clear as you progress through the text.

5-2 SYSTEM ADDRESS BUS

Let us now examine the first of the three buses given in Figure 5-1, the address bus. The three microprocessors presented in this text, 8080, 8085, and Z80, all have an address bus that is 16-bits wide. This means that 16 physical lines are contained in the complete address bus, as shown in Figure 5-2. At this point in the discussion we will assume that the address bus originates at the microprocessor and is output to the system hardware. Later you will see that the address bus does not have to originate at the microprocessor. It may be output by some other hardware in the system.

The function, or job, of the system address bus is to enable or select the path for communication. When communication occurs in the system, the address bus will logically define the hardware that will be sending data or receiving data from the CPU. This is true for all types of hardware communications.

In a typical 8080 microprocessor-based system, the address bus may be

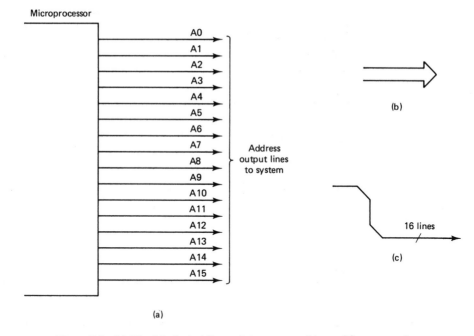

Figure 5-2 (a) The 16 physical lines of the system address; (b) one way these address lines are shown on a schematic; (c) an alternate schematic form of the 16 address lines.

realized as shown in Figure 5-3. In this figure you see that address lines A0–A15 are output directly from the CPU and connect to the system hardware.

The Z80 Address Bus

Like the 8080 microprocessor, the Z80 has 16 physical pins dedicated to the function of address lines A0–A15. These pins are output from the Z80 and connected directly to the external hardware of the system. This is shown in Figure 5-4.

The 8085 Address Bus

The address bus for the 8085 microprocessor is realized physically in a different way than for the 8080 and Z80 microprocessors. For these two microprocessors, a specific pin on the CPU is dedicated to the function of each address line output. No matter in what state the microprocessor is executing, the pin labeled A0 on the microprocessor is always functionally A0. This makes the formation of the 16-bit address bus quite easy. We simply group the 16 bits and pins on the CPU together and designate that group of pins as the system address bus.

However, the 8085 microprocessor does not utilize this technique but

8080

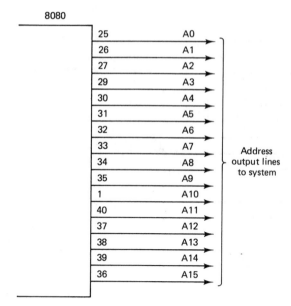

Figure 5-3 Schematic diagram showing the pinout of the 8080 address lines.

instead uses that of time-multiplexed pin functions. This means that the physical pins on the 8085 microprocessor may represent different functions at different points in time. This technique allows more "different" pins in a 40-pin package than would otherwise be available.

The 8 pins of the 8085 CPU that are multiplexed take on the function

Z80

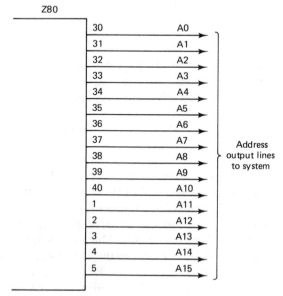

Figure 5-4 Schematic diagram showing the pinout of the Z80 address lines.

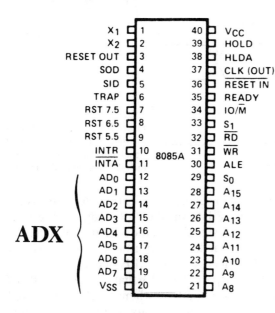

ADX

Figure 5-5 Pin designations for the 8085 microprocessor. Notice that there are pins labeled AD_0–AD_7. These pins are the time-multiplexed data bus and the lower byte of the microprocessor address lines. (Courtesy of Intel Corp.)

either of the data bus or of the lower 8 bits of the address bus. Sometimes they are carrying an address output; at other times they are carrying data in or out of the 8085 microprocessor. These pins are labeled AD0–AD7 (address data) on the 8085 CPU, as shown in Figure 5-5. Referring to Figure 5-5, note that the upper 8 bits of the system address, A8–A15, are exactly the same as other microprocessor address pins—that is, a specific pin on the 8085 microprocessor is designed for use as an address output pin only.

Because the 8085 device uses multiplexed pins for address and data, the formation of the lower byte (8 bits) of the address bus, A0–A7, becomes slightly more complex than that of other microprocessors, such as the 8080 or the Z80. When using multiplexed addressing, we must latch the logical state of the AD0–AD7 pins of the 8085 microprocessor just when they functionally represent the address bus A0–A7. To accomplish this latching we must know the exact moment when the signals on these pins represent the address information. Fortunately, the designers of the 8085 microprocessor made this easy. A special pin (30) on the 8085 CPU is labeled ALE (address latch enable). The ALE signal is normally in the logical 0 state. This signal changes to a logical 1 state when the data on the 8085 AD0–AD7 pins represents address information A0–A7. When the ALE signal reverts from a logical 1 to a logical 0 state, the information on the AD0–AD7 pins should be latched. This is shown in the timing diagram of Figure 5-6.

Figure 5-7 shows how this data can be latched with hardware. Figure 5-7 also shows the complete 16-bit address bus using an 8085 microprocessor. The ALE signal is inverted in Figure 5-7 because of the edge-triggered latch. The specification for the 8085 device requires that the address data be latched on the

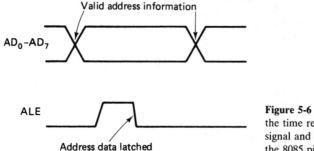

Figure 5-6 Timing diagram showing the time relationship between the ALE signal and the address information on the 8085 pins labeled AD_0–AD_7.

falling edge of the ALE signal. However, specifications for the 74LS374 octal latch show that this device latches data on the rising edge of the clock input. Therefore, we must invert the ALE signal from the 8085 microprocessor. Inverting this signal will allow us to latch the information AD0–AD7 into the 74LS374 octal latches on the falling edge of the ALE signal.

Another type of latch (transparent latch) that is used for this function enables data to pass through to the Q outputs when the clock is in the logical 1 state. When the clock goes to a logical 0, the data on the device output is held fixed. The 74LS373 is an example of this type of latch. When the 8085 address lines are valid, they immediately pass through the latch to the outputs. Figure 5-8 shows the hardware required for using the 74LS373 latches. The difference in timing for the 74LS374 and 74LS373 address latches is shown in Figure 5-9.

Buffering the Address Bus

When interfacing devices to a microprocessor system, the address bus is used. Any device that is connected to the CPU, such as memory, timer chips, floppy disc controller chips, or simple latches will place some electrical load on the address lines. See Figure 5-10.

If the load placed on the address bus by the external devices is excessive, the entire system will malfunction. This is due to the fact that the address output lines of the microprocessor cannot adequately drive the load to a valid logical 1 or logical 0 voltage level. Therefore, in some microprocessor systems the address bus may be constructed with devices called *address buffers*. A block diagram of this type of address bus is shown in Figure 5-11.

The purpose of the address buffers is to increase the current drive capability of the microprocessor address lines. Referring to Figure 5-11, you see that the microprocessor address outputs need only drive the address buffer inputs. All the system load is driven by the buffer outputs. A similar function occurs in an audio system when a power amplifier is installed to drive the heavy load of a speaker system.

Figure 5-12a and b shows how address buffers may be installed in the 8080 and Z80 microprocessor systems. Each line of the address bus is connected to a

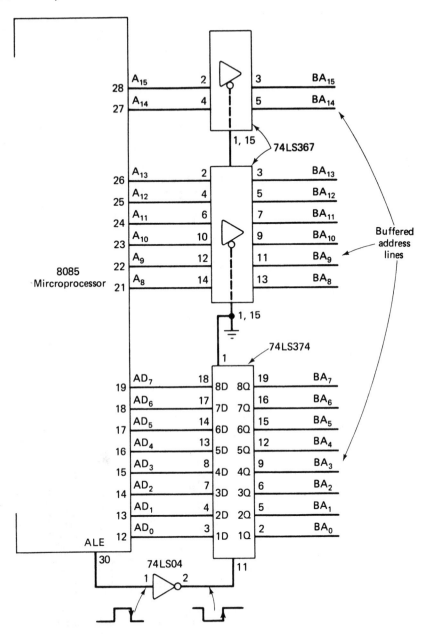

Figure 5-7   Schematic diagram showing how the lower 8 bits of the 8085 microprocessor may be latched with the hardware.

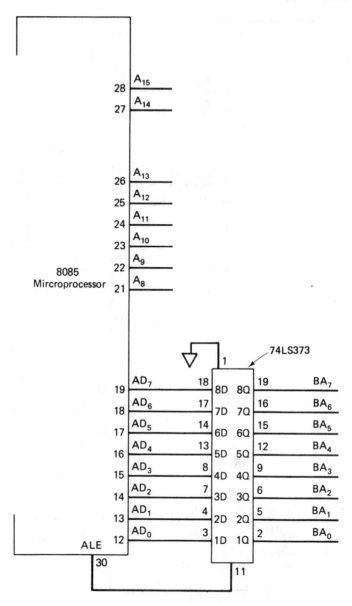

Figure 5-8 An alternate solution for latching the lower 8 bits of the 8085 address bus. This technique uses the 74LS373 latch instead of the 74LS374 latch shown in Figure 5-7.

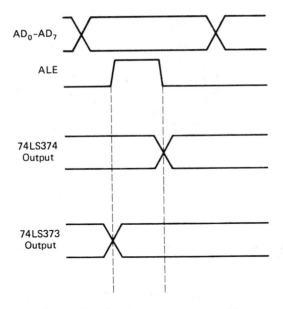

Figure 5-9 Timing diagram showing the relationship of the address output if the 74LS373 latch is used instead of the 74LS374.

buffer. The buffer output becomes the system address bus, which is connected to the rest of the system components.

   Buffering the 8085 address bus is slightly different than buffering the 8080 or the Z80. Since the 8085 address bus is multiplexed, the lower address lines, A0–A7, are already buffered by the use of the 74LS373 or 74LS374 address latch. The remaining address lines, A8–A15, may be buffered in exactly the same way as shown for the 8080 and Z80. Figure 5-13 shows the complete buffered 8085 system address bus.

   The topic of buffering the address bus is introduced at this time because many systems encountered in industry use address buffers. However, address

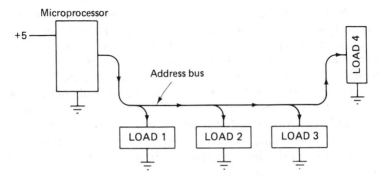

Figure 5-10 The microprocessor system address bus is connected to many loads in parallel. Here the address bus is required to drive four loads.

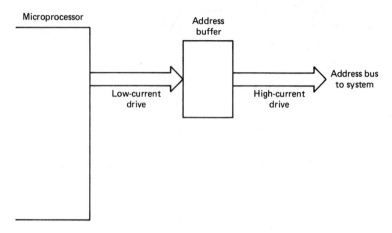

Figure 5-11 Block diagram showing how address buffers fit into the overall concept of generating the address bus.

buffers are not necessary in all microprocessor system applications. The need is dependent on the electrical load that the microprocessor address bus will encounter.

5-3 THE DATA BUS

In the system block diagram given in Figure 5-1, the data bus is used to transfer information from the microprocessor to the system (write) and move the information from the system to the microprocessor (read). The system data bus is not electrically aware of which hardware circuit in the system will be receiving or sending the data for the CPU. That is the job or function of the system address bus. When the data is read into the microprocessor or written out of the microprocessor, the data bus will be the means by which this is accomplished.

Notice that the jobs of the system address and data bus are being separated. Each has a certain function in a communication operation. That is, the address bus can perform its job or function even if the data bus is not doing its function. The converse is also true. Making this assumption allows you to troubleshoot and understand the operation of each bus independently. We discuss this feature of the 3-bus architecture throughout the text.

Figure 5-14 shows the job of the system address bus and the system data bus during a typical system communication. We see in Figure 5-14 that the address bus will select the path, whereas the data bus simply transmitts the data to all paths. Only the path selected by the address bus will make use of the system data bus.

If no data bus buffering is used, the system data bus, like the system address bus, consists of the pins on the microprocessor labeled D0–D7 (for the

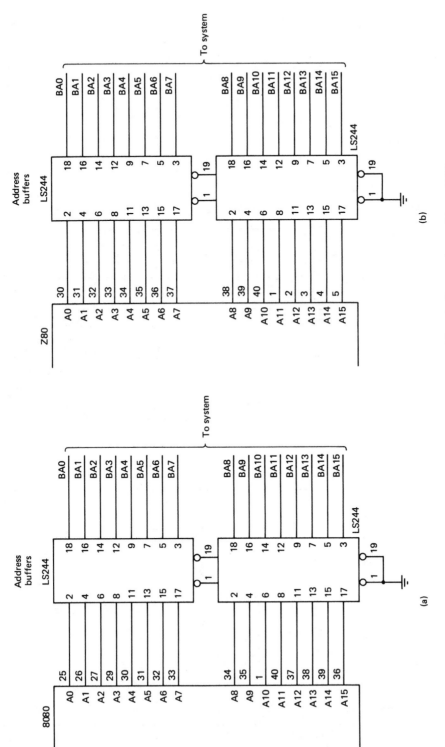

Figure 5-12 (a) Schematic diagram showing how address buffers can be used in an 8080 system; (b) schematic diagram showing how address buffers can be used in Z80 system.

139

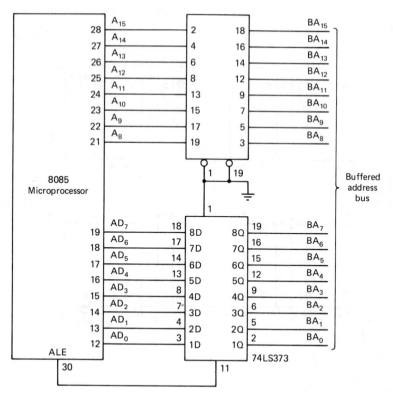

Figure 5-13 Schematic diagram for address buffering in an 8085 microprocessor system. Notice that the lower address lines are already buffered by use of the 74LS373 latch.

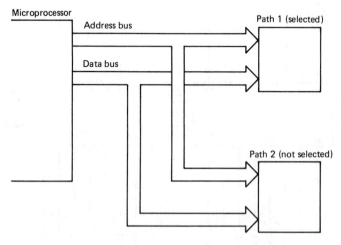

Figure 5-14 During a typical system communication, the address bus will select the communication path and the data bus provides the physical means for the data to be communicated between the CPU and the selected path.

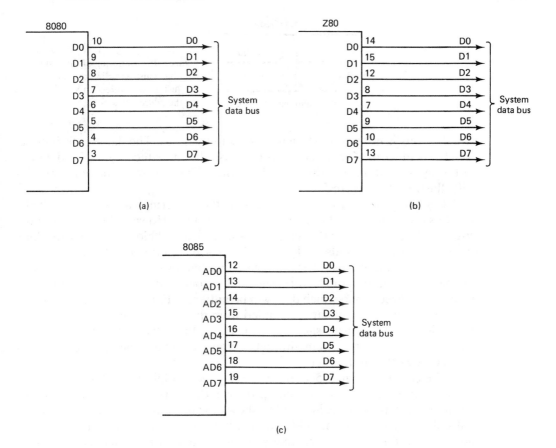

Figure 5-15 (a) Unbuffered data bus for the 8080 microprocessor; (b) unbuffered data bus for the Z80 microprocessor; (c) unbuffered data bus for the 8085 microprocessor.

8080 and Z80) and the pins labeled AD0–AD7 for the 8085. This type of data bus is shown in Figures 5-15a, b, and c for the three different microprocessors.

Buffering of the Data Bus

As mentioned previously, the data bus is bidirectional. That is, information on the data bus sometimes enters the microprocessor (when the microprocessor is in the input mode), and at the other times the information is generated by the microprocessor and is then output to the system (when the microprocessor is in the output mode.) The data bus is often connected to many inputs in parallel in a typical microprocessor-based sytem. This creates the possibility that all these inputs together may load the data bus excessively. This loading is similar to the excessive loading that may occur on the system address bus.

However, an additional factor must be considered for the data bus. That

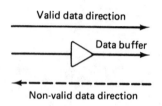

Valid data direction

Data buffer

Non-valid data direction

Figure 5-16 Buffering of the system data bus must occur in both directions, not just a single direction as was needed for the address bus.

is, a data bus is not unidirectional as is an address bus. The data bus is bidirectional. This fact indicates that we connot use the same techniques for data bus buffering as we did for address bus buffering. This new requirement is graphically illustrated in Figure 5-16.

Referring to Figure 5-16 we see that the buffering can work only if the data bus is outputting data to be used by the system. However, if data from the system is being sent to the microprocessor, it will be "blocked" from reaching the microprocessor inputs. In Figure 5-17 we see a bidirectional buffering technique. This technique uses tristate logic to accomplish the bidirectional buffering. Note the addition of a control signal. The logical state of this control signal dictates the direction in which data will be buffered. For example, if the control signal is a logical 1, then data is enabled from the system to the microprocessor input pins. If the control signal is a logical 0, then data is enabled from the microprocessor to the sytem through buffer A. When one tristate buffer is enabled, the other is disabled.

The circuit of Figure 5-17 illustrates a general bidirectional buffering concept that can be applied to the microprocessor data bus. We next discuss each of the three microprocessors in detail to show how this buffering technique can be realized. Keep in mind that bidirectional buffering requires a control signal to indicate data direction. We further show how this control signal can be generated when using the different microprocessrs.

In microprocessor systems that do not require data buffering, the data bus

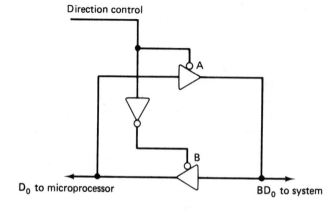

Direction control

A

B

D_0 to microprocessor BD_0 to system

Figure 5-17 A bidirectional buffering technique that uses a control line to determine which direction the buffering will occur.

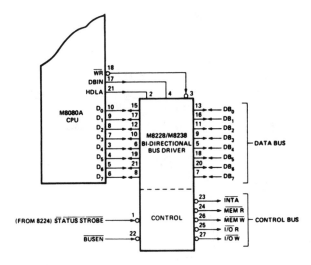

Figure 5-18 Schematic showing how the 8228 system controller connects to the 8080 microprocessor. (The 8228 does all of the data bus buffering and direction control.) (Courtesy of Intel Corp., 1979.)

lines of the system will connect directly to the microprocessor data lines, in Figures 5-15a, b, and c.

The 8080 Buffered Data Bus

When an 8080 microprocessor is used in a system, the buffering of the data bus is not a large design concern. This is due to the fact that the 8080 is made to be used with another special integrated circuit called the 8228 system controller and bus driver.

In brief, the 8228 buffers the data bus and automatically controls the direction of data flow within the system. If we wish to use an 8080 microprocessor, it would also be wise to plan to use an 8228 system controller and bus driver within the system. Figure 5-18 shows the 8228 and 8080 connected together to form the system data bus.

If the system under consideration does not require data bus buffering, then the system data lines are connected directly to the D0–D7 lines on the 8080 microprocessor. This will be true for all microprocessor systems that do not require data bus buffering. However, for an 8080 system, even if data buffering is not required, the 8228 will simplify the design of the system control bus. Therefore, its use with an 8080 is desirable as a minimum amount of hardware.

The Z80 Buffered Data Bus

The Z80 does not have a special system controller similar to the 8228, nor is one required. We will use the 74LS245 device to provide the data bus buffering required for the Z80 microprocessor. The 74LS245 will be enabled (for the present time) by connecting pin 19 to a logical 0. The DIR pin 1 on the 74LS245 device will be connected to the control signal of the Z80 microprocessor, labeled $\overline{RD}$ pin 21 of the Z80 microprocessor. This is shown in Figure 5-19.

When the $\overline{RD}$ is in logical 0, this indicates that the Z80 is in a mode to receive data. Therefore, the 74LS245 must be in a mode to enable data from the system to the Z80 data bus. The 8080 system used the DBIN signal (active logical 1) to control the direction of the data path. In the case of the Z80, the signal output from the Z80 to receive data is active low, but with the 8080 the same signal is active high.

The 8085 Buffered Data Bus

As mentioned earlier, the 8085 has a multiplexed data and address bus. However, this point need not be stressed when dealing with the buffering of the data bus. We again make use of the 74LS245 as the data bus buffer. The $\overline{RD}$ signal pin 32 of the 8085 is active low when the 8085 is in the mode to receive data from the system. This signal will be the DIR signal input to the 74LS245. The complete circuit is shown in Figure 5-20 for the buffering of the 8085 system data bus.

Like the Z80 and 8080, if the system does not require data bus buffering, then the data bus lines are connected directly to the AD0-AD7 pins of the 8085.

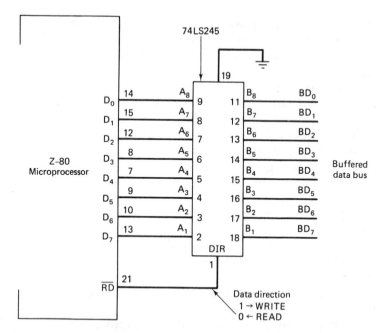

Figure 5-19 Complete schematic for the Z80 data bus buffer using the 74LS245 device.

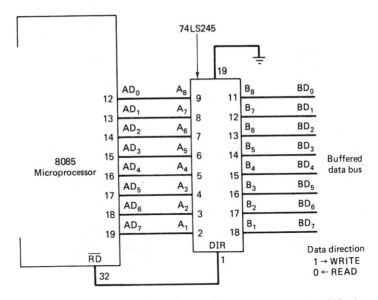

Figure 5-20   Complete schematic for the buffered data bus on the 8085 microprocessor.

5-4 THE CONTROL BUS

Thus far we have discussed the system address bus and the system data bus in the 3-bus architecture. The third bus is the system control bus. This bus has two functions. One function is define the type of communication and the second is to start and stop the communication. Let us examine what is implied in these statements. First, the system control bus electrically defines the type of communication. How many different types of communications are there? Let us list them:

1. Memory read
2. Memory write
3. Input read
4. Output write
5. Interrupt acknowledge
6. Hold acknowledge

For the control bus to define the specific type of activity in the system, there must be a separate control line or a unique method of selection for each one. In some systems, the control bus may not be as straightforward as a single line for each type of communication.

We stated that the second job of the system control bus is to start and stop the communication. This implies that the control lines are timed. They occur at certain points in the clock cycle. This type of information may be obtained from the system specifications.

8080 System Control Bus

When we discussed the 8080 data bus we found that a special device, an 8228 system controller and bus driver, provides all the data bus buffering and direction control required for that system. This same device also generates the four basic system control signals mentioned earlier. This is shown in Figure 5-21, where we see that these four control sigals are labeled $\overline{\text{MEMW}}$, $\overline{\text{MEMR}}$, $\overline{\text{IOR}}$, and $\overline{\text{IOW}}$. These signals are all active logical 0 and are also mutually exclusive. That is, no two of these signals will ever be active logical 0 at the same time.

If we use an 8080 microprocessor for a system controller, then the 8228 device can make system design much easier. However, to appreciate some

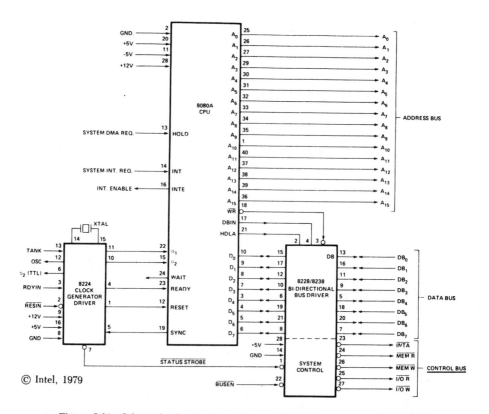

© Intel, 1979

Figure 5-21   Schematic showing how the control bus is an output of the 8228 system controller. (Courtesy of Intel Corp.)

alternatives, we will discuss how the four control signals can be generated by using discrete digital devices. This discussion will enhance our understanding of how an 8080 microprocessor contols the hardware in the system.

The 8080 Status Latch

The 8080 CPU generates a byte on the data bus pins called a *status word*. The status word is generated to let the system hardware know the microprocessor's intentions. The status word is generated at the beginning of the bus cycle. If the 8080 is setting up to read data from memory, the 8080 first generates the status word to indicate this. The system uses the status word to enable the correct hardware logic path for the 8080 to use.

When the status word is present on the data bus, the 8080 informs the system by forcing the sync signal pin 19 of the 8080 to a logical 1. The system then uses the sync signal to enable a strobe to move the data on the data bus into a latch called the *status latch*. This is shown in Figure 5-22.

After the status word is latched, certain bits of the status information are

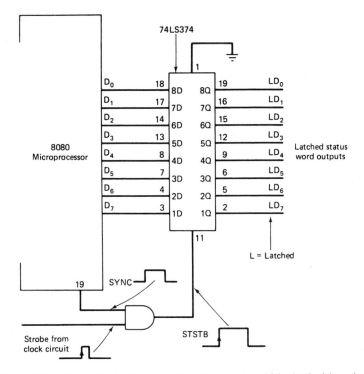

Figure 5-22   Schematic showing how the status word could be latched by using discrete logic. The sync pulse going to logical 1 level enables the status word to be latched into the 74LS374.

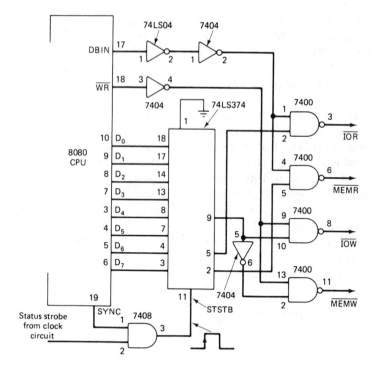

Figure 5-23 Schematic showing how the system control bus for the 8080 could be realized by using discrete logic.

combined logically with the DBIN and $\overline{\text{WR}}$ signals to form the four control bus signals. This is shown in the schematic of Figure 5-23.

We have now realized the system control bus for the 8080 with discrete digital logic. This was done to provide you with a more detailed explanation of how the 8080 actually generates the system control bus. However, if you are using an 8080 microprocessor, it would be wise to consider using an 8228 bus controller for generation of the system control signals and buffering of the system data bus.

TABLE 5-1 LOGICAL CONDITIONS FOR THE CONTROL SIGNALS OUTPUTS OF THE Z-80 MICROPROCESSOR FOR THE SYSTEM FUNCTIONS LISTED

#19 $\overline{\text{MREQ}}$	#20 $\overline{\text{IORQ}}$	#21 $\overline{\text{RD}}$	#22 $\overline{\text{WR}}$		
1	0	0	1	IOR	
1	0	1	0	IOW	These are the system
0	1	0	1	MEMR	functions that the
0	1	1	0	MEMW	four codes designate.

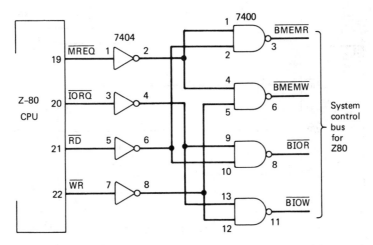

Figure 5-24 Complete schematic showing how the system control bus can be realized with discrete logic for the Z80 microprocessor.

Z80 System Control Bus

Upon examining the signals for control on the Z80, we see signals labeled $\overline{\text{MREQ}}$, $\overline{\text{IORQ}}$, $\overline{\text{RD}}$, and $\overline{\text{WR}}$. A logical combination of these four signals provides the total system control bus. This logical combination is summed up in Table 5-1. We can realize these four system control bus signals as shown in the schematic of Figure 5-24. All the timing of the signals is handled internally by the Z80. Therefore, all that is required is logically to combine the Z80 control lines, as shown in the figure.

8085 System Control Bus

To form the 8085 control bus we use the microprocessor control pins labeled $\text{IO}/\overline{\text{M}}$ pin 34, $\overline{\text{RD}}$ pin 32, and $\overline{\text{WR}}$ pin 31. Table 5-2 shows the logical state of each of these signals listed for the conditions of the control bus functions.

Note from Table 5-2 that the $\text{IO}/\overline{\text{M}}$ pin 34 is logical 1 whenever the system is performing input device read (IOR) or output device write (IOW) opera-

TABLE 5-2 LOGICAL CONDITIONS FOR THE CONTROL SIGNAL OUTPUT PINS OF THE 8085 MICROPROCESSOR FOR THE SYSTEM FUNCTIONS LISTED

#32 $\overline{\text{RD}}$	#31 $\overline{\text{WR}}$	#34 $\text{IO}/\overline{\text{M}}$		
0	1	1	IOR	
1	0	1	IOW	System functions
0	1	0	MEMR	these four codes
1	0	0	MEMW	designate.

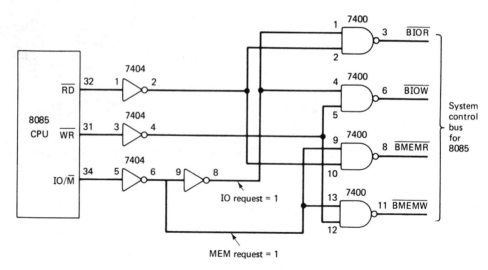

Figure 5-25 Complete schematic showing how the system control bus can be realized with discrete logic for the 8085 microprocessor.

tions. This same pin is a logical 0 whenever the system is performing a memory read (MEMR) or memory write (MEMW) operation.

The $\overline{RD}$ and the $\overline{WR}$ pins, 32 and 31, respectively, are active logical 0 whenever the function designated by the pin label is taking place within the system. That is, when the system is performing a memory read or an input device read, the $\overline{RD}$ signal is a logical 0. When the system is performing a memory write or an output device write operation, the $\overline{WR}$ signal is a logical 0. Figure 5-25 shows the logic required to realize the 8085 system control bus.

5-5 SUMMARY OF THE 3-BUS ARCHITECTURE

We have shown how the 8080, 8085, and Z80 microprocessors can be designed into the 3-bus architecture. We made use of the microprocessor control signals for generation of the system control bus and direction of data flow on the data bus. The nice part is we need not concern ourselves with the timing of the microprocessor control signals. This timing is important, but for now we assume the microprocessor generates correctly timed signals. That is, the microprocessor controls the timing of the signals such as $\overline{RD}$, $\overline{WR}$, and DBIN.

From this discussion it is not a difficult or complex job to design the 3-bus architecture with the microprocessors discussed here. We have not shown all the details yet, such as adding memory to the system or generating the system clocks. These topics are covered in the following chapters.

As a final review of information presented in this chapter, we will show all three microprocessors realized in the 3-bus architecture. These complete schematics are displayed in Figures 5-26a, b, and c. In these figures notice the three

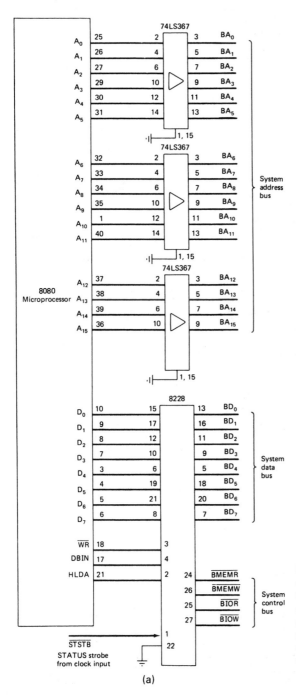

(a)

Figure 5-26 (a) Complete schematic showing the data, address, and control buses for the 8080 microprocessor; (b) complete schematic showing the data, address, and control buses for the 8085 microprocessor; (c) schematic showing the address, data, and control buses for the Z80 microprocessor.

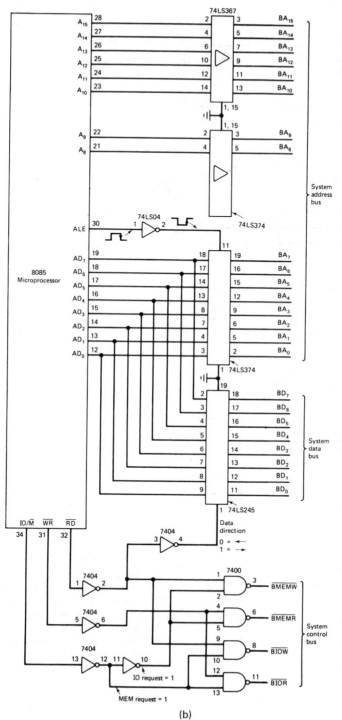

Figure 5-26   (*continued*)

(b)

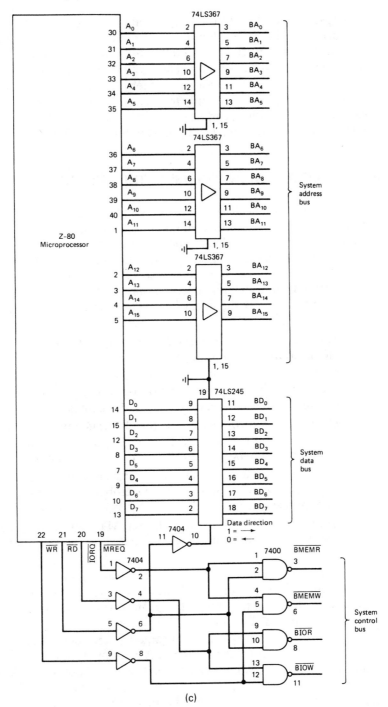

(c)

Figure 5-26 *(continued)*

buses generated: the address bus, the data bus, and the control bus. Also note the similarities in the generation of the three buses on each of the microprocessors.

It should be further stressed that the control bus may or may not be formed in the way we have described. The formation of a control bus depends a great deal upon the type of memory and other ICs used in the system. We present the control bus in this manner only to illustrate similarities between the microprocessors. However, this technique for control bus formation is certainly a valid one. Further, it promotes very easy understanding of the 3-bus architecture.

5-6 GENERAL COMMUNICATION USING THE 3 BUSES

Let us now examine how the 3 buses may be used in a specific sequence to perform a general system communication. From this general communication sequence, we can develop a specific sequence, which may be tailored to a specific microprocessor-based system. This is shown in the next section, where we tailor this general sequence to a specific sequence for a memory read using the 8080, 8085, and Z80 microprocessors.

First let us show the general sequence for a read operation:

1. The address of the read hardware source (memory in this case) is placed on the address bus via the microprocessor.
2. The system data bus is electrically prepared to receive data. That is, the data output buffers in the microprocessor are shut down (tristated). This will allow other hardware to control these lines.
3. The system control bus asserts the READ control line. This action places the data from the read source onto the system data bus and sets the proper direction for the bidirectional buffers on the data bus.
4. Data from the hardware source is now electrically present at the microprocessor data input lines. This data is strobed into an internal register of the microprocessor.
5. After a finite length of time the READ control line is unasserted. This removes the data from the system data bus and terminates the operation.

During this entire discussion there was no mention of time, and each system bus signal was independent. It is much easier to understand what is actually occurring in the system hardware if you view the activity of the system buses as sequences of events.

Let us now take this general sequence and define a specific sequence of events that will occur during a memory read operation for the 8080, 8085, and Z80. This sequence will be repeated in the next chapter, where we learn how to connect read-only memory to the microprocessor. This specific sequence is given here so you will see how a particular microprocessor follows the general sequence of events just described. First, we consider the sequence of events for an 8080 microprocessor.

Memory-Read Sequence for 8080

Let us now discuss a memory-read sequence for a specific microprocessor. The processor chosen for use is the 8080. You should keep in mind that any microprocessor may be used to generate a sequence that will closely follow the general sequence given in the preceeding section. The sequence of events that occurs during a memory read for a typical 8080-based system is as shown. We will expand on this sequence in the next chapter, where connecting to system memory is discussed. For now it is necessary for you to understand that such a sequence of events may be generated. These sequences are obtained from timing diagrams of the microprocessor under study.

Sequence for an 8080 Memory-Read Operation

1. First, the address is output on address lines A0-A15.
2. Status word for a memory-read, op-code fetch, or stack read is placed onto the system data bus.
3. The sync output pin is toggled from a logical 0 to a logical 1 and back to a logical 0 again. This action will provide the means for strobing the status word (STSTB), which will be output to the system hardware.
4. At this time the 8080 internal data bus is electrically prepared to receive data from the system.
5. The DBIN signal from the 8080 is set to a logical 1, which asserts the $\overline{\text{MEMR}}$ system control signal. With this signal asserted the memory places data onto the system data bus. This data is now present at the microprocessor data input pins (D0–D7). During this time the 8080 will strobe data into an internal register.
6. After a finite length of time the 8080 will set the DBIN signal line to a logical 0. When this occurs the $\overline{\text{MEMR}}$ control siganl will become unasssserted. The data is removed from the system data bus and the transfer is complete.

The preceding sequence of events involves many more details, which we did not show at this time. The main point is for you to note that such a sequence does exist. Further, this sequence may be used to understand how the hardware of a microprocessor-based or computer system operates. When you use a sequence of events for different hardware operations, it becomes quite clear how the system hardware involved in the communication must operate. This is our present objective, understanding how the hardware of a system performs the communication.

Let us now take two more microprocessors and show the sequence of events that occurs during a memory-read operation. This is done to show that the operation of almost any microprocessor may be thought of as a sequence of hardware events, which—if followed—will produce reliable communication. This sequence of events is taken directly from manufacturer's timing diagrams.

Sequence for an 8085 Memory-Read Operation

The following is a sequence similar to the 8080 sequence just given. It describes the operations executed each time the 8085 reads data from system memory.

1. First the address is output on address lines A8–A15 and AD0–AD7.
2. ALE is toggled from a logical 0 to a logical 1 and back to a logical 0 again. This action latches the address lines A0–A7 to be output to the system hardware.
3. Next, the IO/$\overline{\text{M}}$ signal goes to a logical 0, indicating a memory communication.
4. At this time the 8085 internal data bus is electrically prepared to receive data from the system.
5. The $\overline{\text{RD}}$ signal from the 8085 is set to a logical 0, which asserts the $\overline{\text{MEMR}}$ system control signal. With this signal asserted, the memory places data onto the system data bus. This data is now present at the microprocessor data input pins (AD0–AD7). During this time the 8085 will strobe data into an internal register.
6. After a finite length of time, the 8085 will set the $\overline{\text{RD}}$ signal line to a logical 1. When this occurs, the $\overline{\text{MEMR}}$ control signal will become unasserted. The data is removed from the system data bus and the transfer is complete.

5-7 SEQUENCE OF EVENTS FOR A Z80 MEMORY READ

In the following sequence of events, we will show how the Z80 CPU would respond during a typical memory-read sequence. This sequence does not include the refresh operations that may occur during a Z80 hardware cycle. If our goal was to understand how the Z80 communicated with dynamic RAM, then we could certainly break the refresh operation into a sequence of events. For now we choose to ignore the refresh operation.

1. First, the address of memory to be read is placed on the system address lines A0–A15. The Z80 does not use a multiplexed address and data bus like the 8085.
2. Next, the $\overline{\text{MREQ}}$ (memory request) output pin is set to a logical 0, indicating a memory operation will occur.
3. The Z80 data bus is electrically prepared to input data from the system data bus.

4. Next, the $\overline{RD}$ signal line from the Z80 is set to a logical 0. This action will assert the $\overline{MEMR}$ system control line. When this line becomes asserted, data from memory is placed onto the system data bus. This data is now present at the Z80 data input pins.
5. During the time data is present at the Z80 input pins, it will be strobed into an internal register.
6. Next, the Z80 sets the $\overline{RD}$ signal line to a logical 1. This unasserts the $\overline{MEMR}$ control line. When this occurs, data from the system memory is removed from the system data bus and the hardware operation is complete.

In this sequence of events for the Z80 memory-read operation, you can see many similarities between it and the 8080 and 8085 sequence. Details of the sequence change due to the differences in microprocessors, but the flow of the sequence does not. Each follows the general sequence given earlier in this section.

5-8 DISCUSSION OF SEQUENCES

The concept of a sequence of events for a hardware operation is simple and easy to understand. Further, it accurately describes the hardware operation of the system during a specific system communication.

We listed six different hardware operations that may occur in a computer system during normal operation:

1. Memory read
2. Memory write
3. Input read
4. Output write
5. Interrupt acknowledge
6. Hold acknowledge (halt)

No matter how complex the software of a microprocessor-based or computer system is, these six operations are the only ones occurring during normal system execution. As an example of this statement, let's take a very complex instruction like CALL. From your knowledge of system software, you know that the CALL instruction does the following:

1. Reads the op-code (memory read).
2. Pushes 8 bit word (A15-A8 return address) on stack (memory write)
3. Pushes 8 bit word (A0-A7 return address) on stack (memory write)
4. Jumps to address of subroutine. (another memory read)

Every instruction may be subdivided into its major hardware events. Therefore, your task is understanding how the system will perform these operations. In the following chapters you will learn how each of these operations occurs in a typical microprocessor system. Further, you do not have to be an expert in system hardware and design to accomplish it. When you approach a new system, keep these sequences in mind to be used as an effective tool to aid in your understanding of the system.

5-9 CHAPTER SUMMARY

This chapter began with an introduction to a microprocessor hardware architecture called the 3-bus architecture. You were shown that microprocessor systems can be thought of as having three distinct buses, address, data, and control. Further, all system communications were performed using these three buses. After this introduction, each bus was presented and its function discussed. During this presentation you were shown how typical 8080, 8085, and Z80 systems use hardware to realize the bus under consideration.

When the system control bus was introduced, it was stated that all system communications fall into the following six catagories regardless of the complexity of the system:

1. Memory read
2. Memory write
3. Input read
4. Output write
5. Interrupt acknowledge
6. Hold acknowledge

After the discussion of the buses, you were shown how general system communication can occur using the 3 buses. You saw the definite function of each bus in the system communication sequence. Further, you were shown how each hardware operaton may be reduced to a specific sequence of events. You were shown the sequence for a memory-read operation with an 8080, 8085 and a Z80.

REVIEW PROBLEMS

1. Define the term bus as it applies to a microprocessor system.
2. What are the buses in a 3-bus system architecture?
3. List seven hardware events that occur in a microprocessor system.

4. Draw a block diagram of a typical microprocessor system showing the buses, ROM, RAM, and I/O.

5. How large is the address bus for each?
   (a) 8080
   (b) 8085
   (c) Z80

6. Which of the three microprocessors discussed in Question 5 uses a multiplexed address bus?

7. Why would you buffer an address bus?

8. How wide is the data bus for each?
   (a) 8080
   (b) 8085
   (c) Z80

9. How does the buffering of the data bus differ from buffering the address bus?

10. What device would you choose for buffering the data bus on an 8080 system?

11. What is the function of the data bus?

12. What is the function of the control bus? There are two functions.

13. Which microprocessor uses a status latch?

14. True or false: In an 8080 system DBIN goes to a logical 1 when the CPU is reading data from memory.

15. What signal on the 8085 indicates the device is reading data? Which logical state is the active state for this signal?

16. Draw a schematic diagram of the Z80 control bus, including the signals for the $\overline{\text{MEMR}}$, $\overline{\text{MEMW}}$, $\overline{\text{IOR}}$, and $\overline{\text{IOW}}$.

17. What is the sequence of events for an 8080 memory-read operation?

6

USING ROM WITH THE 8080, 8085 AND Z80

In this chapter you will be shown how the 8080, 8085, and Z80 microprocessors are interfaced and communicate with ROM. We will start with a general introduction to the operation of ROMs. We will then examine the important aspects of using ROM with the 8080, 8085, and the Z80. By the end of this chapter, you should have a complete understanding of how ROM communicates electrically with these three microprocessors, and you should be able to apply this knowledge to most microprocessor-based systems, regardless of the overall application.

6-1 WHAT IS ROM?

In most microprocessor-based systems there is a certain amount of memory that can store information, and this information will not be lost when power is turned off. This type of nonvolatile memory is called ROM, or read-only mem-

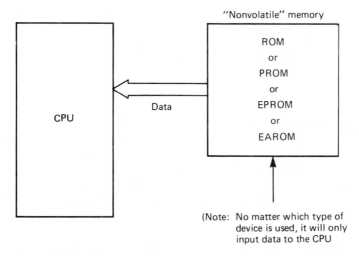

Figure 6-1   A block diagram showing a typical application of a nonvolatile memory device that might be used in a microprocessor system. The nonvolatile ROM usually contains boot-up programs that are executed when the system is first powered up.

ory. Information in ROM can be read out but not altered. ROM is useful in a system because it allows the CPU to initialize all the peripheral hardware to the proper logical states when power is first turned on.

There are several types of nonvolatile memory that can be used in a microprocessor system. These include ROM, programmable ROM (PROM), erasable PROM (EPROM), and electrically alterable ROM (EAROM or EEROM). See Figure 6-1. Normally, in a typical microprocessor system only one type of nonvolatile memory is used. However, since it is possible that a system may require any one of these memory devices, we will now describe them all.

ROM

With ROM the data is placed into the memory device by the manufacturer. ROM is used wherever there is a nonchanging data base of high volume. (*Note:* programming data into a ROM is a very expensive process.)

PROM

With PROM data is programmed into the memory device by the user. High-voltage pulses actually "blow apart" metal strips or polycrystaline silicon inside the integrated circuit, forcing logical 1s and 0s into specific address locations in memory. Once programmed into the device, the data cannot be altered. It is possible to change selective bits of the data. In most PROMs if the data was such that it did not require the metal strip to be blown apart, then you may change the data. Usually this means if the data bit was a logical 1, then it may be programmed to a logical 0. This memory device usually operates at a much faster speed than other programmable memories.

EPROM

With EPROM data is programmed by the user into the memory device, by applying high voltage signals. This is similar to the programming of a PROM. Data can be erased by shining ultraviolet light onto a transparent window that covers the integrated circuit. After a specified time of exposure to the light, all data will be erased and the device can then be reprogrammed with entirely new data. These devices are often used in development work, where the program may undergo many changes.

EAROM, EEROM, or E² ROM (Electrically Erasable)

With EAROM data is programmed into the memory device by the user in a fashion similar to the EPROM. The major difference is that the data in an EAROM can be erased electrically, without the ultraviolet light. Further you can alter single bytes of data.

No matter which nonvolatile memory you are using in your system application, the operating characteristics are all very similar. We will now discusss the important parameters of this group of memories. (*Note:* Throughout this text we will use the word ROM to refer to any of the nonvolatile memory devices that we have just described.)

6-2 IMPORTANT OPERATING CHARACTERISTICS OF ROM

Let's now review some important operating characteristics of ROM. To begin, it is important to note that ROM can only be read (not written) by the CPU; hence, the name read-only memory. Second, information in the memory is fetched whenever an address is applied to the address input lines. The number of address input lines a ROM has depends on the amount and internal organization of the data stored within it.

It is possible to determine the internal organization of data in memory by noting the specification for the ROM. For example, a ROM may be organized as 1024×8, 2048×8, or 4096×8, to name just a few types of memory organizations. The first number in the specification indicates the number of unique address locations contained within the single integrated circuit. The second indicates the number of bits of parallel data that can be read from the ROM at each unique address location.

The following steps help determine the number of address lines available with a particular ROM. If we start with a description of the device, say 2048 × 8, we already know that the 2048 indicates the number of unique address locations, and that this number is equal to the number of different binary combinations existing on the address lines. In other words, $2^X = 2048$, or $X = 11$, where X equals the number of address lines. Therefore, if the memory is described as 4096 × 8, the number of address lines is $2^X = 4096$, or $X = 12$. As you can see, this equation works regardless of the number of address locations in the ROM.

Note that when describing a ROM, the exact number of address locations is usually shortened to the nearest thousand and abbreviated by the single letter K, for Kilo. In other words, a 1024 × 8 ROM is described as a 1K device. A 4096 × 8 ROM is described as a 4K device, and so on.

A final point about ROM operation is that all data is read from ROM in a parallel fashion. In other words, when the microprocessor reads the contents of a particular location in ROM, all the data bits at that location are simultaneously placed on separate conductors of the system data bus. This data is then input to the CPU.

Figure 6-2 displays a functional block diagram of ROM operation. In this figure, we can see an input, called chip-select, that we have not yet discussed. The function of this input line is to turn on and off the data output lines of the ROM device. When the chip-select input line is active, the ROM data outputs are active and are either logical 1 or 0, depending on the data programmed into the device and the values applied to its address input lines. When the chip-select

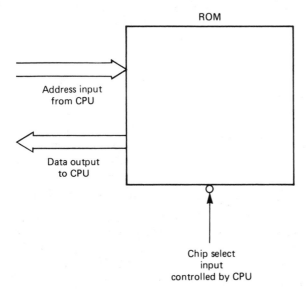

ROM

Address input from CPU

Data output to CPU

Chip select input controlled by CPU

Figure 6-2 A functional block diagram of a typical ROM device. Address lines are input to select the internal data to be output. Data is output when the device is selected with the chip-select line.

line is not active, the data outputs are placed into a tristate, or high-impedance state. In other words, the chip-select input line either electrically enables or disables the outputs of the ROM device. Later in this chapter we will show exactly how this line is used.

6-3 SEQUENCE OF ELECTRICAL EVENTS FOR READING DATA FROM ROM

Let's discuss the block diagram in Figure 6-2. The following sequence of electrical events occurs each time data is to be read from the device. Keep in mind that this general sequence must be followed regardless of the type of CPU used in the system. Let's examine the sequence. (Later on, we will discuss specifically how the 8080, 8085, and Z80 follow this sequence.)

1. An address is input to the ROM from the CPU. This address indicates the internal location of the device from which data will be read. There are thousands of bytes of data stored in the ROM. The address lines select 1 byte of data to be output.
2. The chip-select is made active to enable the data outputs onto the system data bus. At this time the data from ROM is present on the ROM data output pins.
3. The CPU then waits for a finite amount of time, called access time—which is approximately 100–600 nanoseconds, depending on the type of ROM used—thus allowing the memory device to decode the address that is input and then allowing output data to reach the data output lines of the device. From the output lines of the device, the data must return to the CPU data input pins.

During the time the memory data is active on the system data lines, the CPU will strobe the information into an internal register.

4. The chip-select is made inactive to remove the ROM data from the system data bus.

This general sequence is followed each time the CPU reads data from ROM. Figure 6-3 shows a general timing diagram of this sequence.

CPUs are designed to operate within these boundaries. Therefore, in most cases the user does not have to think about this sequence because the microprocessor handles it with the help of internal timing circuits. However, it is important that you understand this sequence, as it makes using ROM—as well as understanding how the circuits operate—much simpler.

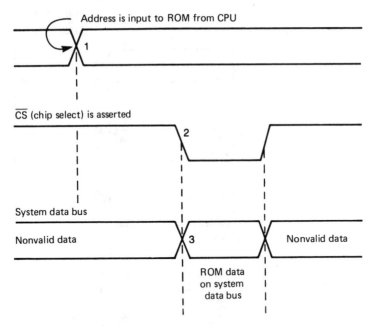

Figure 6-3 A general timing diagram for reading data from a ROM device: (1.) Address is input to the device from the CPU; (2.) the chip-select is asserted; (3.) the ROM data outputs are enabled onto the data bus.

6-4 CONNECTING THE Z80 BUSES TO ROM

Figure 6-4 shows how the Z80 address and data bus may be connected to 2716 ROM.

6-5 CONNECTING THE 8080 BUSES TO ROM

Figure 6-5 shows how the 8080 address and data bus may be connected to a 2716 ROM.

6-6 CONNECTING THE 8085 BUSES TO ROM

Let's now connect the 8085 buses to ROM. Figure 6-6 shows how the address and data bus of the 8085 connect to system ROM.

In Figure 6-6 we see that the 8085 data bus is 8 bits wide. Therefore, the address lines are connected as we may expect, starting with A0. Notice in

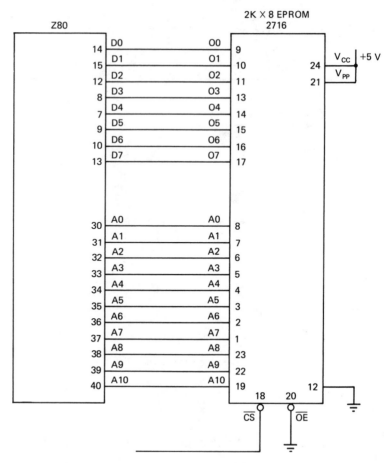

Figure 6-4 Schematic diagram showing how the Z80 data and address buses connect to the system ROM.

Figure 6-6 that the chip-select line of the 2716 EPROM is not connected. This is because we plan first to discuss the topic of address mapping.

6-7 ADDRESS MAPPING

The 8080, 8085, and Z80 have 16 physical address output lines, labeled A0–A15. This means that a total of 2^{16}, or 65,536, storage locations can be accessed directly by these CPUs. A 2716 EPROM has 2048 physical address locations that can be accessed. Therefore, a method for selecting only 2048 locations out

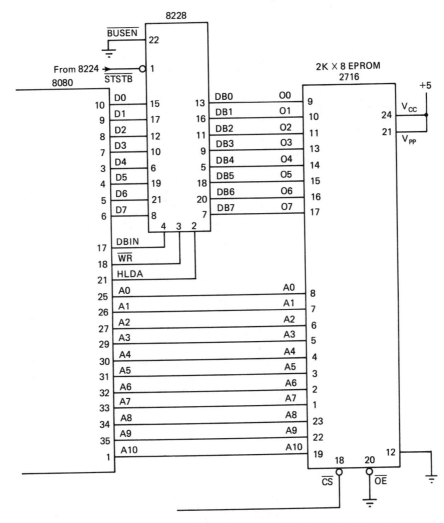

Figure 6-5 Schematic diagram showing how the 8080 data and address buses connect to the system ROM.

of the total 65,536 possibilities is needed. Which 2048 locations do we choose? There are 65,536 divided by 2048, or 32, possible blocks of 2048 locations from which we can select.

Before any hardware is built for a system, the system designer must first construct a memory map. A memory map is used to indicate which address locations are specified for any particular ROM, RAM, input, or output device.

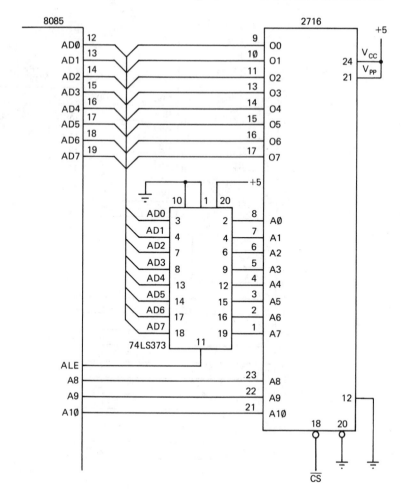

Figure 6-6 Schematic diagram of the connection between the 8085 and one 2716 EPROM. Only one EPROM is required due to the 8-bit data bus of the 8085 CPU.

Figure 6-7 shows a typical memory map for a microprocessor system. We can see in this figure that the entire 65,536 address locations are divided into functional blocks. These blocks indicate the address locations that are reserved for specific ROMs and RAMs.

Note: As shown in Figure 6-7, the lowest block in an 8080, 8085, or Z80 system is usually reserved for ROM. This is because whenever these microprocessors are reset from an external reset switch or from a power-on reset circuit, the processor always expects the first instruction op-code executed after the reset operation to reside at address 0000 hexadecimal.

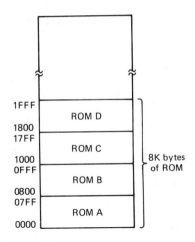

Figure 6-7   Typical memory map for a microprocessor system.

6-8 GENERATING THE CHIP SELECTS FOR ROM

Based on the information in Figure 6-7 we can see that ROM A is enabled whenever addresses are between 0000 and 07FF, inclusive, for an 8080, 8085, or Z80 system. An electrical signal in the system is needed to indicate that the address being output on the address bus is within these limits. Figure 6-8 shows hardware that will accomplish this for the 8080, 8085, and Z80 CPUs.

Let us now discuss how Figure 6-8 operates. The chip-selects that are output from this circuit are labeled $\overline{\text{ROM A}}$, $\overline{\text{ROM B}}$, $\overline{\text{ROM C}}$, and $\overline{\text{ROM D}}$. These signals are connected to the chip-select inputs of the system ROMs. CPU

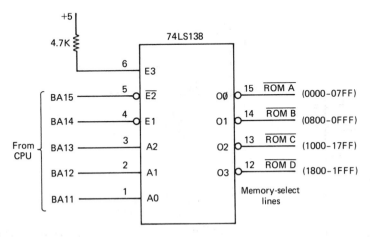

Figure 6-8   Schematic diagram showing how the chip-select lines may be generated with hardware.

			Buffered address lines			
Address range	BA15	BA14	BA13	BA12	BA11	BA10 through BA0
0000–07FF	0	0	0	0	0	XXX XXXX XXXX
0800–0FFF	0	0	0	0	1	XXX XXXX XXXX
1000–17FF	0	0	0	1	0	XXX XXXX XXXX
1800–1FFF	0	0	0	1	1	XXX XXXX XXXX

Figure 6-9   Decoding truth table for Figure 6-8.

output address lines BA0–BA10 are connected directly to the ROM address input lines, as was shown in Figures 6-4, 6-5, and 6-6. The remaining address lines BA11–BA15 are used in the generation of the chip-select lines for the ROM.

Address lines, BA14–BA15 are input to the enable pins of the 74LS138. The 74LS138 is enabled whenever lines BA14–BA15 are 0. This corresponds to a 00XX XXXX XXXX XXXX existing on the system address bus, where X = don't care state of the other address lines. Address lines BA13, BA12, and BA11 will determine which ROM block will be active.

Figure 6-9 shows the address decoding truth table for the circuit of Figure 6-8. The decode circuit of Figure 6-10 works essentially the same way as we described for Figure 6-8. The main difference is that the chip-select lines will become enabled or change at 4K increments instead of 2K increments of the address bus. This would be necessary if 4K × 8 ROM were used in the system. Figure 6-11 shows the address decade truth table for Figure 6-10.

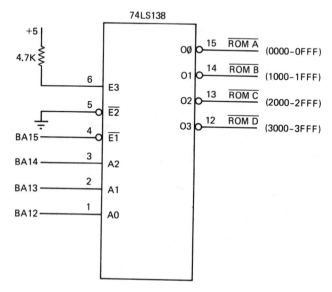

Figure 6-10   Decoding circuit that enables 4K ROM blocks instead of 2K blocks. Note the different address lines used for this decoding circuit.

BA15	BA14	BA13	BA12	BA11–BA0	HEX	Active pin
0	0	0	0	0–0	0000	15
0	0	0	0	1–1	0FFF	15
0	0	0	1	0–0	1000	14
0	0	0	1	1–1	1FFF	14
0	0	1	0	0–0	2000	13
0	0	1	0	1–1	2FFF	13
0	0	1	1	0–0	3000	12
0	0	1	1	1–1	3FFF	12

Figure 6-11   Decoding truth table for 6-10.

6-9 CONNECTING THE CHIP-SELECT LINES

Figure 6-12 shows in block diagram form how the chip-select lines would connect to a typical 8080, 8085, or Z80 ROM system. We see in these figures that when both the $\overline{\text{MEMR}}$ and the ROM select lines are a logical 0, the chip-select input to the 2716 EPROM is active logical 0. Figure 6-13 shows the general timing relationship between important signals involved in the memory-read operation.

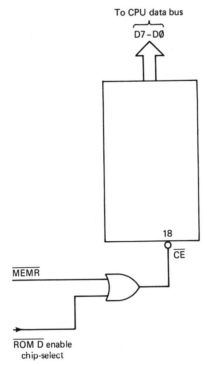

To CPU data bus

D7–DØ

18

$\overline{\text{CE}}$

$\overline{\text{MEMR}}$

$\overline{\text{ROM D}}$ enable
chip-select

Figure 6-12   Block diagram showing how the chip-select lines and the MEMR line are used to enable the system ROM.

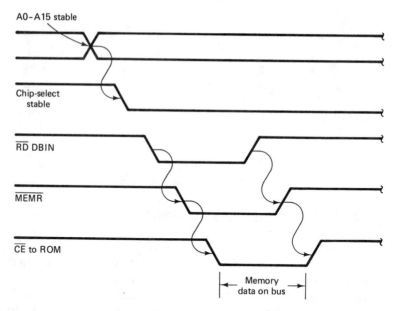

Figure 6-13 Timing relationship between important signals involved in the memory-read operation.

6-10 A VARIATION

Figure 6-14 shows yet another technique that can be used to enable the ROM data outputs onto the system data bus of the microprocessor at the correct time. This technique takes advantage of the internal decoding located inside the 2716

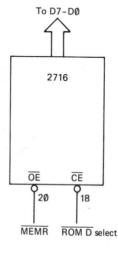

Figure 6-14 A second technique that may be used to enable the ROM data lines onto the system data bus using the MEMR and the chip-select lines. Notice that this technique makes use of the logic contained in the 2716 EPROM devices.

device package. In this application both the $\overline{OE}$ pin 20 and the $\overline{CE}$ pin 18 must be a logical 0 before the data outputs are enabled onto the system data bus. Recall that with the previous technique $\overline{OE}$ pin 20 was always connected to ground or logical 0, which is the active state for this input.

6-11 ADDING MORE ROM

Figure 6-15 shows that it is possible to add more ROM devices to a microprocessor system by using the enabling technique shown in Figure 6-14. This technique allows us to add more physical ROM devices without having to add external gates. Another point to note is that each ROM has a specific address block reserved for its use. However, the microprocessor makes no distinction between blocks when executing the software. The CPU assumes that all the ROM space is available. In other words, a 3-byte instruction may have 1 byte located in the last location of ROM A and the next two bytes located in the first locations of ROM B.

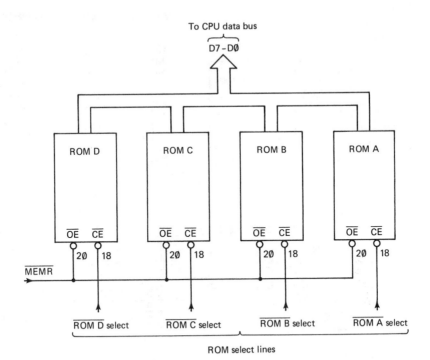

Figure 6-15 By using the enabling technique shown in Figure 6-14, more ROM can be added without the need for additional logic devices.

6-12 MEMORY MAPPING LARGER ROMS

In today's microprocessor systems, ROM chips configured as 4K × 8, 16K × 8, and 32K × 8 are widely used. In this example you will see how ROMs larger than 2K × 8 may be designed into a system. Once you have been shown how this is accomplished, it will be an easy matter to understand how any size ROM can be interfaced to the 8080, 8085, and Z80 microprocessors.

The two ROMs used here are the 2732 (4K × 8) and 2764 (8K × 8). Figure 6-16 shows the pinouts of these two devices.

A point of interest regarding the pinouts of the 2732 and the 2764 is that they can be interchanged in a physical socket even though the pin counts are completely different. This can be accomplished by using 28-pin sockets in the system and connecting Vcc to pins 28 and 26. See Figure 6-17.

These devices are designed so that you can elect to start with 2732 and then later upgrade to the 2764. This allows you to double your available ROM

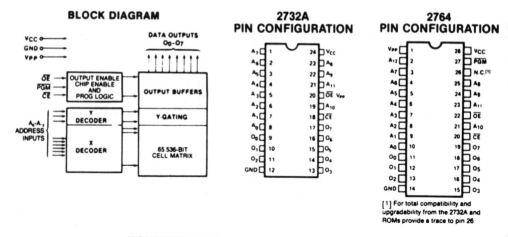

MODE SELECTION

PINS / MODE	CE (20)	OE (22)	PGM (27)	Vpp (1)	Vcc (28)	Outputs (11-13, 15-19)
Read	V_{IL}	V_{IL}	V_{IH}	V_{CC}	V_{CC}	D_{OUT}
Standby	V_{IH}	x	x	V_{CC}	V_{CC}	High Z
Program	V_{IL}	x	V_{IL}	V_{PP}	V_{CC}	D_{IN}
Program Verify	V_{IL}	V_{IL}	V_{IH}	V_{PP}	V_{CC}	D_{OUT}
Program Inhibit	V_{IH}	x	x	V_{PP}	V_{CC}	High Z

x can be either V_{IL} or V_{IH}

PIN NAMES

A_0-A_{12}	ADDRESSES
CE	CHIP ENABLE
OE	OUTPUT ENABLE
O_0-O_7	OUTPUTS
PGM	PROGRAM
N.C.	NO CONNECT

*HMOS is a patented process of Intel Corporation.

Figure 6-16   Pinouts and a block diagram of the 2732 and the 2764 EPROM devices that are widely used in microprocessor applications. (Courtesy of Intel Corp.)

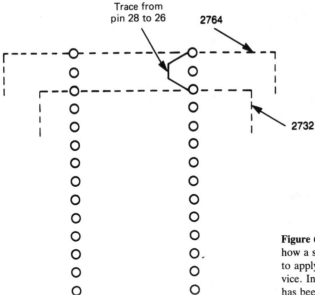

Figure 6-17 A block diagram showing how a single physical socket can be used to apply either the 2732 or the 2764 device. In this application, a 28-pin socket has been used.

space without increasing any physical board space. If the larger memories are chosen (2764), the memory map will change from the one shown in Figure 6-18 (2732) to the map shown in Figure 6-19. Notice that in this new memory map, only one device is required, whereas two were used before.

Figure 6-20 shows the respective hardware required to generate the memory map shown in Figure 6-21.

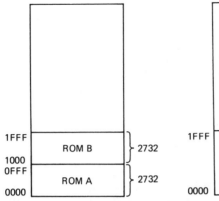

Figure 6-18 Memory map for use with 4K ROMs.

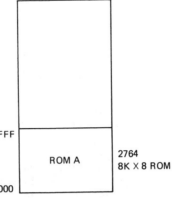

Figure 6-19 Memory map for use with an 8K ROM.

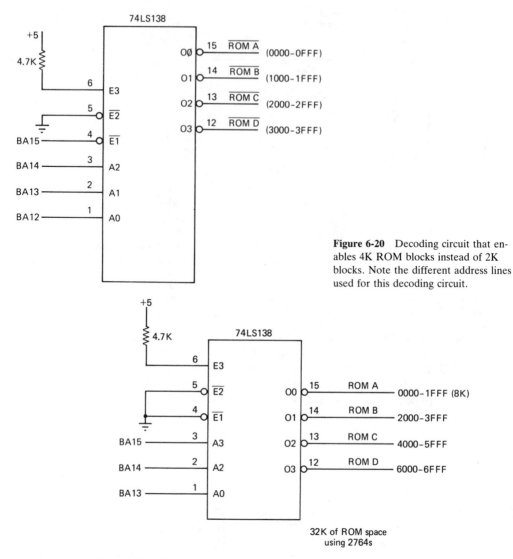

Figure 6-20   Decoding circuit that enables 4K ROM blocks instead of 2K blocks. Note the different address lines used for this decoding circuit.

Figure 6-21   Schematic diagram of chip select decoding for a 32K ROM space constructed of 8K ROM. Hardware for memory map of Figure 6-19.

6-13 ADDING ADDRESS BUFFERS

The ROM system that we have presented is very basic. It is, however, widely used in industry. Note that there is a critical area in this type of system design: the electrical load on the address bus.

In the system designs shown, the load on the address bus has been only a 74LS138 and the inputs to ROMs. Here the LS stands for low-power schottky, and each LS input loads the bus with approximately .4 milliamperes in the logical 0 state and 20 microamperes in the logical 1 state. The ROM devices load the bus only with approximately 10 microamperes in both the logical 1 and the logical 0 state. This combination may not seem too excessive a load for the system address bus. However, the address lines may be connected to many I/O ports, cables or even through edge connectors to another circuit board, all of which add additional capacitance and load and strain the capability of the CPU to provide proper voltages on the address lines.

In these cases and in any applications where the output load on the address lines exceeds the specified rating, address buffers are needed. Address buffers increase the output current drive capability of the address bus from the CPU.

Keep in mind that not all applications require address buffering. The use of buffers is dependent on the type and amount of electrical load being driven by the system address lines. If address buffering is required, it may be accomplished using the method outlined in Chapter 5.

6-14 MEMORY DATA BUFFERING

In some system applications, the data outputs from the system ROM do not have enough current drive capability to control the system data bus lines adequately. In such cases, it is necessary to use memory data buffers. Memory data buffers perform a function similar to the CPU data buffers discussed in the previous chapter.

The memory data buffers must be capable of tristate operation. A tristate operation is necessary because the memory data must be electrically removed from the system data bus when the microprocessor is not electrically requesting it. Figure 6-22 presents a schematic diagram showing how memory data buffers can be installed in a typical 8 bit microprocessor ROM system.

In this figure, the ROM data outputs need only drive the data buffer inputs. The outputs of the memory data buffers will drive the entire load of the system data bus.

Like the address buffers, data buffers may not have to be used in your system. The use of data buffers is dependent on the amount and type of electrical load the memory data must drive on the system data bus. Since the memory data buffers must be enabled whenever any one of the ROMs is accessed, extra circuitry must be added to generate the buffer enable signal. This signal is composed of the logical AND of all the ROM select lines, logically ORed with the $\overline{\text{MEMR}}$ signal to preserve proper signal timing.

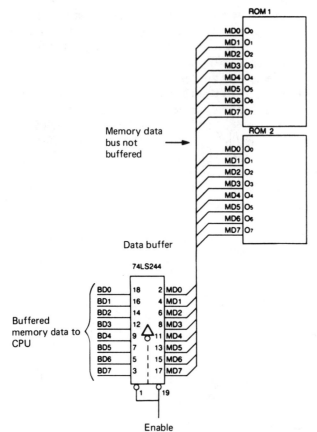

Figure 6-22   Schematic diagram showing one way the memory data can be buffered prior to connecting to the system data bus. The buffer enable line will be asserted whenever the CPU is reading data from this system address space.

6-15 A COMPLETE ROM SYSTEM

The schematic shown in Figure 6-23 is a complete ROM system for the 8080, 8085, or Z80 microprocessor. This system is based on the concepts presented in this chapter. The memory map for Figure 6-23 is shown in Figure 6-24.

6-16 TIMING CONSIDERATIONS FOR ROM

We have shown how the ROM devices may be connected to the 8080, 8085, and Z80 with all the peripheral circuits, such as address buffers, chip-selects, and memory data buffers. However, we have not discussed what timing effects these peripheral circuits or the ROM itself have on the overall system speed. That is our objective in this section. The discussion will start with a block

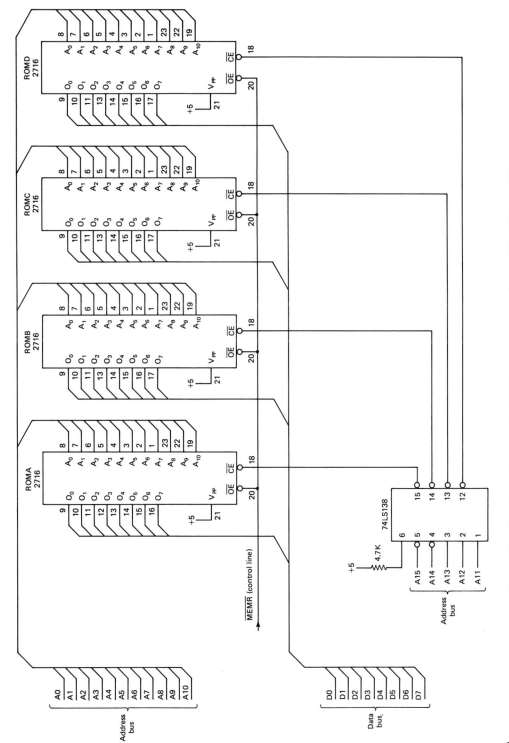

Figure 6-23 Complete schematic of a ROM system for use with an 8080, 8085, or Z80 microprocessor using the 3-bus architecture.

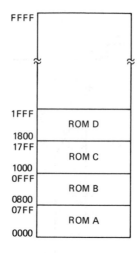

Figure 6-24   Memory map for the ROM system shown in Figure 6-23.

diagram of the ROM interface. From this block diagram we will present a typical timing diagram of 8080, Z80, and 8085 read cycles.

In this diagram, important timing parameters will be shown. At the conclusion of this section you will have been shown the important timing aspects of using a ROM with a microprocessor.

Figure 6-25 shows a block diagram of a microprocessor-based system which uses

1. Address latches/buffers
2. Chip-select logic

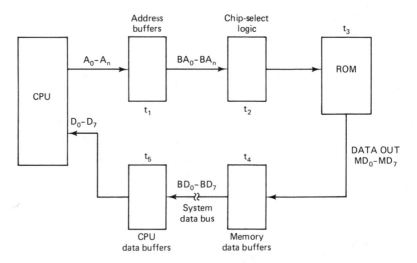

Figure 6-25   Block diagram of a microprocessor ROM system showing all delays.

3. Memory data buffers
4. CPU data buffers

These components are typical in many microprocessor systems, and each has an effect on the system timing. When a microprocessor outputs an address on the system address bus, a certain length of time is given by the CPU to allow the memory data to be input to the CPU data input pins. This concept is shown in Figure 6-26.

The time allowed by the CPU is dependent on the clock frequency of the system. This is due to the fact that all timing in the system is derived from the clock edges. You usually wish the CPU to operate at the fastest speed possible. However, you do not want the operation to be so fast that reliable communication is not achieved with the system ROM.

There are input pins to the microprocessor that will allow the system to

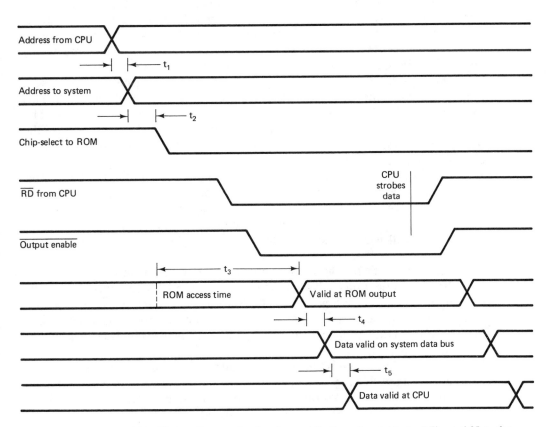

Figure 6-26 Timing diagram showing the contribution of each block of Figure 6-25 to the overall round-trip delay time.

slow down the CPU during a memory access. This will let the CPU run as fast as possible except when accessing memory. The pin for this function is called *ready* on the 8080, 8085, and *wait* on the Z80.

For this discussion we will assume that the ready line is not used. The objective is to show the maximum delay expected from the system ROM during any memory cycle. With this information you may then know the maximum clock speed at which the CPU will operate and still communicate reliably with ROM.

In Figure 6-25 we see that each block in the figure will cause a certain delay time. This delay time must be accounted for in the overall system operating speed. When the CPU first outputs an address, it will have to pass through the address latch/buffers before the reaching the system address bus. This time is labeled as t1. For a 74LS373 device, the delay time is approximately 18 nanoseconds, whereas for a 74LS244, the delay time is 12 nanoseconds.

From the address latches and buffers, the address lines will then be input to the ROM and chip-select circuits. We will concentrate on the delay caused by the chip-select circuits because these circuits will cause a delay of the signal applied to the ROM. Figure 6-23 shows a typical chip select circuit. The total delay of Figure 6-23 will be the delay of the 74LS138 decoder. This time will be approximately 26 nanoseconds; it is labeled t2 in Figure 6-25.

Once the address has been applied and the chip is selected, it will take a finite amount of time for the ROM device itself to access the internal data. This time will vary from ROM to ROM, depending on the speed of the device. The ROMs we will use have an access time of 450 nanoseconds. If we add up the total delay thus far, we see that it takes 18 + 26 + 450 nanoseconds before the data is output from the ROM to the system data bus.

When the data is output from the ROM, it must be input to the CPU. This will be along another path, generating yet another delay. The first delay encounted by the output data will be the memory data buffers. Figure 6-25 shows this time as t4; it is dependent on the type of data buffer used. For this example we will use a 74LS244, which has a delay time of approximately 12 nanoseconds.

After the data has gone through the memory data buffer, it is placed onto the system data bus. The data bus will then apply the memory data to the CPU data bus buffer. We will use a 74LS245 device, which has a propagation delay of 12 nanoseconds, for this function. Adding up the return delay time for the ROM data we have 12 nanoseconds + 12 nanoseconds, which equals 24 nanoseconds.

The total delay time is the addition of all system delays, or address delay + data return delay; this is 523 nanoseconds + 24 nanoseconds, which is equal to 547 nanoseconds. Using this information you can set the system clock speed to a frequency consistent with the overall ROM speed.

6-17 CHAPTER SUMMARY

In this chapter we have presented useful information regarding the use of ROM with the 8080, 8085, and Z80 CPUs. We have presented basic information concerning ROM operation and described the fundamental elements that give a reliable connection between these microprocessors and ROM. The examples given in this chapter have shown that using ROM with a microprocessor can be a relatively simple task. There are certain guidelines, however, that must be followed. This chapter presented these guidelines.

REVIEW PROBLEMS

1. What is a ROM?
2. List four different types of memory devices that may be referred to as a ROM.
3. Whay would you need ROM in a system?
4. Assume a ROM is organized as 32K × 8.
   (a) How many address inputs are used?
   (b) How many data lines are needed?
5. What is the sequence of events necessary for reading data from ROM.
6. Draw a schematic diagram showing how an 8080 address bus would connect to a 4K × 8 ROM. Assume no address buffering is used.
7. What is meant by the term address map?
8. Draw an address map with 8K of ROM space that is realized with 2K × 8 ROM devices.
9. For what is the chip-select pin on a ROM used?
10. Use a timing diagram to show what is meant by access time.

7

RANDOM ACCESS
MEMORY (RAM)

We continue our study of the 8080, 8085, and Z80 microprocessors by learning how to use them with static *random access memory,* or RAM. In a microprocessor system, RAM is used for temporary storage of programs, data, or variables. Unlike ROM (discussed in chapter 6), RAM is volatile, and information is lost when power is turned off. There are systems that provide battery backup, which keeps the information in RAM after power is lost. These are special cases, and we do not discuss them in this chapter. Our RAM system will lose information when power is turned off.

The electrical connections between the 8080, 8085, Z80, and static RAM are not difficult to realize. There are certain guidelines, however, that should be followed. In this chapter we examine these guidelines and discuss all the important topics relating to static RAM. We also display and discuss a popular static RAM system that is used in industry. This system comprises 8192 × 8 bit memories, which are given the industry device number of 6264.

7-1 OVERVIEW OF STATIC RAM COMMUNICATION

We begin our discussion by explaining how a semiconductor read and write memory is communicated with in general. Once you understand this process, you will find that it is an easy and straightforward jump to understanding how the microprocessor communicates with RAM.

To start, a RAM in a microprocessor system is capable of having data written to it and read from it by the CPU. Specific electrical signals are necessary for performing these two types of communication. Figure 7-1 shows the major electrical signals involved.

Let's first examine the power-supply connection. A RAM must be powered up, and the most common voltages are +5 and ground. (We suggest that you refer to the manufacturer's data sheet for the exact voltages.) For our purposes, however, we assume that the RAMs require +5 volts and ground.

In Figure 7-1, the lines labeled "Data In" are the physical wires that allow the electrical information to be written into the RAM. The lines labeled "Data Out" are the wires that allow the information stored in the RAM to be read out

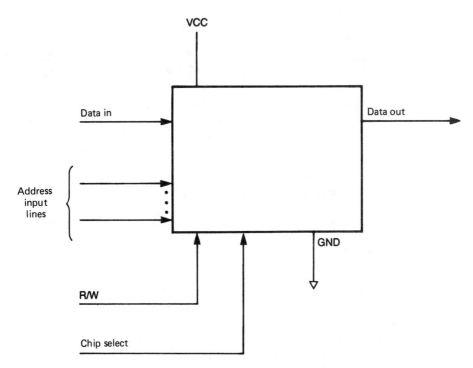

Figure 7-1   A block diagram showing the major electrical signals of a typical RAM device.

electrically. A RAM must have at least one data input line and one data output line. The exact number depends on the internal organization of the device. Recall from the discussion of ROM that only data-out lines were required, since information was read from the ROM but not written into it.

In Chapter 6 we described the internal organization of ROM. Much of the same information and reasoning can be used to describe the RAM. Like ROM, much information can be obtained from the data sheet description of a RAM. For example, a RAM may be described as being a 256 × 1, a 256 × 4, a 1024 × 8, and so on. The second number indicates the number of data input and output lines contained in the memory. In other words, every time the memory is electrically written to or read from, the second number in the description determines the number of parallel data bits that are written or read.

Let's now examine the address input lines shown in Figure 7-1. (We previously examined these lines in Chapter 6 during our discussion of ROM.) Address lines of the static RAM have the electrical function of selecting the internal location of the memory device from which data will be read or written to. It is possible to determine the number of unique storage locations available in a RAM by simply counting the number of address lines. The converse of this is also true. That is, you can determine the number of address lines in a device by knowing the stated number of unique storage locations. For example, a 1024 × 4 static RAM will have 1024, or 1K, unique storage locations available, and at each storage location, 4 bits of data will be written or read at the same time.

The final signal line in the figure (labeled R/W) electrically determines whether the RAM will have data written to it or read from it.

Let's now turn our attention to the general sequence of events that occurs whenever any semiconductor memory is written to or read from, that is, whenever the microprocessor communicates with RAM. Let's first examine the events that occur during a read operation.

7-2 SEQUENCE OF EVENTS FOR A RAM READ

1. First, the address lines are input to the memory. At this time the internal location from which the data will be read is logically decoded by the RAM.
2. The R/W control line is placed in the correct logical state for a memory read. On some memories this is a logical 1; on others it is a logical 0. (The manufacturer's data sheets will help you determine the exact logical condition for a memory read for your system.)
3. The system must wait for a certain length of time, called *read access time,* to allow the internal circuits of the memory chip to decode the address and select the addressed data.
4. After the wait time, the data is available on the memory output lines and can then be read by the system microprocessor. If the microprocessor

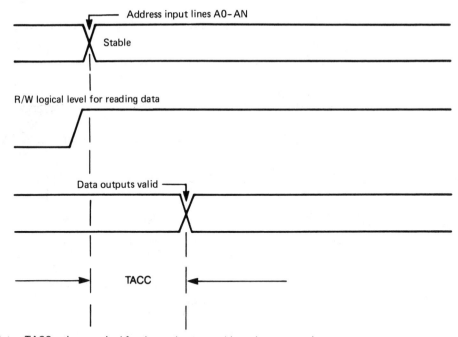

Address input lines A0– AN

Stable

R/W logical level for reading data

Data outputs valid

TACC

(Note: TACC = time required for data to become stable at the memory data output
lines during a memory read operation)

Figure 7-2   A timing diagram showing the general sequence of events for reading
data from system memory.

reads the data too soon—that is, if it does not wait for the read access
time—invalid data may be read from the device.

A timing diagram of this sequence appears in Figure 7-2. We shall refer
back to this sequence when we examine communication between the microprocessor and the semiconductor RAM. The sequence for reading data from memory is exactly the same for RAM as it was for ROM. In fact, the microprocessor
system does not electrically distinguish between reading data from ROM and
reading it from RAM.

Let's now discuss the sequence of events that occurs during the writing of
data to RAM.

7-3 SEQUENCE OF EVENTS FOR A RAM WRITE

1. First, the address input lines to the memory are set to the logical conditions of the internal memory location to which data will be written.
2. Next, the data that will be written into the memory are placed on the data
   input lines.

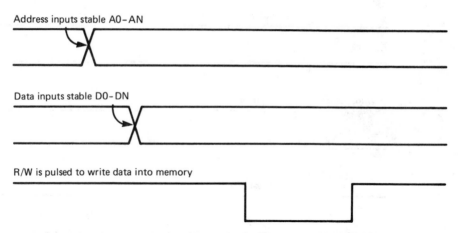

Address inputs stable A0–AN

Data inputs stable D0–DN

R/W is pulsed to write data into memory

Figure 7-3 A timing diagram showing the sequence of events necessary when writing data to a typical RAM device.

3. The system must wait a finite amount of time, called *write access time* (usually less than a few hundred nanoseconds), to allow the internal decoding circuits contained in the RAM to stabilize.

4. After the wait time, the R/W control line of the memory is set to the correct logical level or pulsed to allow the data present at the data input lines to be written into the RAM.

Figure 7-3 shows a timing diagram representation of this sequence of events. The hardware required to follow this sequence varies with each microprocessor.

7-4 A REAL MEMORY DEVICE

Let's now examine a typical RAM memory that is widely used in industry. We begin by examining the important specifications of the device. We then show how data is read from it and written to it. It is essential, when using RAMs, to understand the electrical specifications of the device. However, once you understand the working of one RAM device in detail, it is easy to understand the workings of other RAMs used in other microprocessor systems and applications.

The RAM that we have chosen for this discussion is the 6264, an 8192 × 8, static, common I/O RAM. The term *common I/O* refers to the electrical configuration of the data input and output lines of the RAM. A RAM may have either separate or common I/O. *Separate I/O* means that there are unique data input and output lines for the RAM (as shown in Figure 7-4). Common I/O means that the data input and output lines of the device are the same physical pin (see Figure 7-5).

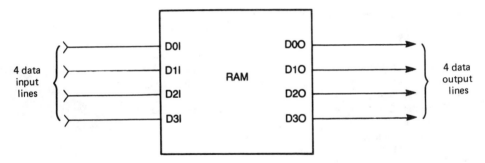

Figure 7-4   A block diagram of a RAM with separate input and output lines.

Since RAM can only be written to and read from, only data input and output lines are needed. Further, these two functions are mutually exclusive: You can never read and write to memory at the same time. Thus the input and output lines are never used at the same time. This fact allows for a reduction in the number of hardware pins needed to perform the read and write functions in RAM. It also allows for a common I/O, a condition in which the data input and output lines are time-multiplexed. This means that during a memory-read operation, the data I/O lines are used as data outputs and during a write operation, they are used as inputs. The logical state of the R/W control line determines the definition of the data I/O pins at any given time for the memory. Figure 7-6 shows a pinout and functional block diagram for a 6264 memory device.

Let's now examine the important parameters of this memory device. (*Note:* The information given in this section may be applied to most semiconductor memories, even if they do not use common I/O.)

For this example, we consider the 6264 at the user level only. Further, we assume that the memory access time is consistent with the overall system speed and that we won't need to generate any wait states for slow memory.

Figure 7-7 shows a partial data sheet and general timing diagrams for

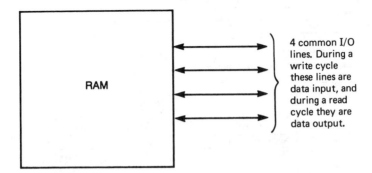

Figure 7-5   A block diagram of a RAM with common input and output lines.

HM6264P-10, HM6264P-12, HM6264P-15

8192-word x 8-bit High Speed Static CMOS RAM

■ FEATURES

- Fast access Time   100ns/120ns/150ns (max.)
- Low Power Standby   Standby: 0.1mW (typ.)
  - Low Power Operation   Operating: 200mW (typ.)
- Single +5V Supply
- Completely Static Memory..... No clock or Timing Strobe Required
- Equal Access and Cycle Time
- Common Data Input and Output, Three State Output
- Directly TTL Compatible: All Input and Output
- Standard 28pin Package Configuration
- Pin Out Compatible with 64K EPROM HN482764

(DP-28)

■ BLOCK DIAGRAM

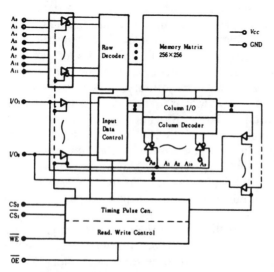

■ PIN ARRANGEMENT

NC	1	28	V_{CC}
A_{12}	2	27	$\overline{WE}$
A_7	3	26	CS_2
A_6	4	25	A_8
A_5	5	24	A_9
A_4	6	23	A_{11}
A_3	7	22	$\overline{OE}$
A_2	8	21	A_{10}
A_1	9	20	$\overline{CS}_1$
A_0	10	19	I/O_8
I/O_1	11	18	I/O_7
I/O_2	12	17	I/O_6
I/O_3	13	16	I/O_5
GND	14	15	I/O_4

(Top View)

■ ABSOLUTE MAXIMUM RATINGS

Item	Symbol	Rating	Unit
Terminal Voltage *	V_T	−0.5 ** to +7.0	V
Power Dissipation	P_T	1.0	W
Operating Temperature	T_{opr}	0 to +70	°C
Storage Temperature	T_{stg}	−55 to +125	°C
Storage Temperature (Under Bias)	T_{bias}	−10 to +85	°C

* With respect to GND.   ** Pulse width 50ns: −3.0V

■ TRUTH TABLE

$\overline{WE}$	$\overline{CS}_1$	CS_2	$\overline{OE}$	Mode	I/O Pin	V_{CC} Current	Note
X	H	X	X	Not Selected	High Z	I_{SB}, I_{SB1}	
X	X	L	X	(Power Down)	High Z	I_{SB}, I_{SB2}	
H	L	H	H	Output Disabled	High Z	I_{CC}, I_{CC1}	
H	L	H	L	Read	Dout	I_{CC}, I_{CC1}	
L	L	H	H	Write	Din	I_{CC}, I_{CC1}	Write Cycle (1)
L	L	H	L		Din	I_{CC}, I_{CC1}	Write Cycle (2)

X : H or L

 HITACHI

Figure 7-6   Pinout and functional diagram of the 6264, 8K X 8 static RAM. (Courtesy Hitachi America, Ltd. Semiconductor and IC Sales and Service Division.)

• WRITE CYCLE (1) (OE clock)

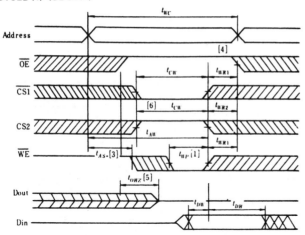

• WRITE CYCLE (2) (OE Low Fix)

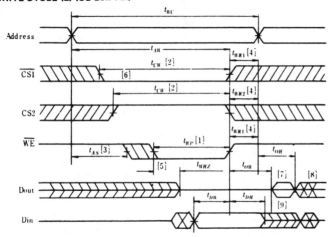

NOTES: 1) A write occurs during the overlap of a low $\overline{CS1}$, a high CS2 and a low $\overline{WE}$. A write begins at the latest transition among $\overline{CS1}$ going low, CS2 going high and $\overline{WE}$ going low. A write ends at the earliest transition among $\overline{CS1}$ going high, CS2 going low and $\overline{WE}$ going high. t_{WP} is measured from the beginninng of write to the end of write.
2) t_{CW} is measured from the later of $\overline{CS1}$ going low or CS2 going high to the end of write.
3) t_{AS} is measured from the address valid to the beginning of write.
4) t_{WR} is measured from the end of write to the address change.
t_{WR1} applies in case a write ends at $\overline{CS1}$ or $\overline{WE}$ going high.
t_{WR2} applies in case a write ends at CS2 going low.
5) During this period, I/O pins are in the output state, therefore the input signals of opposite phase to the outputs must not be applied.
6) If $\overline{CS1}$ goes low simultaneously with $\overline{WE}$ going low or after $\overline{WE}$ going low, the outputs remain in high impedance state.
7) Dout is the same phase of the latest written data in this write cycle.
8) Dout is the read data of next address.
9) If $\overline{CS1}$ is low and CS2 is high during this period, I/O pins are in the output state. Therefore, the input signals of opposite phase to the outputs must not be applied to them.

(a)

Figure 7-7 (a) Partial data sheet showing timing diagrams for a write cycle of the 6264 device; (b) partial data sheet showing timing diagrams for a read cycle of the 6264 device. (Courtesy Hitachi America, Ltd. Semiconductor and IC Sales and Service Division.)

● **READ CYCLE**

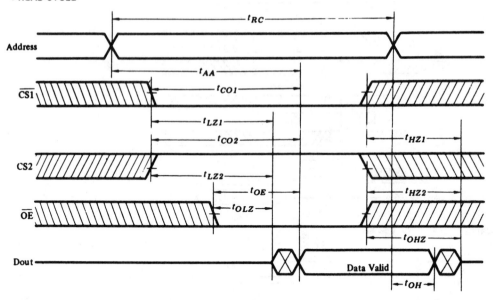

NOTE: 1) $\overline{WE}$ is high for Read Cycle

● **WRITE CYCLE**

Item		Symbol	HM6264P-10		HM6264P-12		HM6264P-15		Unit
			min	max	min	max	min	max	
Write Cycle Time		t_{WC}	100	--	120	–	150	–	ns
Chip Selection to End of Write		t_{CW}	80	–	85	–	100	–	ns
Address Setup Time		t_{AS}	0	–	0	–	0	–	ns
Address Valid to End of Write		t_{AW}	80	–	85	–	100	–	ns
Write Pulse Width		t_{WP}	60	–	70	–	90	–	ns
Write Recovery Time	$\overline{CS1}, \overline{WE}$	t_{WR1}	5	–	5	–	10	–	ns
	CS2	t_{WR2}	15	–	15	–	15	–	ns
Write to Output in High Z		t_{WHZ}	0	35	0	40	0	50	ns
Data to Write Time Overlap		t_{DW}	40	–	50	–	60	–	ns
Data Hold from Write Time		t_{DH}	0	–	0	–	0	–	ns
$\overline{OE}$ to Output in High Z		t_{OHZ}	0	35	0	40	0	50	ns
Output Active from End of Write		t_{OW}	5	–	5	–	10	–	ns

(b)

Figure 7-7   (*continued*)

electrical communication with the 6264 memory. Let's first discuss writing data to the memory. These events follow the general sequence described in Section 7-3. However, now we will describe the process in more detail and include the actual specifications of the memory under consideration.

7-5 SEQUENCE OF EVENTS FOR WRITING DATA TO THE 6264

1. Address inputs are set to the address to which data will be written.
2. Data is applied to the data I/O lines of the device. These lines are tristated. If this were not the case, there would be a conflict between the data being applied to the memory and the data stored in the memory; that is, each would try to control the I/O line. Figure 7-8 shows a diagram of such a conflict.
3. Next, the $\overline{WE}$ and $\overline{CS}$ inputs to the memory device are asserted at approximately the same instant in time. This is shown in Figure 7-9.

Now that we are familiar with the events that occur each time the 6264 memory has data written to it by the CPU, let's go on to explore the events that occur when the 6264 memory has data read from it.

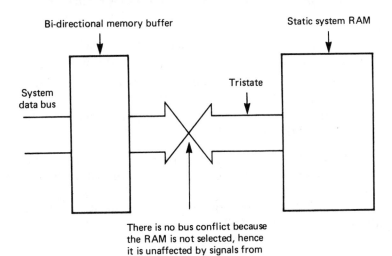

There is no bus conflict because
the RAM is not selected, hence
it is unaffected by signals from
the buffer.

Figure 7-8 A block diagram of a possible bus conflict between the data output lines and the memory data buffer outputs. This conflict can be avoided through proper control of the memory data buffer direction control line.

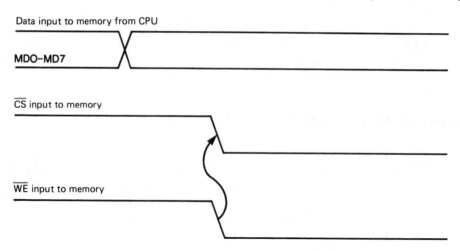

No bus conflict will result and memory will operate correctly.

Figure 7-9 A timing diagram showing the correct way to control the memory data buffers. In this case, the bus conflict shown in Figure 7-8 will be avoided.

7-6 SEQUENCE OF EVENTS FOR READING DATA FROM THE 6264

1. The address is input to the memory from which data will be read.
2. Next, the $\overline{CS}$ is asserted. At this time the $\overline{WE}$ input will be a logical 1.
3. The $\overline{OE}$ (output enable) line is set to a logical 0. During the write operation we assumed that this line was a logical 1.
4. At this time the data I/O lines output the data that has been previously stored at the address location.

Remember, reading of data from static RAM works in the same way as the reading of data from ROM. That is, all the same timing and buffering rules apply.

Now that we understand how the 6264 device operates, let's connect it to the 8080, 8085, and Z80 CPUs.

7-7 CONNECTING THE ADDRESS LINES TO THE 8080, 8085, AND Z80

To illustrate how the 8080, 8085, and Z80 communicate with RAM, we now present a complete design of a general-purpose RAM system. As we examine this system, we will explore potential problem areas that can arise when using

RAM with these CPUs. Although the information presented here is general, it is important information for anyone who wishes to use or understand how static RAM is used in a microprocessor system application.

We can see how the 8080, 8085, or Z80 address bus connects to the RAM system by examining the partial system memory map shown in Figure 7-10. This memory map shows where in the total 65,536 unique address locations the system RAM will reside. The map of Figure 7-10 shows that the system RAM will reside between the addresses 8000 and BFFF. This is a total of 16,384 bytes.

To realize the memory system design, the first thing we must do is to determine the number of memory devices needed to construct the entire RAM space. This can be calculated in the following way. The available RAM space is equal to 16,384 bytes. Each 6264 memory device stores up to 8192 bytes; therefore, for each 8192 bytes of storage using 6264 memories, we need one physical device. Thus we need two 6264 memory devices to realize the entire RAM space.

For this example, our system will require only 16,384 bytes of RAM. We will assume, however, that the system was designed to address up to 32,768 bytes of RAM space. (*Note:* It is normally a good idea to include more RAM space in your design than you feel you will use, if your system permits it. This allows you easily to expand the memory in the future if your application requires it. Bear in mind that adding memory to a system that has not been designed with that concept in mind can be a difficult task.) A memory map that shows 32,768 bytes of RAM memory space is shown in Figure 7-11.

We know that our system requires two 6264 memories to realize the entire

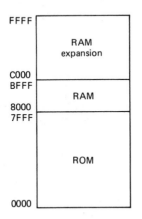

Figure 7-10 System memory map showing that the system RAM exists between memory addresses 8000 and BFFF, inclusive.

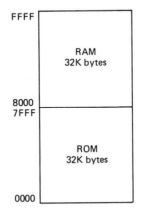

Figure 7-11 In this system memory map the RAM space is from 8000 to FFFF, which is equal to 32,768 bytes. This space can be physically realized with four 6264 memory devices.

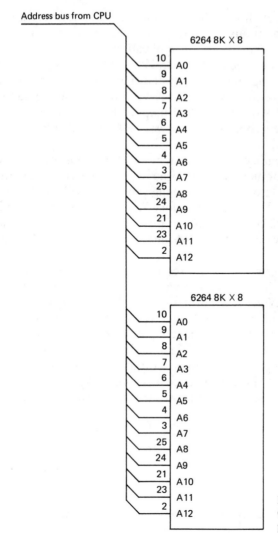

Figure 7-12 The address lines of the CPU connect directly to the address inputs of the 6264 device.

memory space. Each 6264 device has 13 address input lines, A0–A12; all these lines must be connected together in a parallel fashion; that is, A0 of all memory chips must be connected; A1 of all devices must be connected, and so on.

Address line A0 of the CPU is connected to address line A0 of the memory devices. For the 8085 we assume that the address has been latched using the ALE signal, as described in Chapter 5. Figure 7-12 shows the system address bus connected to the system RAM devices.

The remaining system address lines, A13–A15, will be used in the generation of the memory-select lines. These upper address lines will be used to

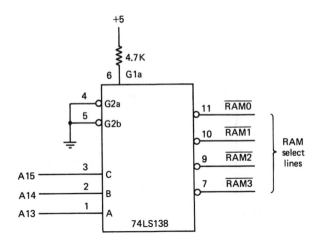

A15	A14	A13	A12		
1	0	0	X	8000–9FFF	$\overline{RAM0}$
1	0	1	X	A000–BFFF	$\overline{RAM1}$
1	1	0	X	C000–DFFF	$\overline{RAM2}$
1	1	1	X	E000–FFFF	$\overline{RAM3}$

Figure 7-13 Decode logic for generating the memory-select lines for the memory space from 8000 to FFFF.

decode the RAM address space from the entire memory space 0000–FFFF. This is exactly the same problem as was described in Chapter 6 for decoding the ROM space.

There are several different techniques that can be used to decode address lines for any particular application. The circuit shown in Figure 7-13 is a typical one. Figure 7-13 is provided to give you a general idea of how logical memory selection can be accomplished. Although this circuit does work, it is not meant to be the ultimate solution. All solutions must be tailored to the system application and normally depend on the system speed, bus loading, and final memory map.

7-8 CONNECTING THE DATA LINES—NONBUFFERED

We will now show you how to connect the memory data outputs to the microprocessor system data bus. For this example we will not use data buffers; we will assume that the loading on the data bus lines does not exceed the specified output drive capability of the memory data I/O lines. Thus we will simply connect the 6264 memory outputs in a parallel fashion and then connect the resulting eight data I/O lines directly to the CPU data bus. The connection is straightforward. The process is shown in Figure 7-14.

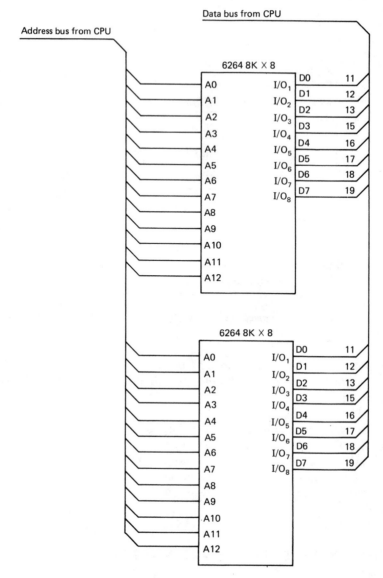

Figure 7-14   When connecting the memory data bus to the system data bus in a nonbuffered mode, the data I/O lines of the memory are connected directly to the CPU data I/O lines.

It is important to recall that the I/O lines of the 6264 memory system are controlled by the $\overline{\text{WE}}$, $\overline{\text{CS}}$, and $\overline{\text{OE}}$ inputs to the device. If your application is such that data buffers are not required, a simple direct connection from the memory output lines to the system data bus is all that is needed.

7-9 GENERATING THE MEMORY READ AND WRITE CONTROL LINES

When the microprocessor communicates electrically with memory, there are physical output lines that are used to generate the memory-control lines labeled $\overline{\text{MEMR}}$ and $\overline{\text{MEMW}}$. These physical lines are different for each of the three microprocessors. The logical decoding for the $\overline{\text{MEMR}}$ and $\overline{\text{MEMW}}$ control lines was presented in Chapter 5. It is repeated here for your convenience. Figure 7-15a, b, and c, shows a typical logical circuit that generates a system memory read and write signals for the 8080, 8085, and Z80 microprocessors.

Using the two system signals $\overline{\text{MEMW}}$ and $\overline{\text{MEMR}}$, let's now discuss how the system connects to the 6264 $\overline{\text{CS}}$, $\overline{\text{OE}}$, and $\overline{\text{WE}}$ input lines. Figure 7-16 shows the complete connection of the 6264 to the CPU system buses. Remember that each of these system buses is constructed differently for each microprocessor. The following describes how the circuit of Figure 7-16 operates.

When the system address is output, the memory-decoding circuits shown in Figure 7-13 assert the $\overline{\text{CS}}$ input line to the proper 6264 device. Next the CPU generates the $\overline{\text{MEMR}}$ or $\overline{\text{MEMW}}$ system-control signal, depending on the operation taking place. The $\overline{\text{MEMW}}$ control line is connected to the $\overline{\text{WE}}$ input pin of all 6264s. Only the device that has its $\overline{\text{CS}}$ input low will actually perform the write operation.

When the CPU is reading from the system memory, the $\overline{\text{MEMR}}$ control line will become asserted. The $\overline{\text{MEMR}}$ control line is connected to the $\overline{\text{OE}}$ input pin all 6264 devices. Only the 6264 device that has its $\overline{\text{CS}}$ input low will output data to the CPU.

We can see from Figure 7-16 that connecting the 6264 devices to the 8080, 8085, or Z80 in an unbuffered mode is not a complicated task.

7-10 USING BUFFERED DATA LINES WITH STATIC RAM

In Chapter 6 we discussed the use of a bidirectional data bus buffer for the system data bus. This buffer was installed whenever the electrical requirements of the data bus exceeded the capabilities of the CPU. If the CPU requires a data bus buffer, then there is a high probability that the system memory will require a data bus buffer. This is due to the fact that the memory data outputs have approximately the same output drive capability as the CPU data outputs.

Let's assume that our RAM system does require a data buffer. We cannot use the same unidirectional buffering that was shown in Chapter 6 for ROM because information on the system data bus must be written into RAM as well as read from RAM. Therefore, bidirectional buffering is needed for RAM.

Figure 7-17 shows one way to install data buffers on a static RAM system. Note that the 6264 memory I/O lines are all connected as before; however they are not connected directly to the CPU data lines. Instead, they are connected to

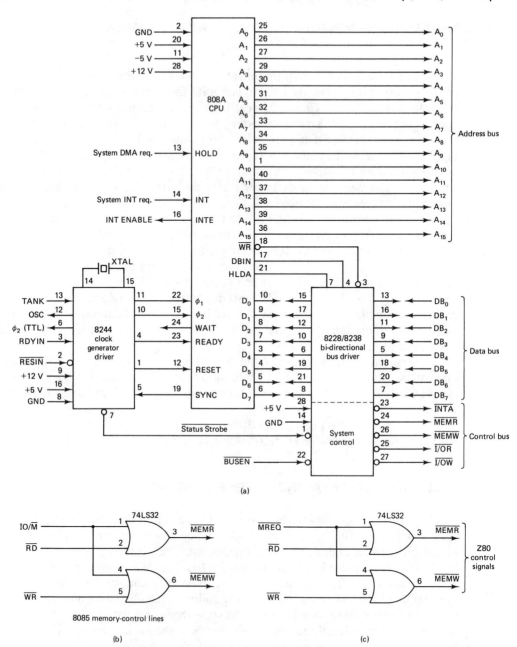

(a)

(b)

8085 memory-control lines

(c)

Figure 7-15 (a) Schematic showing how the control bus is an output of the 8228 system controller; (b) schematic diagram showing how the memory-read and memory-write control lines are generated for the 8085 CPU; (c) schematic diagram showing how the memory-read and memory-write control lines are generated for the Z80 CPU.

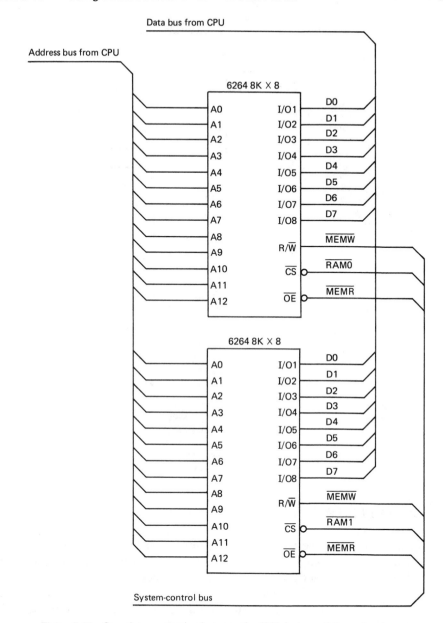

Figure 7-16   Complete connection between the 6264 device and the system buses.

one side of a bidirectional buffer. The direction control for the data buffer is dependent on the logical state of the $\overline{\text{MEMR}}$ signal. Whenever this signal is a logical 0 and the memory space is enabled, the memory data buffers will then buffer data from the RAM outputs to the system data bus.

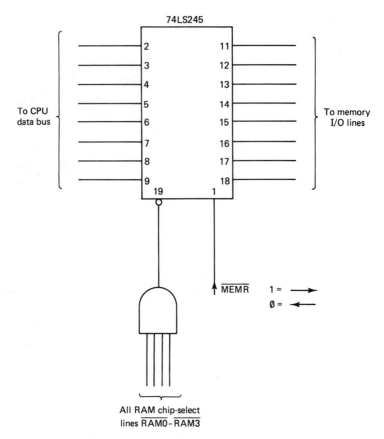

Figure 7-17 The 74LS245 device may be used to provide bidirectional buffering for the data I/O lines of the static RAM.

7-11 CHAPTER SUMMARY

In this chapter we have presented important information regarding the use of static RAM with the 8080, 8085, and Z80 microprocessors. Our discussion began with a description of a general sequence of events for a memory read and write operation. We then examined a typical RAM system comprised of 6264 8k × 8 static RAMS. Important memory parameters were presented, and details for reading and writing to memory were shown. We then went on to examine the complete connection between the microprocessor and the 6264 device.

You can see from this chapter that using static RAM devices with a microprocessor is not a difficult task. Starting with the system memory map, the number of address lines and the correct number of memory chips are calculated. Once these are determined, the chip-select lines are generated using combinational logic circuits. Next, the data and control lines are connected to the CPU. As a final point, it is usually better to design more RAM memory space into your system than you think you will use. This makes it easy to expand should you need more memory later.

REVIEW PROBLEMS

1. List two characteristics of static RAM.
2. Draw a block diagram of the 6264 RAM showing input and output lines.
3. List the sequence of events for reading data from static RAM.
4. List the sequence of events for writing data to static RAM.
5. Define read access time for a static RAM.
6. Explain the difference between common and separate I/O as it relates to static RAM.
7. Why is it a good idea to include more RAM space in your system than you think you will use?
8. The 6264 static RAM requires _____ address input lines.
9. If your system was designed to accommodate 32K bytes of static RAM, how many 6264 devices would this require to realize?
10. Show the address decoding if the 32K bytes of static RAM existed in the upper 32K address space.
11. Draw a schematic showing decoding and connections for the memory system of problem 10.

8

MICROPROCESSOR INPUT AND OUTPUT

Two major hardware operations that a microprocessor performs electrically are reading data from an input device and writing data to an output device. In this chapter, we learn how this is accomplished with the 8080, 8085, and Z80. We also construct a general I/O port using discrete logic devices.

We have chosen to discuss I/O at this point in the text because electrical communication with I/O is similar to that of static RAM. As you read through the chapter, you will notice many similarities between these two functions. In fact, some microprocessor systems are designed using an I/O architecture called *memory-mapped I/O*, which treats I/O and static RAM the same way.

Many of the remaining chapters in this text build on the information presented in this chapter. Interfacing to input and output devices is an important topic. Understanding this concept will allow you to understand how many different types of peripheral chips and devices are interfaced to these three popular microprocessors.

8-1 WHAT ARE INPUT AND OUTPUT?

The general architecture of a microprocessor system is shown in Figure 8-1. We see in Figure 8-1 that the microprocessor will communicate with ROM, RAM, and I/O. ROM and RAM are lumped together, forming the quantity called system memory. For the 8080, 8085, and Z80 the system memory has valid addresses from 0000H to FFFFH. The I/O block shown in Figure 8-1 is not included in this memory space.

Some microprocessors must allocate some of the available memory space for I/O. Examples of these types of microprocessors are the 6800, 6502, 6809, and 68000, to name a few. Microprocessors that use some of the memory space for I/O are said to use memory-mapped I/O.

In an 8080-, 8085-, or Z80-based system, you do not have to use memory-

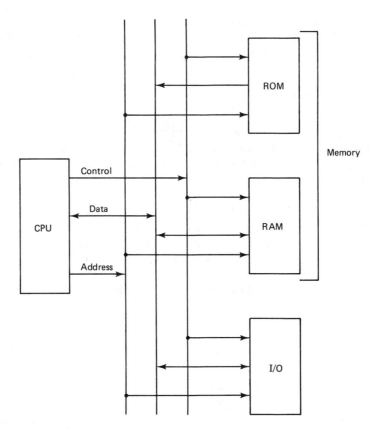

Figure 8-1   The diagram shows the general 3-bus architecture for a microprocessor system.

mapped I/O. These microprocessors have a separate I/O architecture. This means all the system memory space can be used for memory. System I/O will have its own address space. An I/O architecture like this is called *I/O mapped I/O*. It should be noted that although these microprocessors have a separate I/O space, they can use memory-mapped I/O in any application that requires it.

An I/O operation may be defined as follows:

Input An input operation is reading data from a source which is not system memory.

Output An output operation is writing data to a destination that is not system memory.

We stated earlier in this text that the system-control bus defines the type of communication that occurs. If the system uses I/O mapped I/O, then there are separate control lines for the I/O system and memory system. The memory system uses control lines labeled memory read (MEMR) and memory write (MEMW), whereas the I/O system uses control lines labeled input read (IOR) and output write (IOW). This is shown in Figure 8-2.

8-2 I/O ADDRESSING

In a typical microprocessor system, there are usually several I/O *ports*. A port is a unique place to read or write data that is not system memory. A port is very similar to a unique location in the system memory from which data will be read or to which it will be written during a memory operation.

Each port in the I/O system is given a unique address. The address of the I/O port is called the *port-select code*. See Figure 8-3. To generate the port-select code, the system makes use of address lines A0–A7. These lines are decoded and allow the I/O to respond to a specific address combination. For example, a port may be have a port-select code of 57H. This is shown in Figure 8-4.

Although the I/O system and the memory system are completely separate,

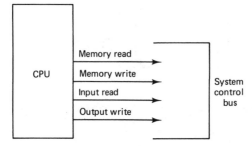

Figure 8-2 The system control signals for memory are memory read and memory write. The corresponding control signals for I/O are input read and output write.

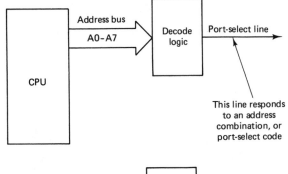

Figure 8-3 Each I/O port in the system is designed to respond electrically to a unique combination on the system address bus.

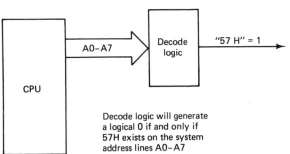

Figure 8-4 This I/O port will respond electrically to system address 57H.

the systems use the same address lines. This means you may have a memory address of 00F4H and an I/O address of F4H. Distinction is made between the two systems by the lines contained in the system-control bus. That is, memory will use the MEMR and MEMW control lines to communicate with the selected address, whereas the I/O system will use the IOR and IOW control lines.

8-3 WHAT IS AN I/O DEVICE?

An I/O device may be defined as any hardware that the system controls. The device may have one or more I/O ports or I/O addresses associated with it. This is shown in Figure 8-5. Examples of I/O devices are special LSI (large-scale

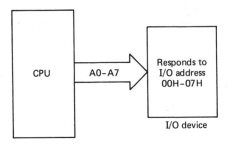

Figure 8-5 Block diagram showing an I/O device that responds to several I/O addresses. In this example the I/O device responds to I/O addresses between 00 and 07H, inclusive.

integration) chips such as a floppy disc controller or timer chip. We will be discussing some of these LSI devices in the later chapters of this text. Other examples are

Digital voltmeters
A/D converters
Position sensors
Motor controllers
Solid-state or electromechanical relays

8-4 INPUT AND OUTPUT INSTRUCTIONS

To read data from an input port, the microprocessor will execute the IN instruction. This instruction is written in mnemonic form like this:

```
IN port number
```

The IN instruction transfers a byte to the accumulator from an input port address. As you learned in Chapter 2, the IN instruction is encoded as shown in Figure 8-6. We see in Figure 8-6 that the first byte has the op-code, and the second byte shown in Figure 8-6 is an 8-bit port address. Since there are only 8 bits, this gives a total of 256 (00–FF) unique I/O ports that can be accessed with this instruction. During the execution of the IN instruction, the 8-bit address is placed on address bus lines A0–A7.

The Output Instruction

To transfer data from the CPU to an output port, the OUT instruction is used. This instruction has a similar mnemonic format to the IN instruction:

```
OUT port number
```

This is a 2-byte instruction with the first byte being the op-code; the second is the 8-bit address to which to write data.

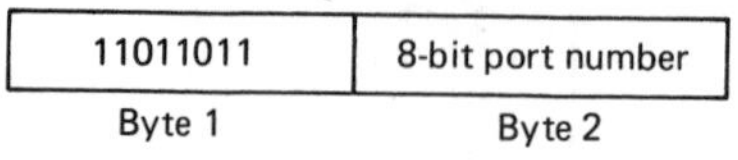

11011011	8-bit port number
Byte 1	Byte 2

Figure 8-6 Diagram showing the two bytes of the IN instruction.

8-5 DECODING THE PORT ADDRESS

Now that we have had an introduction to I/O operations, let's examine the system hardware required for input and output. The hardware schematics given are very general and are designed to allow you to understand exactly how the system hardware and software operate together in performing I/O operations. We start our discussion with the topic of decoding the port address.

Recall from an earlier discussion that I/O and memory in a microprocessor system use the same address lines. It is the microprocessor's job to separate electrically the memory requests from the I/O requests. Each I/O port in a microprocessor system will respond to a unique combination of 8 bits on the system address lines A0–A7. The address combination to which a port will respond electrically is known as the *port address,* or *port-select code.* We will assume the port address is equal to FFH. This is an arbitrary choices of a port number. Figure 8-7 shows a schematic diagram of a circuit that will detect this port address.

In the diagram of Figure 8-7 we can see that output pin 8 of the 74LS30 is a logical 0 if and only if all input pins are a logical 1. Note that output pin 8 of the 74LS30 (labeled *port-select line*) is active logical 0 whenever the system address bus is logically equal to the unique select code of the port. In this case the port-select code equals FF.

It is important to remember that the port-select line can become active even when there is no I/O communication occurring in the system. This is because system memory uses the same address lines. As an example, let's assume that the microprocessor is reading data from memory address XXFF and the port-select line suddenly becomes active. This can occur because the address lines A0–A7 are equal to the port-select code. In other words, if the port-select line in your system begins to signal that it is being selected at a time when software instructions are not indicating system communication with the port, do not be alarmed—the condition is valid.

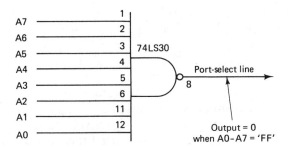

Figure 8-7 Schematic diagram showing the decoding of the port number FF.

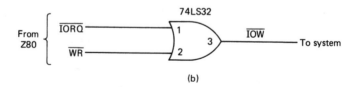

(a)

(b)

For 8080 $\overline{\text{IOW}}$ signal, see Figure 7-15a

Figure 8-8   Schematic diagram showing how the $\overline{\text{IOR}}$ and $\overline{\text{IOW}}$ control lines can be generated for the 8085 and Z80 microprocessors. Refer to Figure 7-15 to see how these signals are generated for the 8080 CPU.

8-6 GENERATION OF THE IOW AND IOR CONTROL LINES

In Chapter 5 you were shown how the microprocessor generates the $\overline{\text{IOR}}$ and $\overline{\text{IOW}}$ control lines. Whenever the CPU executes an IN instruction, the $\overline{\text{IOR}}$ control line is active. When an OUT instruction is executed, the $\overline{\text{IOW}}$ control lines becomes asserted. The schematic diagram for generating these control lines with the 8085 and Z80 is shown in Figure 8-8 for your convenience. In an 8080 system the 8228 system controller generates the $\overline{\text{IOW}}$ and $\overline{\text{IOR}}$ control signals. This was shown in Figure 7-15a.

8-7 THE PORT-WRITE STROBE

The port-write signal for a microprocessor system is defined as the write-enable strobe for a selected output port. This signal is generated whenever the port-select line and the $\overline{\text{IOW}}$ control signal are both active. The port-write signal provides the active digital signal that specifies that the data from the microprocessor is to be latched or written to an output port.

The timing diagram in Figure 8-9 shows the general timing sequence that occurs when data is written to an output port. In this figure, all the signals except $\overline{\text{IOW}}$ can be thought of as static logic levels. The signal voltage levels remain stable for the entire hardware operation. We have used this concept and the timing diagram in Figure 8-9 to create the schematic in Figure 8-10. This schematic shows one way the port-write strobe can be generated using hardware.

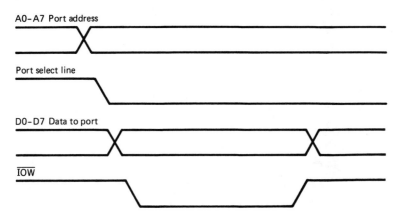

Figure 8-9   General timing diagram of the sequence of events that occur during a typical output operation.

Let's examine this diagram and see how the circuit operates. Understanding this operation will help you to better understand how each signal is used on the different microprocessors. (Remember, however, there are other ways that a port strobe line can be generated.)

The electrical objective of the circuit in Figure 8-10 is to provide an active logical 0 strobe whenever the CPU is writing data to the specified output port. In this figure, the system address lines A0–A7 are input to an 8-input NAND gate. This is the same circuit that was shown in Figure 8-7.

When all the address lines (A0–A7) are a logical 1 (which is the port-select code FF), the output pin 8 of the NAND gate is a logical 0. This port-select line is connected to one input of the 74LS32 OR gate. The other input of the OR gate is connected to the $\overline{\text{IOW}}$ strobe signal (displayed in Figure 8-8).

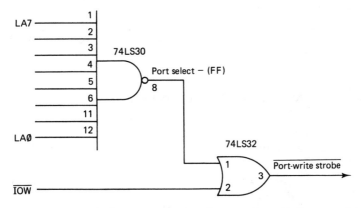

Figure 8-10   Schematic diagram showing how the port-write strobe is generated for use during the output instruction.

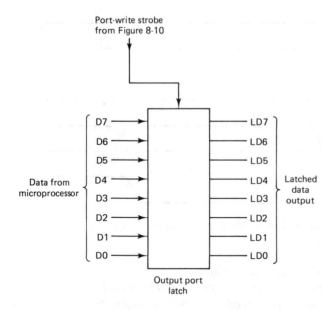

Figure 8-11   Block diagram of a typical output port latch. The data is strobed into the latch by the port-write strobe generated in Figure 8-10.

When the $\overline{\text{IOW}}$ strobe is a logical 0, it is indicating electrically that the CPU is performing the function of outputting a byte to a system output port.

Therefore, if the $\overline{\text{IOW}}$ signal is a logical 0 and the port-select line is a logical 0, the CPU must be electrically outputting data to the selected port. In effect, we have electrically qualified the $\overline{\text{IOW}}$ signal with the port-select line. The result is a unique active logical 0 strobe that occurs if and only if the CPU is performing an output operation to the system output port FF. This resulting strobe signal is called the *port-write strobe*. It is used to strobe data into the selected output port. This is shown in Figure 8-11. The action of writing data into an output port is very similar to the action that occurs during a memory write cycle with the CPU.

Note that this is only one way that an output port strobe can be realized. The sequence of electrical events and the concept of qualifying the $\overline{\text{IOW}}$ strobe are common to most microprocessor systems.

8-8 GENERATION OF THE PORT READ SIGNAL

Let's now examine how the microprocessor reads data from an input port. The timing diagram displayed in Figure 8-12 shows a general sequence of electrical events that occurs during an input port-read operation. In this diagram, the port-select and $\overline{\text{IOR}}$ signals are similar to the port select and $\overline{\text{IOW}}$ signals used for an output operation.

The electrical effect of the $\overline{\text{RD}}$ signal for the 8085 and Z80 (DBIN signal for the 8080) is shown in Figure 8-13. $\overline{\text{RD}}$ or DBIN is the timed control signal from the CPU that generates the $\overline{\text{IOR}}$ signal for the input port hardware. This

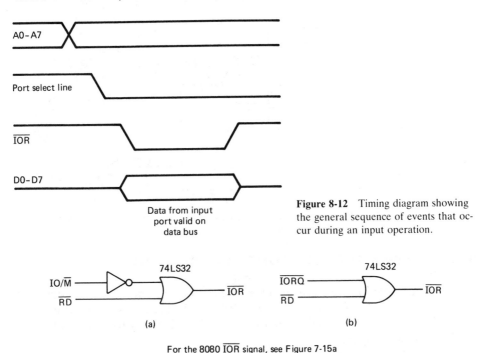

Figure 8-12 Timing diagram showing the general sequence of events that occur during an input operation.

For the 8080 $\overline{\text{IOR}}$ signal, see Figure 7-15a

Figure 8-13 (a) Schematic diagram showing how the $\overline{\text{IOR}}$ signal may be generated for the 8085 microprocessor; (b) $\overline{\text{IOR}}$ signal for the Z80 CPU.

signal electrically starts the data transfer. When $\overline{\text{RD}}$ goes to a logical 0 or DBIN goes to a logical 1, the data from the input port is placed on the system data bus and is strobed by the CPU into an internal register. At a later point in time, RD or DBIN will become unasserted. When this occurs the input port data will be electrically removed from the system data bus, and the hardware transfer will be complete.

Figure 8-14 shows one of several ways a port-read signal can be generated, based on the logical conditions given in Figure 8-12. We see in Figure 8-13 that this circuit is very similar to the logic shown in Figure 8-10. Indeed, the only difference between the two circuits is that one uses the $\overline{\text{IOW}}$ signal and the other uses the $\overline{\text{IOR}}$ signal. Apart from that, they are identical. The port-read strobe becomes active if and only if the CPU is electrically inputting data from the addressed port.

8-9 A COMPLETE SCHEMATIC FOR AN I/O PORT

Figure 8-15 shows a complete schematic for an input and output port for the 8080, 8085, and Z80 microprocessors configured into the 3-bus architecture that was presented in Chapter 5.

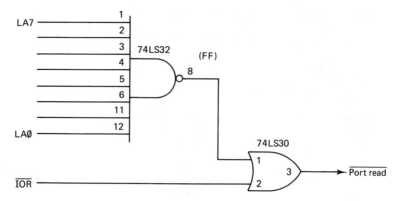

Figure 8-14 Schematic diagram showing how the port-read signal is generated for use during an input instruction.

8-10 SEQUENCE OF EVENTS FOR AN OUTPUT WRITE

In this section we go over the sequence of electrical events that occurs during an output write operation with the 8080, 8085, and Z80 CPUs. At each step in the sequence you should refer to the schematic of Figure 8-15 and note how the hardware responds.

8080 Sequence of Events

1. For an output write, the address lines A0–A7 are first set to the desired output address under control of the CPU. At this time, the output address lines A0–A7 are decoded by the port-select hardware.
2. The status word for an output write is placed on the data bus. For an 8080 the status word for an output write is 0001000.
3. The 8080 then sets sync output equal to 1.
4. The sync output is reset equal to 0 (status word is now latched).
5. Next, the 8080 outputs the electrical data to be written to the output port on the data bus lines, D0–D7. Notice in Figure 8-15 that the data is now valid at the input to the 74LS374 octal latch. At this point in the operation, all the static decoding from the 8080 output signals has occurred.
6. Next, the $\overline{WR}$ timed control output line from the 8080 is set to a logical 0. This section asserts the $\overline{IOW}$ system-control line. With $\overline{IOW}$ going to a logical 0, the port-write strobe also goes to a logical 0. Next, the $\overline{WR}$ output from the 8080 goes to a logical 1. At this point the data that was present on the D0–D7 data lines is written to the 74LS374. The signal is now a logical 1, and the output operation is complete.

Steps 2, 3, and 4 are accomplished inside the 8228 controller chip shown in Figure 5-26a.

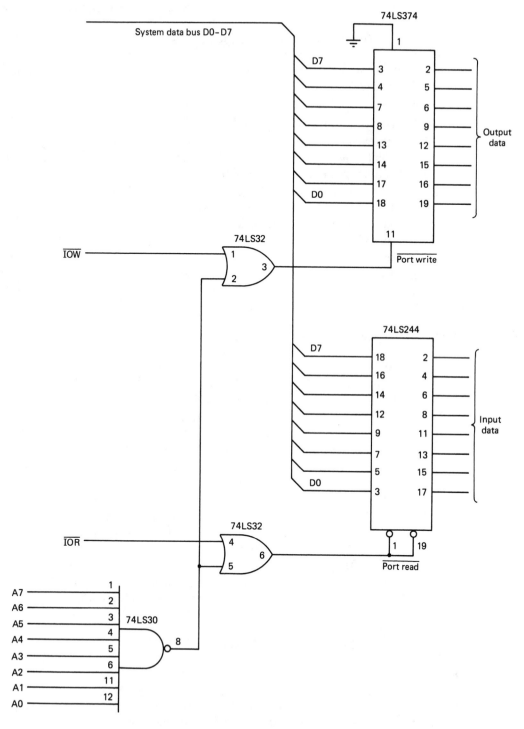

Figure 8-15 Complete schematic diagram for a general I/O port decoded at address FF.

Sequence of Events for an 8085 Output Write

1. For an output write, the data lines AD0–AD7 are set to the desired output address under control of the CPU.
2. The 8085 the sets the ALE output equal to 1.
3. The ALE output is reset to 0 (address is now latched).
4. Next, the 8085 outputs the electrical data to be written to the output port on the data bus lines, D0–D7. Notice in Figure 8-15 that the data is now valid at the input to the 74LS374 octal latch.
5. Next, the IO/$\overline{\text{M}}$ goes to a logical 1 for an 8085, indicating that an I/O operation will be performed. At this point all the static decoding from the 8085 output signals has occurred.
6. Next, the $\overline{\text{WR}}$ timed control output line from the 8085 is set to a logical 0. This action asserts the $\overline{\text{IOW}}$ system control line. With $\overline{\text{IOW}}$ going to a logical 0, the port-write strobe also goes to a logical 0. Next, the $\overline{\text{WR}}$ output from the 8085 goes to a logical 1. At this point the data that was present on the D0–D7 data lines is written to the 74LS374.

Sequence of Events for a Z80 Output Write

1. The address lines A0–A7 are set to the desired output address under control of the CPU. At this time, the output address lines A0–A7 are decoded by the port-select hardware.
2. Next, the Z80 outputs the electrical data to be written to the output port on the data bus lines, D0–D7. Notice in Figure 8-15 that the data is now valid at the input to the 74LS374 octal latch.
3. Next, the $\overline{\text{IORQ}}$ output goes to a logical 0, indicating that an I/O operation will be performed. At this point all the static decoding from the Z80 output signals has occurred.
4. Next, the $\overline{\text{WR}}$ timed control output line from the Z80 is set to a logical 0. This action asserts the $\overline{\text{IOW}}$ system control line. With $\overline{\text{IOW}}$ going to a logical 0, the port-write strobe also goes to a logical 0.
5. Next, the $\overline{\text{WR}}$ output from the Z80 goes to a logical 1. At this point the data that was present on the D0–D7 data lines is written to the 74LS374. The signal is now a logical 1, and the output operation is complete.

8-11 SEQUENCE OF EVENTS FOR AN INPUT-PORT READ OPERATION

Let's now discuss the sequence of events performed by the 8080, 8085, or Z80 when reading data from an input port (as shown in Figure 8-15). The major events that occur in an I/O read are the same as those in an I/O write, except that the $\overline{\text{RD}}$ or DBIN timed control output line is used rather than the $\overline{\text{WR}}$ control line.

8080 Sequence of Events for an Input Read

1. For an input read, the address lines A0–A7 first set to the desired output address under control of the CPU. At this time, the output address lines A0–A7 are decoded by the port-select hardware.
2. The status word for an input read is placed on the data bus. For an 8080, the status word for an input read is 01000100.
3. The 8080 the sets sync output equal to 1.
4. The Sync output is reset to 0 (status word is now latched).
5. Next, the 8080 tristates its output lines to prepare electrically to read data from the system. At this point in the operation all of the static decoding from the 8080 output signals has occurred.
6. Next, the DBIN timed control output line from the 8080 is set to a logical 1. This action asserts the $\overline{\text{IOR}}$ system-control line. With $\overline{\text{IOR}}$ going to a logical 0, the port-read line also goes to a logical 0. At this time the data from the input port is placed onto the system data bus and is present at the 8080 data input pins. The data is then strobed into the A register of the CPU.
7. Next, the DBIN output from the 8080 goes to a logical 0. At this point the data that was present on the D0–D7 data lines is electrically removed from the system data bus and the transfer is complete.

Sequence of Events for an 8085 Input Read

1. For an input read, the data lines AD0–AD7 are set to the desired output address under control of the CPU.
2. The 8085 the sets the ALE output to 1.
3. The ALE output is reset to 0 (address is now latched).
4. Next, the 8085 tristates its data outputs to prepare the CPU to read data from the input port.
5. The $\text{IO}/\overline{\text{M}}$ goes to a logical 1 for an 8085, indicating that an I/O operation will be performed. At this point all the static decoding from the 8085 output signals has occurred.
6. Next, the $\overline{\text{RD}}$ timed control output line from the 8085 is set to a logical 0. This action asserts the $\overline{\text{IOR}}$ system control line. With $\overline{\text{IOR}}$ going to a logical 0, the port-read strobe also goes to a logical 0. At this time the data from the input port buffer is placed onto the system data bus and is present on the CPU data input lines.

7. The $\overline{RD}$ output from the 8085 goes to a logical 1. At this point the data that was present on the D0–D7 data lines is electrically removed from the system data bus.

Sequence of Events for a Z80 Input Read

1. The address lines A0–A7 are set to the desired output address under control of the CPU. At this time, the output address lines A0–A7 are decoded by the port-select hardware.
2. Next, the Z80 tristates its data lines, preparing to read data from the system input port.
3. Then the $\overline{IORQ}$ output goes to a logical 0, indicating that an I/O operation will be performed. At this point all the static decoding from the Z80 output signals has occurred.
4. Next, the $\overline{RD}$ timed control output line from the Z80 is set to a logical 0. This action asserts the $\overline{IOR}$ system-control line. With $\overline{IOR}$ going to a logical 0, the data from the selected input port is placed onto the system data bus and is present at the CPU input pins. The CPU now strobes the data into the A register.
5. Next, the $\overline{RD}$ output from the Z80 goes to a logical 1. At this point the data that was present on the D0–D7 data lines is removed from the system data bus and the operation is complete.

8-12 CHAPTER SUMMARY

In this chapter we presented the essential points of electrical communication with I/O for the 8080, 8085, and Z80 microprocessors. We started by presenting the essential points of input and output and discussed the two I/O instructions, IN and OUT. After these instructions were reviewed, the sequences of events for output and input operations were discussed. From this discussion we determined the hardware required to realize a complete 8-bit I/O port. In examining this hardware we discussed the sequence of events required for reading and writing to an I/O port. As the sequence was given, the hardware response to each step was explained.

The information given in this chapter lays the foundation for topics covered in many of the remaining chapters of the text. In those chapters we will connect several LSI I/O devices to the three CPUs and write the software to control these special I/O chips.

REVIEW PROBLEMS

1. Write a general definition of an input port and an output port.
2. How many I/O ports will the 8080 CPU support? Assume that only the lower half of the address bus is used.
3. What is the main difference between I/O mapped I/O and memory-mapped I/O architecture?
4. Which type of I/O architecture is the 8080 designed to use?
5. What are the control lines used in the I/O architecture given in Problem 4?
6. Give three examples of output devices.
7. Give three examples of input devices.
8. Give three examples of I/O devices.
9. Draw a schematic that will decode the port number 36H.
10. What is the function of the port-select line?
11. What is the function of the port-write strobe?
12. What is the function of the port-read signal?
13. Draw a complete schematic of an I/O port with the select code equal to F7H.

9

TROUBLESHOOTING A MICROPROCESSOR SYSTEM USING STATIC STIMULUS TESTING

In this chapter, we discuss a hardware debugging and troubleshooting technique, called *static stimulus testing,* or SST. This technique offers a simple and inexpensive way to check out the hardware of your system without using software. You can use SST to verify the operation of ROM, RAM, and all peripheral hardware of a microprocessor system.

SST can also be used when you are expanding an existing system. SST will allow you to determine if the hardware of the interface is operating correctly. The problem may be one of hardware or software. After using the SST to verify the hardware, you can concentrate your efforts on making the software operate correctly. By using SST, you can deal with the hardware-software question quite easily.

SST can be used for repairing microprocessor-based systems that fall into two main categories:

1. Systems that are completely nonfunctional and cannot execute any software. These are systems that once worked but now have a hardware problem.

2. Systems that are in a prototype mode and have never worked. These are systems that have not yet been debugged or systems for which the software is not yet complete and cannot be used in the diagnostic process. With SST you can debug a prototype system as it is being built. You do not have to wait until the complete system is finished to determine if a section of the hardware is operational.

In both cases, classical hardware troubleshooting techniques such as logic state analysis and signature analysis are less effective than SST. This is because both of these techniques require that the system be capable of executing some software in order for the technique to be useful. Although techniques have been developed for signature analysis that tend to eliminate this software dependence, your system may not have been designed for use with signature analysis. We discuss signature and logic state analysis in Chapter 11 of this text. After you have been introduced to these techniques, you will be able to choose for yourself when and where each technique will be of the most value to you in your own troubleshooting.

An important question with which we will be dealing for a malfunctioning microprocessor system is, Where and how do you start to troubleshoot a system that is completely inoperative? That is, how do you repair and troubleshoot a system that will not execute any system software?

This is the situation in which SST will do the job easily. With SST it is possible to debug an inoperative system in a direct and orderly fashion. SST will get the system to a point where the software diagnosis can be run. You want the defective system to help you find the problem area. SST allows you to accomplish this objective in an efficient and straightforward manner, using simple and inexpensive instruments.

A very important point of SST is that it does not depend on software for its operation. This means that a person with little training in system software can use SST. Also, if a person is skilled in software and has little hardware training, the basics of SST can be learned quite easily.

When you start to use SST to verify each section of the hardware, you will begin to gain experience in examining exactly what occurs during a system communication. Using the SST will give you a better understanding of how the microprocessor communicates electrically in a system environment. We will be presenting the SST for the 8080, 8085, and Z80, but the concept of SST applies to most microprocessors.

9-1 GETTING STARTED WITH SST

The main idea behind SST is that the electrical communication of a microprocessor system is a static operation. That is, there are two voltage levels, representing 1s and 0s. A microprocessor system alternates between these two static states. The electrical events occur in rapid succession, but they do not have to.

There is an upper limit to how fast a system can operate, but there is usually no lower limit.

In a microprocessor system, the signal lines are performing a unique electrical function at any given instant in time. For example, during a memory operation the system address lines are logically pointing to a particular memory location. This occurs regardless of the logical state of any other system signals. Using SST, each signal line in the system can be treated as an independent logic signal.

Each signal *always* has a point of origin and a point of destination. Using SST as the stimulus, you essentially inject a logic level that will force a single signal line to a desired logical condition at the point of origin. You can then statically trace the electrical response of the line. The element of time-dependent signals is eliminated with SST.

Using standard digital troubleshooting techniques and SST, you can debug the hardware of an entire microprocessor system. Regardless of the complexity of the system hardware, once the dynamic situation has been transformed into a static one, problem areas can be found much easier.

SST will also work well in systems where special LSI devices are used, such as PIOs and USARTS. This is due to the fact that SST operates on the premise that all microprocessor communication within the system is static.

One area that is not static is dynamic RAMs. Although this statement is basically true, it must be qualified. The only part of the dynamic RAM system that is dynamic is the storage cell of the RAM chip. All address inputs, RAS, CAS, REFRESH, and MUX signals can be treated as static lines. You cannot test the memory cell in a static fashion, but all the peripheral signals for the dynamic RAM may be verified statically. This point will become much clearer when you study dynamic RAM operation in Chapter 15.

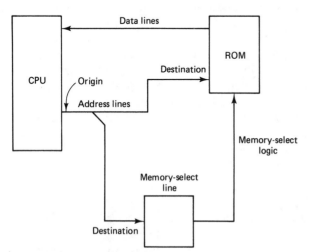

Figure 9-1 Block diagram showing major signal paths during a memory read cycle.

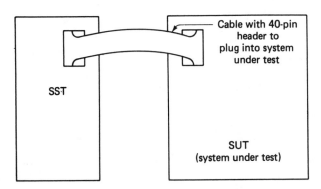

Figure 9-2 Pictorial diagram showing how the SST is connected to the system under test. The microprocessor is removed from the system under test, and the SST cable is installed in the vacated socket.

To illustrate how SST operates in a system, let's take an example. In this example we will check the address inputs to the system ROM. (*Note:* We will be giving expanded examples in later sections of this chapter. This example is meant only to illustrate the main points of how SST works.)

Figure 9-1 shows a block diagram of how ROM electrically exists in a microprocessor system. Also shown are the origin and termination for the address inputs for the ROM. In this example we will verify proper hardware operation of these address-input lines. To accomplish this, the following steps are used;

1. First, we remove the microprocessor from the system and install a cable that is connected to the SST switch panel, as shown in Figure 9-2. You do not have to remove the microprocessor physically from the system. It is possible to remove the device electrically and place the SST cable in parallel with the existing microprocessor pins. This is done by asserting the input pin which tristates the CPU address, data, and control line. See the data sheets for the pin on the CPU which does this.

2. We are using an 8080 SST with a switch for each address line A0–A15. A switch can force one address line to a logical 1 or a logical 0. This is shown in Figure 9-3. We use 16 switches to set the address lines to any of the over 65,000 different possible logical conbinations.

3. The combination that is set on the address switches can be left indefinitely (static). You can then verify the inputs to the address buffers or system

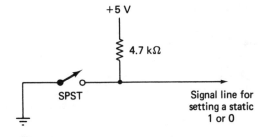

Figure 9-3 Schematic diagram showing how a SPST switch is used statically to control a signal line.

ROM using static DC measurement techniques. This allows you to determine if the signal is correct at every point along the address path.

4. You verify the output of the memory-select lines for proper decoding of the address space. All decoding and address inputs to the ROM are static. A system troubleshooter can take all the time required to examine a particular digital line.

In this general example we have examined only the address lines. Each line may be treated as a separate digital signal because the SST creates an independence of system signals. Further, there has been no mention of absolute time. With SST the emphasis is on the sequence of electrical events, and you can take as long as needed to verify a signal line from its origin to its destination. You are applying your standard digital troubleshooting skills to a microprocessor-based system.

Prior to presenting specific examples of using SST with an 8080-based system, let's cover the basics of the hardware used to realize the technique. As you will see, the hardware required for SST is inexpensive and can be constructed quite easily.

9-2 SST HARDWARE

The SST is a very simple hardware device. It is well suited for educational purposes and industry. The idea behind the instrument operation is that the user has static control of the logic level of any system signal line that the microprocessor would normally control. For the 8080 these are

```
A0-A15
D0-D7
DBIN
WR
SYNC
HLDA
```

The remaining 8080 device pins are inputs with which we need not be concerned at this time.

It is important to remember that the SST will break all feedback loops in the system. An example of a feedback loop is address to memory from CPU, data from memory back to CPU. The address lines and data lines can be verified separately with no interdependence.

Let's design an SST, and you will see how simple the hardware really is. We start with the address stimulus.

Address Output Lines for the SST

Figure 9-4 shows a block diagram of the hardware required for the address stimulus for the SST. The logical value of the address output is determined by the physical position of the DIP switches. The output of each switch is inverted.

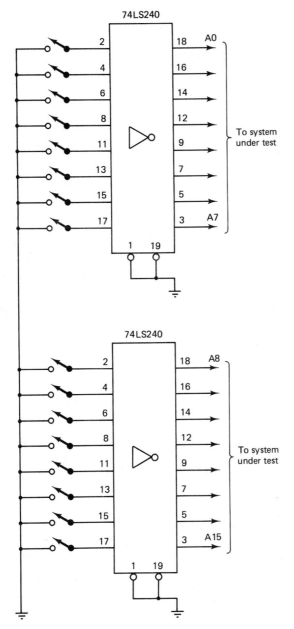

Figure 9-4. Schematic showing how address stimulus is generated.

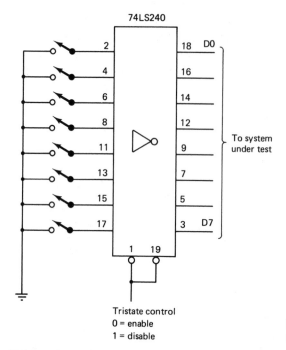

Figure 9-5 Schematic showing how data stimulus is generated.

When the switch is open (off), the input to the inverter will float. This is equivalent to a logical 1, which will force a logical 0 on the output of the inverter. When the switch is closed (on), the input to the inverter is a logical 0. This forces a logical 1 on the output. Therefore, when the switch is open the address is 0. When the switch is closed, the address is 1.

Data Stimulus

Figure 9-5 shows the hardware required for realizing the data stimulus for the SST. You should be aware that the data stimulus must turn off when the SST is simulating the read mode for the 8080. This is the situation where the system, not the SST (CPU), controls the data bus.

The outputs of the buffers labeled D0–D7 are used for the generation of the data lines.

9-3 GENERATING THE DBIN, $\overline{WR}$, SYNC, and HLDA

We now show how to generate the following control signals on the SST:

```
DBIN
WR
SYNC
HLDA
```

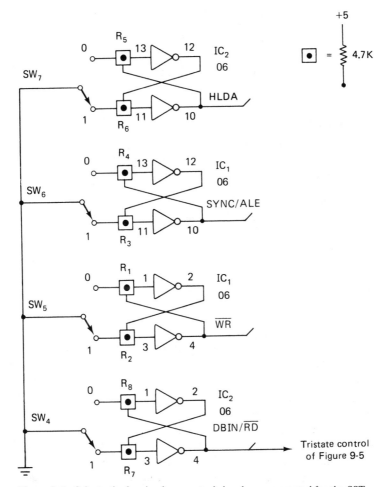

Figure 9-6   Schematic showing how control signals are generated for the SST.

These signals are presented at the same time because the hardware required for each is identical.

All these signals control output lines of the 8080 to the system. To generate these signals, single-pole–double-throw (SPDT) toggle switches are used. Outputs of the switches are input to debounce circuits. Figure 9-6 shows the schematic for generation of these signals. The schematic in Figure 9-6 makes use of the 74LS05 IC. This device contains six open-collector inverters.

9-4 DISPLAYING THE SYSTEM DATA BUS

We have now discussed all the circuits of the SST hardware that output signals. Let us now show the circuit that allows you visually to monitor the data inputs. This visual examination can be accomplished by using eight LEDs to display the

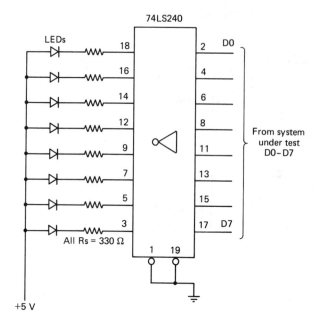

Figure 9-7 Schematic diagram for driving data bus LED's.

logical condition of each line of the system data bus. Figure 9-7 shows the schematic of this display. In this figure the microprocessor (or SST) data bus is connected to the inputs of the 74LS240 inverters. The outputs of these devices drive the LEDs. A logical 0 on the inputs of the 74LS240 turns the LEDs off, whereas a logical 1 turns them on.

The LEDs show the logical condition of the signal lines at the microprocessor device pins D0–D7. This allows you to see the logical value of the data being input to or output by the microprocessor. If two data lines are shortened or if the correct data is not being input to the microprocessor input pins, you will quickly note it on the LED display.

In a write operation you can see what data the 8080 or SST is outputting to the system hardware. Notice that this display shows data directly at the CPU device pins. The complete schematic for an 8080 SST is given in Figure 9-8.

9-5 USING THE SST

The SST will be used to check or debug the system hardware described in Chapters 5, 6, 7, and 8 for the CPU controller, ROM and RAM. The basic system to be debugged is shown in Figures 9-9a, b, c, and d. Using the SST and a logic probe, oscilloscope, or voltmeter, we will verify the operation of each section of system hardware in an orderly, systematic, and effective way. The techniques that are being presented have been used in actual practice in industry and have proven to be effective.

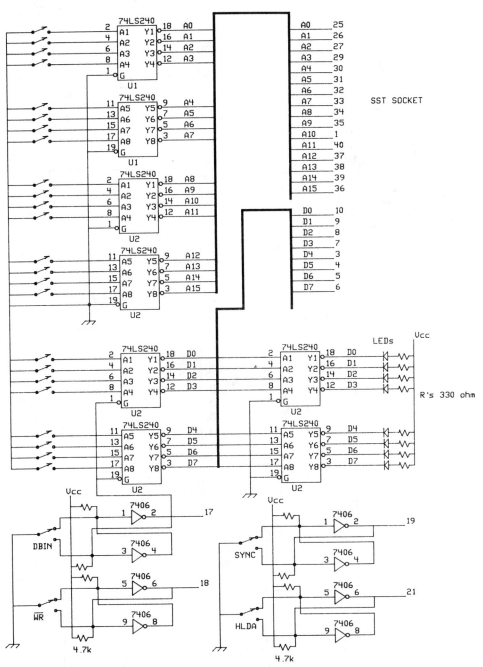

Figure 9-8 8080 SST schematic.

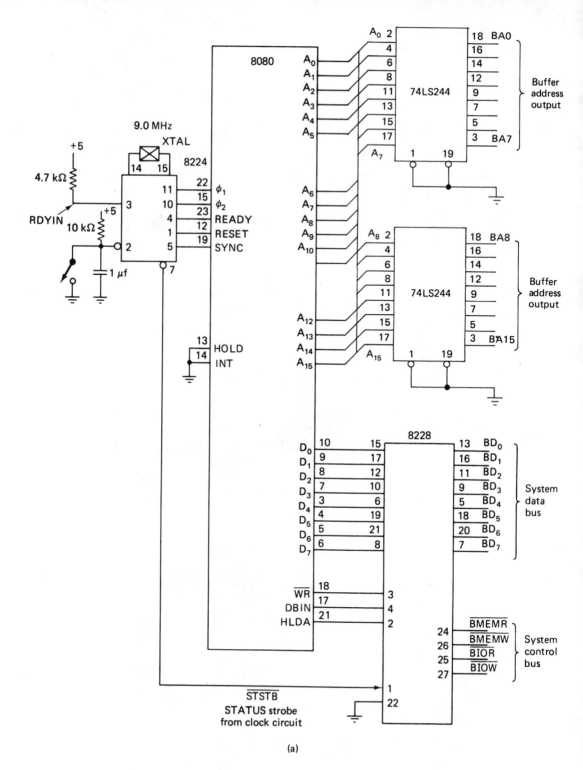

Figure 9-9 (a) Complete schematic for the 8080 as a system controller; (b) complete schematic of a ROM system for use with an 8080, 8085, or Z80 microprocessor using the three bus architecture; (c) schematic of a 16K X 8 RAM system; (d) schematic of an I/O port with address = FF.

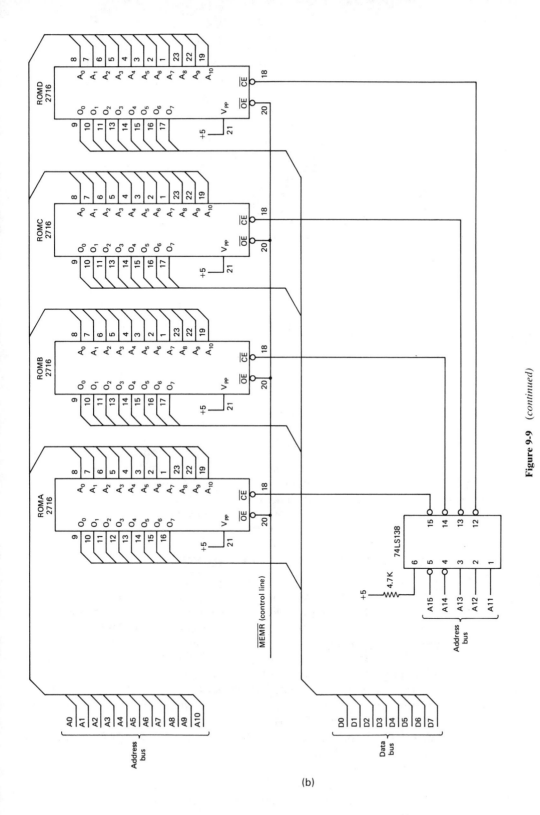

Figure 9-9 (continued)

(b)

231

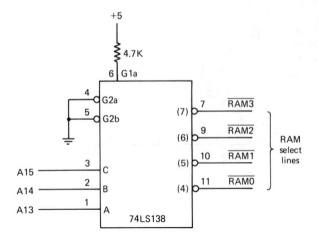

A15	A14	A13	A12		
1	0	0	X	8000-9FFF	$\overline{RAM0}$
1	0	1	X	A000-BFFF	$\overline{RAM1}$
1	1	0	X	C000-DFFF	$\overline{RAM2}$
1	1	1	X	E000-FFFF	$\overline{RAM3}$

(c)

Figure 9-9 (*continued*)

These techniques are guidelines to enable a beginner to have a low-cost, easily understood means of debugging microprocessor hardware. A more experienced microprocessor user might wish to apply the SST in different ways. This is certainly to be encouraged. The guidelines that we will present may be modified to suit an individual system or application.

9-6 CHOOSING A STARTING POINT FOR DEBUGGING

No matter what type of microprocessor system we are using, the system requires some type of power supply. The power supply is a good place to start our system checks. With a voltmeter, we check to ensure that all power supplies are operating correctly. When we have made certain that all power supplies are correct (within the specifications of the system), we turn our attention to the clocks. We should monitor these clocks with an oscilloscope to determine if they are within specifications. The system specifications give minimum and maximum limits for all clock timing and clock-voltage levels. When we have verified that all voltages and free-running clocks are operating correctly, then we may go on to the next step. We must always be certain to verify that the power and clocks are valid at the device pin.

The details of the next step in the troubleshooting process vary from microprocessor to microprocessor. We will discuss the troubleshooting details for the 8080 microprocessor. When we have been through an exact discussion of

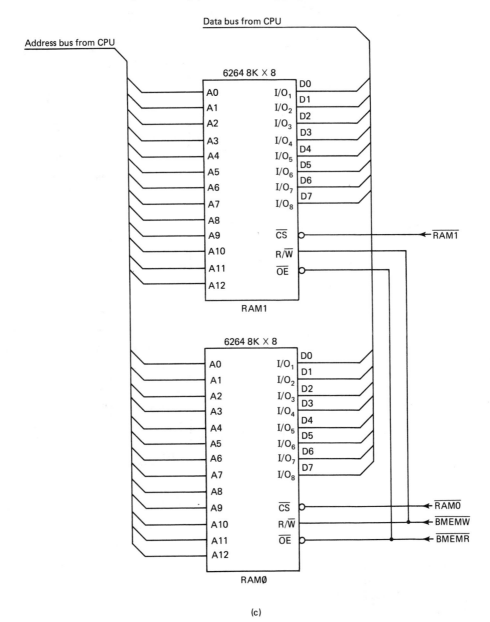

(c)

Figure 9-9(c)   (*continued*)

one microprocessor, we can easily adapt the details to a different microprocessor. We have discussed the operation of the system to be debugged in detail. We know that the microprocessor will be operating as a system controller; that is, the microprocessor will be executing a program continually. The microprocessor will not be interrupted, or put in a wait state, nor will any other device

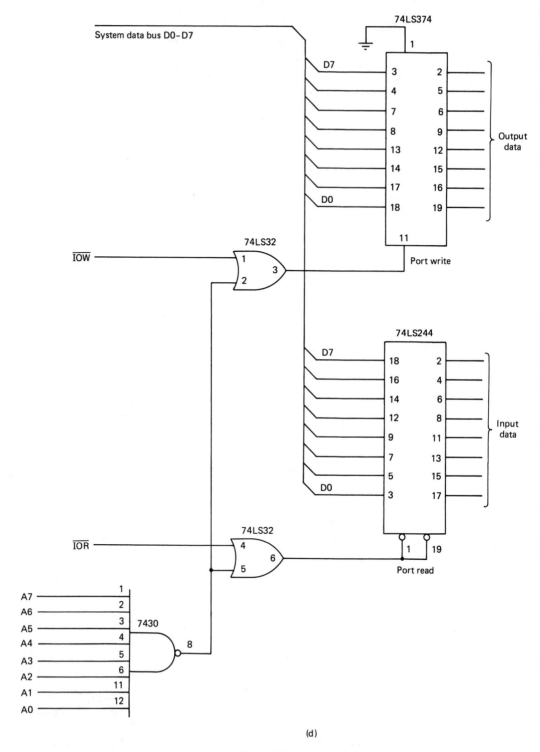

(d)

Figure 9-9 (*continued*)

perform a DMA on the system. Bearing this fact in mind we notice that the input pins to the microprocessor that perform these functions are put into the nonassertive state; that is, they are disabled. These are the inputs that we should verify next in the debugging procedure. For the 8080 the INT (interrupt input) pin 14 should be a logical 0. This is the unasserted state of the interrupt input.

The READY input pin 23 of the 8080 is controlled by the 8224 clock generator. For the READY input to be in the active mode (no wait states), it should be a logical 1. This means the RDYIN input pin 3 of the 8224 should be in the logical 1 state.

Finally the HOLD input pin 13 should be in the logical 0 state. This state indicates that no other hardware in the system will perform a DMA operation.

The actual logic state and numbers of the input control pins vary from microprocessor to microprocessor. We should become familiar with the details of each individual microprocessor pin configuration and control logic state.

We have now determined that all signals provided by the system are in a valid state to permit the microprocessor to execute normally. If we have not already done so, the actual microprocessor is removed from the system. DO NOT REMOVE ANY LOGIC DEVICE OR PC BOARD FROM A SYSTEM WITHOUT FIRST TURNING OFF THE SYSTEM POWER. If power is not turned off first, damage to the integrated circuits may result. The troubleshooting or debugging procedure can hardly be termed effective if in the process we cause more hardware failures.

When the microprocessor is removed, we connect the SST to the system, as shown in Figure 9-2. We note here that to connect the SST, the microprocessor must be installed in a socket within the system. If your microprocessor is not installed in a socket, you have two options.

1. Take the trouble to install the microprocessor in a socket. This may or may not be feasible in a production environment. However, in a development application, this may be a quite reasonable option.
2. The second option is to use a special logic clip that will clamp on the top of the 40-pin microprocessor chip. When you do this you must ensure that the HOLD pin 13 is set to a logical 1, active state. This will place the microprocessor into a state where the busses and control lines are unasserted or tristate. This action will electrically remove the microprocessor from the circuit.

If you do not electrically remove the microprocessor from the circuit, you could cause damage to the CPU by connecting the SST.

Let us assume the SST is connected to the system hardware. Now what do we do? In the discussion that follows we show a very methodical step-by-step approach to checking a microprocessor-controlled system. We discuss each major section of the system ROM, RAM, and I/O. We start with the ROM section, which was discussed in Chapter 6; the schematic is shown in Figure 9-9b.

9-7 TROUBLESHOOTING ROM USING SST

The first item of hardware that we will check is the system address bus. We will ensure that all address lines are connected to every bus point in the system. Further, we will ensure that there is no decoding problem either from defective hardware or from some oversight in the system. If the system once worked, then the decoding problem is from defective hardware. If the system is in a development stage, then the decoding problem may be either defective hardware or an oversight in the design.

Checking the System Address Bus

The address input to the system originates from the microprocessor address line output pins, as shown in Figure 9-9a. The following procedure is meant only as a guideline for checking the address bus and is complete to the extent of being ultraconservative. However, it will accurately check the entire system address bus. As you gain experience in troubleshooting microprocessor hardware, you can shorten this procedure by using certain quick checks that you will learn to make on a particular system in question.

Procedure

1. Set all address lines at the microprocessor output to logical 0 via the address switches on the SST.
2. Next, set address line switch A_0 on the SST to a logical 1 state. This will set pin 25 on the 8080 socket to logical 1.
3. Monitor the output pin 18 on the address buffers to ensure this output is a logical 1.
4. If the output pin 18 of the 74LS244 is a logical 1, then set the address switch A_0 to a logical 0 and recheck pin 18 for a logical 0. Make certain that output pin 18 of the IC will follow a logical state of the address switch A_0 on the SST.
5. If output pin 18 of the 74LS244 does follow the logical state of the address switch A_0, then the 74LS244 device is operating correctly and the connection from the microprocessor to the system is correct. Go directly to Step 7.
6. If the output pin 18 of the 74LS244 does not follow the logical state of the address switch A_0, then it is not clear whether the 74LS244 is defective or

the output is being loaded excessively. Remember, the outputs of the address bus go to many parallel points in the system. Any one of these points may be causing the problem. We must determine the cause of this failure. However, you have reduced your troubleshooting problem from a system failure to one of finding a single "stuck" line failure. This is a standard digital troubleshooting problem.

    a. If output pin 18 of the 74LS244 device does not repond correctly, make certain that input pin 2 does respond to changes in the logic level from output pin 25 of the microprocessor.

    b. If the input to the 74LS244 pin 2 does respond correctly but the output does not respond correctly, then the troubleshooter can use standard digital troubleshooting techniques to determine the cause.

7. Assume that pin 18 of the 74LS244 does respond as it should. We next proceed to every parallel point in the system to which this output is connected and ensure that each point can and does respond correctly. This includes each A_0 input pin on the ROM chips. In this way buffered address A_0 is sure to be connected and bused to every point within the system.

    a. In this system we put our monitoring device first on the ROMA pin 8, then on ROMB pin 8, on ROMC pin 8, and finally on ROMD pin 8. At each point toggle the address switch A_0 on the SST to ensure that each parallel point in the system is operating correctly.

8. Repeat Steps 2–7 for all address line outputs A_1–A_{15} on the 8080 CPU.

It may become clear at this point that by using the SST we have provided a means by which standard digital troubleshooting techniques can be employed to debug a very complicated microprocessor system. We do not have to learn a new set of troubleshooting skills. Rather, we learn a new application of old skills.

Checking the Memory-Select Lines

Next we will check the memory-select decoding lines shown in Figure 9-9b. The memory-select lines are generated by the 74LS138. Inputs to the 74LS138 are address lines A_{11}–A_{15}. The output of the 74LS138 provides the memory-select signal for the system ROM chips. Examining the circuit diagram, we can easily determine the decoding for the memory-select lines:

A_{15}	A_{14}	A_{13}	A_{12}	A_{11}	ACTIVE OUTPUT LINE
0	0	0	0	0	pin 15—ROMA
0	0	0	0	1	pin 14—ROMB
0	0	0	1	0	pin 13—ROMC
0	0	0	1	1	pin 12—ROMD

We set the address switches A_{11}–A_{15} to the different logic states shown for enabling ROMA, ROMB, ROMC, and ROMD. We check to ensure that the outputs of the 74LS138 will respond correctly to changes in the address inputs.

The outputs of the memory-select decoder are connected to different points within the system. For example, pin 15 of the 74LS138 is connected to pin 18 of ROMA. Monitor the chip-select pin at each memory chip to ensure it is operating correctly with the changing address inputs.

After the memory-select inputs are verified, we have determined that the entire address bus system is operating correctly. We next check the system control bus.

Checking the Control Bus

We can start the process of determining if the control bus is operating correctly using the following procedure. Again, this procedure is meant only as a guideline and may be modified to suit any special system applications.

Procedure

1. Set the control line switches on the SST to the unasserted position. This is the same logic level to which the 8080 would set these control lines during normal system operation.

$$
\begin{array}{rcl}
\text{DBIN} & = & 0 \\
\overline{\text{WR}} & = & 1 \\
\text{HLDA} & = & 0 \\
\text{SYNC} & = & 0
\end{array}
$$

2. Next set the status word on the SST data bus lines to the correct value for a memory read operation. To obtain the status word for a memory read, refer to the status word chart in the appendix of this text. For your convenience the following four status words will be used:

$$
\begin{array}{rcl}
\text{MEMR} & = & 82\text{H} \\
\text{MEMW} & = & 00\text{H} \\
\text{IOR} & = & 42\text{H} \\
\text{IOW} & = & 10\text{H}
\end{array}
$$

3. After the status word is set to 82H for a memory-read operation, the sync line is set to a logical 1. This condition enables the status strobe to the 8228 system controller, which latches the status strobe.

4. Next the sync line is set to a logical 0, which disables the status strobe. This is exactly what the 8080 does automatically when latching the status word in a correctly operating system. The SST is simulating 8080 activity on a clock-by-clock basis.

5. Now that the status word is latched, the DBIN signal is set to a logical 1. This action asserts the MEMR control line to a logical 0.

6. Verify that the $\overline{\text{MEMR}}$ control line is a logical 0. This should be verified at the ROM output enable input lines pin 20.

7. Set the DBIN signal to a logical 0 level, which unasserts the $\overline{\text{MEMR}}$ control line.

The preceding sequence of events verifies that the $\overline{\text{MEMR}}$ control circuits work correctly. We can verify the proper operation of the other three control lines by following a similar procedure.

Procedure for checking the $\overline{\text{IOR}}$ control line

1. Set the status word on the system data bus to 42H.
2. Set the sync line to a logical 1.
3. Set the sync line to a logical 0. The status word is now latched.
4. Set the DBIN output line from the SST to a logical 1 level. The $\overline{\text{IOR}}$ control line, pin 25 of the 8228 system controller, will be a logical 0. It is asserted.
5. Set the DBIN output line from the SST to a logical 0 level. The $\overline{\text{IOR}}$ control line will be unasserted.

At this time you have determined that all circuits that generate the $\overline{\text{IOR}}$ control line are operating correctly.

Procedure for checking the $\overline{\text{MEMW}}$ control line

1. Set the status word on the system data bus to 00H.
2. Set the sync line to a logical 1.
3. Set the sync line to a logical 0. The status word is now latched.
4. Set the $\overline{\text{WR}}$ output line from the SST to a logical 0 level. The $\overline{\text{MEMW}}$ control line, pin 26 of the 8228 system controller, will be a logical 0. It is asserted.
5. Set the $\overline{\text{WR}}$ output line from the SST to a logical 1 level. The $\overline{\text{MEMW}}$ control line will be unasserted.

At this time you have determined that all circuits that generate the $\overline{\text{MEMW}}$ control line are operating correctly.

Procedure for checking the $\overline{\text{IOW}}$ control line

1. Set the status word on the system data bus to 10H.
2. Set the sync line to a logical 1.

3. Set the sync line to a logical 0. The status word is now latched.
4. Set the $\overline{WR}$ output line from the SST to a logical 0 level. The $\overline{IOW}$ control line, pin 27 of the 8228 system controller, will be a logical 0. It is asserted.
5. Set the $\overline{WR}$ output line from the SST to a logical 1 level. The $\overline{MEMW}$ control line will be unasserted.

At this time you have determined that all circuits that generate the $\overline{IOW}$ control line are operating correctly.

When we have checked all control bus signals in the system for proper operation, we have made certain that all control bus signals will perform as expected within the system architecture.

Reading Data Back from the ROM

Using the SST you can actually read different locations from the system ROM space. This data will be reflected on the LEDs of the SST. As an example, suppose the first data bytes of the ROM program were C3 00 01. This corresponds to a JMP 0100 instruction. This is a perfectly valid instruction, which may jump the system to location 0100 whenever the system is reset.

Using the SST we can read this data in the following way.

1. Set the system address on the SST to the address location of the data to be read. In this case let us assume that these data bytes are at locations 0000, 0001, and 0002.
2. Latch the status word for a memory read. This is accomplished by the procedure outlined in the previous section.
3. Set the DBIN signal to a logical 1 level. At this time the data from the system ROM will be enabled onto the system data bus. The data will be reflected at the SST LEDS.

   If the data is correct you know that the memory-read circuits are working properly and that some of the system data lines will operate correctly. We say *some* of the data lines because not all these lines were set to a logical 1 or logical 0 when reading the first data byte. (Only D7, D6, D1, and D0 were set to a logical 1 when the data C3 is read from the ROM.)
4. Set the DBIN signal to a logical 0 level. This removes the data from the system data bus.
5. Change the address on the SST address switches. Note that you do not have to latch the status word again as it is still valid. In this way you are simplifying the memory-read process. During system operation the 8080 would latch a new status word even if it were performing a memory-read operation again.

6. Set the DBIN signal on the SST to a logical 1 state. The data at memory address 0001 will be enabled onto the system data bus and reflected on the LEDs of the SST.

If the data is not correct, you can leave the SST set in this condition. With the SST set in this condition, you can statically verify all the logical conditions that exist in the system.

You have reduced the complicated dynamic problem of a memory-read operation into a simpler static one. The element of time dependence has been removed from the troubleshooting process.

With the SST you can read specific data bytes from any memory space in the system and determine if the system data bus is operating correctly. That is, you can decide if any of the data lines are shorted together.

ROM Troubleshooting Summary

You can see from the preceding discussion of troubleshooting the system ROM that the SST provides a very methodical method for debugging. Further, the SST removes the time dependence from the troubleshooting process. You can apply your standard digital troubleshooting skills to debugging microprocessor systems. All the system logic responsible for the ROM data being read by the 8080 may be checked easily with the SST.

You may safely say that if the microprocessor system does not operate by using the SST, then it will not operate at system speed. It has been determined by actual practice that the probability of the system operating at system speed once it has been checked statically is extremely high.

9-8 TROUBLESHOOTING RAM USING SST

In the preceding section you were shown how to troubleshoot a memory-read operation with the 8080 SST. We now wish to present the details of writing to the system memory with the SST. We undertake this discussion to show the flexibility of the SST. This discussion will further aid in the development of your skills in analyzing and troubleshooting a microprocessor system from a static point of view.

Let's start by reviewing the sequence of events required to write data to system memory under the control of an 8080 microprocessor:

1. Address bits A0–A15 are output on the system address bus. This specifies the memory location where data will be written to memory.
2. The data bus bits D0–D7 are output on the system data bus. This specifies the correct status word for a memory-write operation. In the 8080 system

there are two valid status words for the 8080 to write data to system memory. These status words are 00_{16} (memory write) and 04_{16} (stack write). We will use status word 00 for our memory-write operation.

3. The sync bit is made to equal logical 1. This action latches the status word by enabling the status strobe signal.
4. The sync bit is made to equal logical 0. This action disables the status strobe.
5. The system data bus D0–D7 outputs the data to be written into memory.
6. The $\overline{WR}$ bit from the 8080 is set to a logical 0. This action asserts the $\overline{MEMW}$ system control line. The memory that is enabled will have the data written into it.
7. The $\overline{WR}$ bit is set to a logical 1. This action terminates the memory-write sequence.

Let's examine each of these steps in detail as we describe the hardware activity of the circuit shown in Figure 9-9c. When the memory address A0–A15 is applied, the lower address lines A0–A12 are connected to the two memory devices in a parallel fashion. The upper address lines A13–A15 are input to the 74LS138 device to select the correct memory chip. Let us assume that the address lines A15–A13 are equal to 1 0 0.

With this input combination the $\overline{RAM0}$ output line from the 74LS138 will be active logical 0. This output is connected to the $\overline{CS}$ input of the RAM0 device. Using a static probe you can verify the correct chip-select decoding with the memory-address lines output by the SST.

Next the status word is output on the data bus and the sync line is toggled to latch the status word into the 8228 system controller. This is exactly the same operation that occurred in the memory-read operation.

The next step in the memory-write sequence is that SST applies the data lines D0–D7 to the system data bus. At this time the data bus outputs will be set to 1 0 1 0 1 0 1 0. You can verify the logical conditions at the input of the RAM chips.

Next, the $\overline{WR}$ control bit from the SST is set to a logical 0. This asserts the $\overline{MEMW}$ system control line. The $\overline{MEMW}$ control line is connected to the R/$\overline{W}$ input of the RAM chip. You can verify that this pin is a logical 0.

Finally, the $\overline{WR}$ control bit is set to a logical 1. This terminates the write sequence. The data should be written into memory. You can verify that the write operation occurred successfully by performing a memory read operation from the same memory address.

Data is read from the system RAM exactly the same way as for system ROM. When the data is read from the RAM, it will be displayed on the LEDS of the SST. By reading data from the memory location using the SST, you can determine if the RAM is capable of electrical communication with the 8080.

Let's look now at how the SST can be applied to troubleshooting an input or output port.

9-9 TROUBLESHOOTING AN I/O PORT USING THE SST

In this section you will learn how to apply SST techniques to troubleshooting a defective input or output port. The I/O port that we will be troubleshooting is shown in Figure 9-9d. To troubleshoot this I/O port, we will follow the general sequence of events for an input read and output write operation for the 8080. As each event is performed with the SST, system hardware will be checked statically to ensure proper operation. If at any point in the sequence the hardware fails to respond correctly, then you stop and examine the problem further to determine the cause.

Troubleshooting an Input Port

Let's start the sequence of event for an input read operation and examine the hardware response for each step in the sequence.

1. First, the 8080 outputs the port address on address lines A0–A7. Recall that only the lower address lines are used for I/O operations. For our input port the address is FF. When this address is output on the SST, verify that pin 8 of the 74LS30 is a logical 0. This is the active mode of the port-select line.
2. Next, the 8080 outputs the status word for an input read on the data lines. The status word for a memory read is 42H. This is the same step in the sequence for any hardware operation performed by the 8080.
3. The sync line is set to a logical 1. This action enables the status strobe for latching the status word.
4. The sync line is set to a logical 0.
5. Next, the 8080 sets the DBIN line to a logical 1. This action will assert the $\overline{\text{IOR}}$ control line. Verify that pin 4 of the 74LS32 is a logical 0. Further, the port-select line is a logical 0 and the port-read line should be active. This is verified by examining the logic state of pins 1 and 19 of the 74LS244.

    During the time the DBIN line is a logic 1, the data on the system data bus is strobed into the 8080. When using the SST, this data will be displayed on the LEDs. The source of data is from the addressed input port. In this example the source of data is the input pins of the 74LS244 shown in Figure 9-9d.
6. The last step in the sequence is the DBIN line is set to a logic 0. At this time the $\overline{\text{IOR}}$ system control line is set to a logical 1. With the $\overline{\text{IOR}}$ line going to a 1, the port-select line also goes to a 1. These conditions may be verified using a static probe on the correct pins of the circuit shown in Figure 9-9a.

Notice that the troubleshooting process for each type of hardware operation performed by the microprocessor is essentially the same in concept. The sequence of events for the hardware activity is followed using the SST. At each step in the sequence, the hardware is examined for proper operation.

Let's now examine the sequence of events and troubleshooting of an output port using the SST.

Troubleshooting an Output Port

The output port to troubleshoot is shown in Figure 9-9d. We will follow the sequence of events for an output write operation and determine if the hardware is responding correctly at each step.

1. The first step in the sequence is that the port address is output on the 8080 address lines A0–A7. The port-select address is FF. Notice that this is the same hardware that was used to select the input port. We have already verified that this hardware will respond to the correct address. Therefore, we need not duplicate our efforts.

2. Next, the status word for an output write is placed onto the system data bus. The status word for an output write is 10H.

3. The 8080 now sets the sync line to a logical 1, enabling the status strobe to latch the status word.

4. The 8080 sets the sync line to a logical 0.

5. Now the 8080 places the data to be written to the output port on the system data bus. We will write the data 55H to the port FF. With the data 55 on the system data bus, you can verify that the data is reaching the output port latch. In our output port the data is input to the 74LS374 input pins.

6. With the data on the system data bus, the 8080 next sets the $\overline{\text{WR}}$ control line to a logical 0. This action will assert the $\overline{\text{IOW}}$ system control line. When the $\overline{\text{IOW}}$ control line goes to a logical 0, the port-write line also goes to a logical 0 level. This is due to the fact that the port select line is active. You should verify that the port-write line is a logical 0 by examining the logic state of the clock input pin 11 of the 74LS374 shown in Figure 9-9d.

    Notice at this time that the data is not latched on the 74LS374 outputs. This is due to the fact that the 74LS374 clocks the data on the 0 to 1 transition of the clock input pin.

7. The final step in the sequence is the $\overline{\text{WR}}$ line is set to a logical 1 by the 8080. The system $\overline{\text{IOW}}$ is unasserted and the port-write line goes to a logical 1. This is the active transition of the clock for the 74LS374 output port latch. At this time the data that was on the system data bus has been written to the output pins of the 74LS374. This data is static and may be verified by using a static probe.

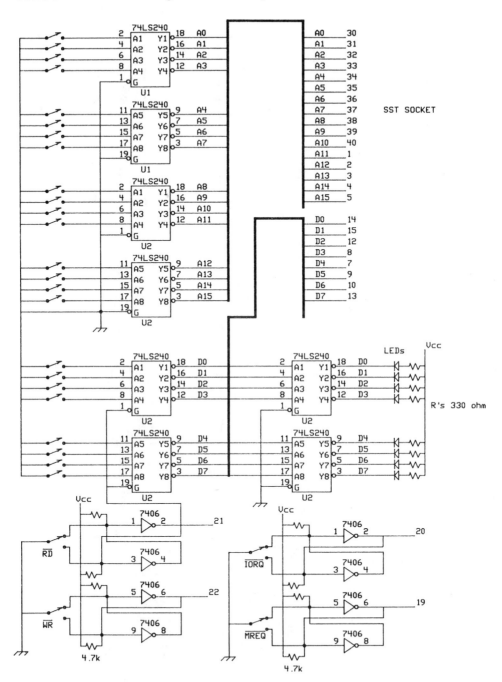

Figure 9-10   Z80 SST schematic.

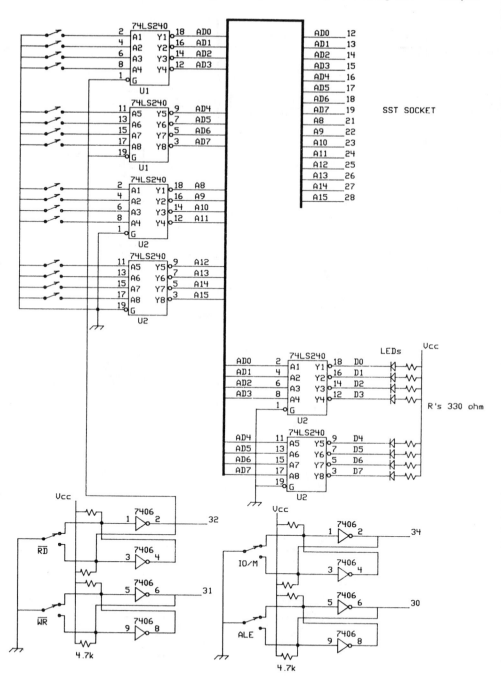

Figure 9-11   8085 SST schematic.

9-10 SCHEMATICS FOR THE 8085 AND Z80 SSTS

Figures 9-10 and 9-11 show the hardware required to realize the SSTs for the 8085 and Z80 CPUs. Using this hardware you can troubleshoot 8085- and Z80-based systems in the same way as for the 8080. All you need to do is follow the sequence of events of these microprocessors for the hardware activity that you are troubleshooting.

9-11 CHAPTER SUMMARY

In this chapter we have presented the concept of static stimulus testing or SST. You were shown that the hardware operations that occur in a microprocessor-based system are static. Building on this, an example of testing the system address bus was shown. In this example, it was assumed that you could set any of the address lines to a logical 1 or a logical 0 using a switch. This is exactly what the SST does.

The hardware of the SST was then discussed. From the schematics you were shown that the hardware for an SST is very simple and easy to use. These devices can be constructed to suit your own personal system. Finally you were shown two detailed examples of troubleshooting using the SST.

SST can be applied to most microprocessor-based systems. Using this tool will provide another method for debugging computer systems. Further, the more you use it, the better you will understand how the system hardware operates. The deeper your understanding of how the system hardware operates, the more effective you will become in using all troubleshooting tools and repairing defective systems.

REVIEW PROBLEMS

1. Define the troubleshooting technique of SST.
2. List the two main categories of malfunctioning systems.
3. True or false: SST allows the troubleshooting of dynamic RAM. Explain your answer.
4. List the 8080 outputs that the SST must control.
5. Draw the circuit for the address stimulus section of the SST.
6. Why does the $\overline{WR}$ signal require a debounce circuit when used with the SST?
7. Why would you wish to monitor the system data bus lines with LEDs during a memory write operation?
8. Draw a flowchart of the steps required to test the address inputs of a ROM.
9. Draw a flowchart of the steps required to test the memory-select lines in the decoding logic of a CPU system.

10. Why do you need to know the status word when troubleshooting an 8080 system using an SST?

11. When you set the DBIN line to a logical 1 during a memory read operation and the $\overline{\text{MEMR}}$ control line does not become asserted, list four probable causes.

12. When reading data back from ROM, you need to read at least two different bytes to completely check the system data lines. Why?

13. If you know how to read data from ROM using SST, you also know how to read data from RAM. Is this a true statement? Explain.

14. During a memory-write operation, what section of the system is in control of the system data bus?

15. If an I/O port was not responding correctly when the system was running, how would you determine if the port-select lines were being decoded properly using SST.

16. Draw a flowchart of the steps required to test the input lines of an input port with a port address of 45H.

17. What is the main difference in using an SST for an 8080, a Z80, or an 8085?

10

CLOCKS AND INTERRUPTS FOR THE 8080, 8085, AND Z80

This chapter discusses the general topic of system clocks and interrupts for the 8080, 8085, and Z80 microprocessors. We start by presenting the specifications for the microprocessor free-running clock inputs. After this the concept of interrupts is given. From here we show how each different type of interrupt is electrically handled with the 8080, 8085, and Z80. As each new interrupt type is discussed, examples of software will be presented to explain in detail what occurs in the CPU during the operation.

10-1 THE 8080 SYSTEM CLOCKS

The 8080 microprocessor requires two high-voltage (12-volt) clock inputs. These clock inputs are called phase 1 and phase 2. The timing relationship of the clock inputs to each other is shown in the timing diagram of Figure 10-1. This includes the specifications and limits for all appropriate clock levels.

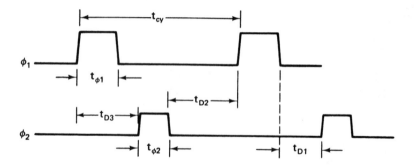

AC Characteristics

$T_A = 0°C$ to $70°C$, $V_{DD} = +12 V \pm 5\%$, $V_{CC} = +5 V \pm 5\%$, $V_{BB} = -5 V \pm 5\%$, $V_{SS} = 0V$, unless otherwise noted

Symbol	Parameter	Min.	Max.	Unit
$t_{CY}{}^{(3)}$	Clock period	0.48	2.0	μ sec
t_r, t_f	Clock rise and fall time	0	50	n sec
$t_{\phi 1}$	ϕ_1 Pulse width	60		n sec
$t_{\phi 2}$	ϕ_2 Pulse width	220		n sec
t_{D1}	Delay ϕ_1 to ϕ_2	0		n sec
t_{D2}	Delay ϕ_2 to ϕ_1	70		n sec
t_{D3}	Delay ϕ_1 to ϕ_2 leading edges	80		n sec

Figure 10-1 Timing diagram of phase 1 and phase 2 clocks for the 8080. Below is the excerpted clock specification for the 8080 clocks. (Courtesy of Intel Corp., 1979.)

As can be seen from the timing diagram and specifications listed in Figure 10-1, the generation of these clock waveforms is a significant hardware task. Fortunately, a special integrated circuit has been designed to provide the 8080 with properly timed and voltage-limited phase 1 and phase 2 clock inputs. This device is called an 8224 clock generator. In addition to providing the 8080 with phase 1 and phase 2 clocks, the 8224 provides certain other functions, such as reset and status strobe, that make its use very convenient for the 8080 microprocessor.

It is not worth the extra effort for a designer using an 8080-based system to design a clock-generator circuit for the 8080. If you are using an 8080 as a system controller, then you should plan on using an 8224 clock generator also. Package count reduction and system reliability will most certainly be enhanced by using the 8224 device.

The block diagram and partial specification for the 8224 clock generator are shown in Figure 10-2a and b. Basically, all that is needed to operate the 8224 device is a satisfactory DC power supply and an external crystal. The required frequency of oscillation for the external crystal is nine times as great as the Tcy (period of time for one microprocessor cycle) that we want at phase 1 and phase 2 clock inputs to the 8080. For example, suppose we want a Tcy of 1 microsecond, which corresponds to a frequency of 1 megahertz. The crystal frequency will be equal to 9 megahertz, or nine times the desired Tcy.

Figure 10-3 shows how the 8224 device interfaces to the 8080 microprocessor. Notice the output signal from the 8224 called $\overline{\text{STSTB}}$. This is the status strobe signal we mentioned in Chapter 5 for strobing the status word into the status latch. This signal is active when status information is present on the 8080 data bus. If you are uncertain about the function of the status strobe in the system, refer to Chapter 5. You can see that by using an 8224 clock generator, there is little difficulty in providing the correct phase 1 and phase 2 clock inputs

PIN CONFIGURATION   **BLOCK DIAGRAM**

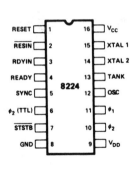

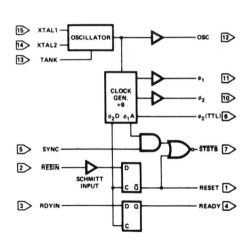

PIN NAMES

$\overline{\text{RESIN}}$	RESET INPUT		XTAL 1	CONNECTIONS
RESET	RESET OUTPUT		XTAL 2	FOR CRYSTAL
RDYIN	READY INPUT		TANK	USED WITH OVERTONE XTAL
READY	READY OUTPUT		OSC	OSCILLATOR OUTPUT
SYNC	SYNC INPUT		ϕ_2 (TTL)	ϕ_2 CLK (TTL LEVEL)
$\overline{\text{STSTB}}$	STATUS STB (ACTIVE LOW)		V_{CC}	+5V
			V_{DD}	+12V
ϕ_1	8080		GND	0V
ϕ_2	CLOCKS			

(a)

Figure 10-2   (a) Block diagram and pinout for the 8224 clock generator for use with the 8080 microprocessor; (b) specifications for the 8224 clock generator. (Courtesy of Intel Corp., 1979.)

D.C. Characteristics

$T_A = 0°C$ to $70°C$; $V_{CC} = +5.0V \pm 5\%$; $V_{DD} = +12V \pm 5\%$.

Symbol	Parameter	Limits			Units	Test Conditions
		Min.	Typ.	Max.		
I_F	Input Current Loading			-.25	mA	$V_F = .45V$
I_R	Input Leakage Current			10	μA	$V_R = 5.25V$
V_C	Input Forward Clamp Voltage			1.0	V	$I_C = -5mA$
V_{IL}	Input "Low" Voltage			.8	V	$V_{CC} = 5.0V$
V_{IH}	Input "High" Voltage	2.6 2.0			V	Reset Input All Other Inputs
$V_{IH} \cdot V_{IL}$	REDIN Input Hysteresis	.25			mV	$V_{CC} = 5.0V$
V_{OL}	Output "Low" Voltage			.45	V	(ϕ_1, ϕ_2), Ready, Reset, $\overline{STSTB}$ $I_{OL} = 2.5mA$
				.45	V	All Other Outputs $I_{OL} = 15mA$
V_{OH}	Output "High" Voltage ϕ_1 , ϕ_2 READY, RESET All Other Outputs	 9.4 3.6 2.4			 V V V	 $I_{OH} = -100\mu A$ $I_{OH} = -100\mu A$ $I_{OH} = -1mA$
I_{SC}[1]	Output Short Circuit Current (All Low Voltage Outputs Only)	-10		-60	mA	$V_O = 0V$ $V_{CC} = 5.0V$
I_{CC}	Power Supply Current			115	mA	
I_{DD}	Power Supply Current			12	mA	

Note: 1. Caution, ϕ_1 and ϕ_2 output drivers do not have short circuit protection

CRYSTAL REQUIREMENTS

Tolerance: .005% at $0°C - 70°C$
Resonance: Series (Fundamental)*
Load Capacitance: 20-35pF
Equivalent Resistance: 75-20 ohms
Power Dissipation (Min): 4mW

*With tank circuit use 3rd overtone mode.

(b)

Figure 10-2 (*continued*)

to the 8080 microprocessor. It is highly recommended that this device be used for system control when designated with an 8080 microprocessor.

10-2 THE 8085 SYSTEM CLOCK

The 8085 clock generator circuit is contained entirely within the 8085 microprocessor. We simply connect a crystal between pins labeled X_1 (pin 1) and X_2 (pin 2) of the 8085 microprocessor. The crystal frequency can be any value between

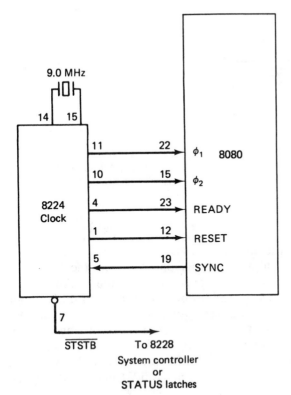

Figure 10-3   Schematic showing how the 8224 clock generator is connected to the 8080 microprocessor.

1 and 6 megahertz. This frequency is then divided by 2 and used as the Tcy of the 8085 microprocessor. Figure 10-4 shows how simple it is to provide correct timing for the 8085 microprocessor.

10-3 THE Z80 SYSTEM CLOCK

The Z80 microprocessor requires a single clock input to function correctly. This clock input may be generated in several ways. One method of providing a free-running clock input to the Z80 is shown in Figure 10-5. Here we see that the frequency of oscillation is set by the crystal X_1. However, in order for the circuit to work, the R_1C_1 product must be greater than the period of oscillation. For example, assume that we want a frequency of 1 megahertz. This represents a period of 1×10^{-6} seconds. This means that the R_1C_1 product must be greater than 1×10^{-6}. R_1 should remain constant at approximately 330 ohms. This gives a value of C_1 greater than or equal to $1 \times 10^{-6}/330$ (farads), or approximately 3000 picofarads. This RC product allows enough phase shift at the frequency of oscillation for the circuit to oscillate. Specifications for the Z80 clock are given in Figure 10-6.

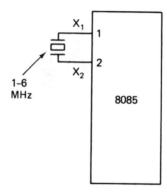

Figure 10-4 Schematic showing how the crystal is connected to the 8085 for proper operation. Note that the 8085 will operate at one-half the crystal frequency.

10-4 SYSTEM CLOCK SUMMARY

For proper operation of the microprocessors we have presented (except for the 8085), we are required to input at least one phase, and for the 8080 we need two phases of clock inputs. These clock inputs are free running. And, regardless of the state of the microprocessor, the proper clock input to the microprocessor must be present. A simple and wise thing to do when troubleshooting a microprocessor-controlled system is first to ensure that all power supply levels are valid and then make certain that all input clocks are timed correctly and are valid at the correct voltage levels. The microprocessor will have no chance to operate correctly if the system clocks are not valid. Checking the system clocks

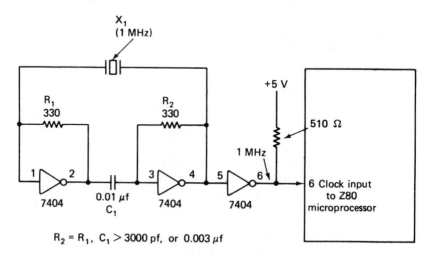

Figure 10-5 Schematic for a possible clock generator that can be used with the Z80 microprocessor. The clock output is not TTL. The 510 pull-up resistor will increase the clock output voltage level to near +5 volts. Refer to Figure 10-6 for clock specifications.

Timing measurements are made at the following
voltages, unless otherwise specified:

	"1"	"0"
CLOCK	V_{cc} –.6V	.45V
OUTPUT	2.0 V	.8 V
INPUT	2.0 V	.8 V
FLOAT	ΔV	± 0.5 V

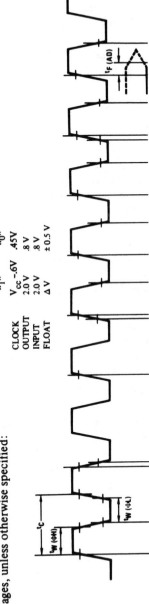

$T_A = 0°C$ to $70°C$, $V_{cc} = +5V \pm 5\%$, Unless Otherwise Noted.

Signal	Symbol	Parameter	Min	Max	Unit	Test Condition
Φ	t_c	Clock Period	.4	[12]	μsec	
	$t_w(\Phi H)$	Clock Pulse Width, Clock High	180	[E]	nsec	
	$t_w(\Phi L)$	Clock Pulse Width, Clock Low	180	2000	nsec	
	$t_{r,f}$	Clock Rise and Fall Time		30	nsec	

[12] $t_c = t_{w(\Phi H)} + t_{w(\Phi L)} + t_r + t_f$

Figure 10-6 Excerpted specifications for the Z80 microprocessor clock input. Reproduced
by permission. 1986 Zilog, Inc. This material shall not be reproduced without the written
consent of Zilog, Inc.)

should be one of the first things we do when troubleshooting a microprocessor-controlled system.

It is important to understand how the system clocks operate in order to know how fast the system will respond and how fast the external hardware will operate. By knowing the speed of the system clock, you can calculate the speed of operation for any type of instruction executed by the microprocessor. This is easily accomplished by examining the data sheets to determine how many clock cycles any given instruction takes to execute.

In the remaining sections of this chapter we will discuss the important topic of interrupts. We start by introducing the concept of an interrupt.

10-5 DEFINITION OF INTERRUPT

An example of an interrupt for micrprocessors systems may be seen by examining a human system, with which most of us are familiar. Picture this: You are having a conversation with one other person. A third person walks up and speaks your name. This is an indication to you that this third person wishes to get your attention. You may think of this third person as external interrupt request.

This person wishes to interrupt your conversation. The following is a list of possible actions you may take to this external request.

1. You can completely ignore the third person and keep talking as though that person were not there.
2. You can get to a convenient stopping place in your conversation and then turn your attention to the third person. At this time you will be conversing with the third person and not the original person.
3. You can immediately stop your conversation with the person to whom you are talking and start talking with the third person.

When you are finished talking with the third person, you will want to go back to the same place in the conversation with the original person to whom you were speaking before the external interrupt request occurred.

The preceeding scenario may seem simplistic, but it accurately presents the concept of interrupts for microprocessor systems. Here is how.

Think of yourself as the CPU. The person with whom you are talking initially is a main program being executed. A third person is an external interrupt request. That is, some hardware in the system wishes the CPU to give it attention. The CPU must handle this request in some fashion. There are various ways to handle the interrupt request; the three mentioned in the previous example are the most common.

With this general introduction to an interrupt, let us concentrate on the details of interrupts for the 8080, 8085, and Z80 microprocessors.

10-6 THE EXTERNAL INTERRUPT REQUESTS

In a microprocessor system there are many hardware components that comprise the entire system. These can be printers, floppy or hard discs, CRTs, timers, motors, or DACs, just to name a few. Some of the external hardware will need the attention of the CPU only at certain times. At all other times the hardware can function on its own.

For example, suppose there was a system time base (clock) as an external hardware device. The time of day was displayed in a corner of the CRT screen. As the screen was viewed, the clock digits were updated each second. That is, once each second the clock hardware would require the CPU to read the time and print it to the CRT screen. All other times the CPU could do any other tasks required of it.

The external hardware of the clock does not need the CPU's attention all of the time. Therefore, the clock hardware will electrically request the CPU to read only when required. One way to accomplish this is through the CPU's interrupt system. Once each second the CPU will be given an external, electrical, interrupt request from the clock hardware. At that time the CPU will stop the task presently in progress and read the clock time. After the clock has been read, the CPU will resume the program it was executing prior to the clock interrupt.

This is a simple example of interrupts, but we may extend it to any external hardware issuing an electrical interrupt request to the CPU. The answer to the initial question of the origin of the interrupt requests is that they come from the microprocessor system hardware.

10-7 RESETTING THE 8080, 8085, AND Z80

The first type of interrupt input to the CPU that we will discuss is the RESET input. This interrupt forces the microprocessor to begin executing a program from a specific location in system memory. The input is asserted when the system is first powered up or whenever we wish the CPU to restart the program from the beginning. The following is a presentation of how each of the three microprocessors responds during a reset operation.

Resetting the 8080

When the 8080 is initially powered up there must be a way to force the CPU to start executing a program at a certain address location each time rather than at random. To accomplish this we must reset the 8080 when power is first turned on or at any time that we wish the 8080 to return and start executing the system program from that same special location in memory. To reset the 8080, we connect components to the $\overline{\text{RESIN}}$ input of the 8224 system clock generator pin

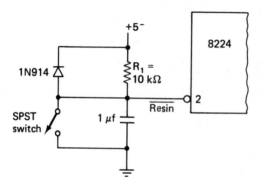

Figure 10-7   Schematic showing a power-on reset circuit for the 8080 microprocessor when using an 8224 clock generator.

2, as shown in Figure 10-7. When the 8080 is reset, it will begin executing from address location 0000. This always allows the programmer to put any initialization software at a fixed address space.

Referring to Figure 10-7 we see that the RC network will provide a power-on reset signal. Prior to V_{CC} being applied to the system, the capacitor will be in a discharge mode. Upon the application of power, the capacitor charges toward V_{CC} via the resistor R_1. When the capacitor voltage reaches a certain predetermined value, the reset will be disabled and the system will start execution from address 0000.

The switch shown in Figure 10-7 will allow the system to be reset at any time. Pressing the switch will discharge the capacitor and the 8080 will again reset until the capacitor voltage charges to a specified value. The 8224 clock generator device has a Schmidt trigger input that will monitor the voltage between the capacitor and resistor and at the same time provide very little loading of the RC network.

Resetting the 8085

To reset the 8085 microprocessor, we connect the reset input pin 36 to the same components we discussed for the 8224 clock-generator reset input in Figure 10-7. With the 8085 connected in this fashion, it will be reset when the switch is pressed or when power is first turned on. The 8085 will begin executing program code from address 0000. This is exactly the same as for the 8080. Figure 10-8 shows the reset circuit connected to an 8085 microprocessor.

Resetting the Z80 CPU

For resetting the Z80 microprocessor, we can use a circuit as shown in Figure 10-9. This circuit works like the reset circuit we discussed for the 8080 and 8085. As with the other microprocessors, when the Z80 is reset it begins program execution from address location 0000.

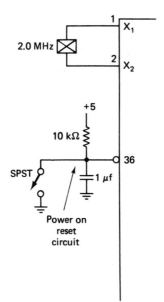

Figure 10-8   Reset circuit for an 8085 CPU.

10-8 COMMENTS ON INTERRUPTS

The process of interrupting the microprocessor consists of two parts, hardware and software. Hardware is used to request the interrupt electrically. Software is used to control and perform the action that must occur as a result of the interrupt request. The following presentation of interrupts covers both hardware and software. First the hardware for requesting interrupts is given and then software routines that are used for interrupts are explained. These software routines are given the names *interrupt service routines*.

When an interrupt request is applied to the microprocessor it is done so in

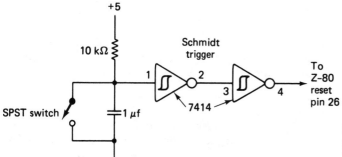

Figure 10-9   Schematic showing a power-on reset circuit for use with the Z80 microprocessor.

an *asyncronous* mode, which means that the input can be applied at any point during program execution regardless of the state of the system clock.

In most microprocessor literature, a term called *vector interrupts* is used. The term *vector* means *pointer*. This vector shows the microprocessor which memory address to use to start the interrupt program from the point in time that an interrupt signal is input. So the term vector is really nothing more than an address location in memory. The term *vectored interrupt* means that interrupts will be occuring within a microprocessor system and a memory address, or vector, will be the starting address for the interrupt routine.

On some microprocessors there are two types of interrupts: (1) maskable interrupts and (2) nonmaskable interrupts. In the case of a nonmaskable interrupt, when an interrupt signal is applied, the microprocessor will always respond to that interrupt input. The microprocessor must respond to a nonmaskable interrupt whenever that occurs during a program.

For a maskable interrupt, the software can enable or disable that particular input from ever being recognized by the microprocessor. We will not discuss all of the applications for maskable and nonmaskable interrupts in a microprocessor system. What we will do is show how these two functions can be realized with software and hardware. From that starting point, you will be able to use the interrupt system of the microprocessor for your own special applications.

10-9 EXAMPLE HARDWARE FOR ASSERTING THE INTERRUPT

To perform the hardware function of interrupting microprocessors, we will use the same circuit for all three. This will allow you to concentrate on the details of how the interrupt is handled by the CPU rather than on the actual external interrupt circuit. The circuit to force the external interrupts is given in Figure 10-10; it applies the hardware interrupt signal to each of the microprocessors. By doing this, we may be able to understand how each of the processors responds to an interrupt request being applied by external hardware. The microprocessor will then communicate with the circuit of Figure 10-10 in some fashion. This is depicted by the arrow at bottom right coming from the microprocessor to the clear input of the 74LS74 flip-flop.

Let us discuss briefly how the circuit shown in Figure 10-10 operates. The first section at the left shows the switch debounce circuit, which comprises two 7406 open-collector inverters and a SPDT momentary pushbutton switch.

When switch S_1 is in the resting position (NC), the center terminal of the switch is connected to the normally closed pole. The normally closed pole is connected to pin 3 of IC1. Pin 3 of IC1 is a logical 0. Pin 4 of IC_1 is a logical 1. This forces a logical 1 on the clock input pin of the 74LS74. The 74LS74 is an edge-triggered flip-flop. When the clock input of the 74LS74 goes from a logical 0 level to a logical 1 level, data at the D input of the 74LS74 will be transferred

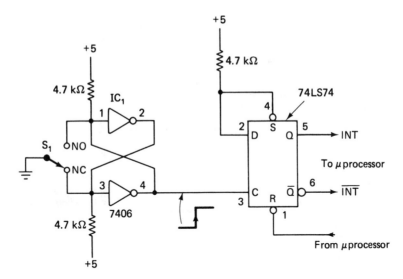

Figure 10-10   Schematic of hardware used to implement or force the interrupt input to a microprocessor.

to the Q and $\overline{Q}$ outputs. In Figure 10-10 we see that when the circuit is in the normally closed position, the clock input to the 74LS74 is logical 1.

When we press switch S_1, the center terminal of S_1 is connected to the normally open pin of the switch. This forces a logical 0 on pin 1 of IC_1 and a logical 1 at output pin 2 of IC_1. The result is that the clock input of the 74LS74 goes to a logical 0 when the SPDT switch S_1 is pressed. This action will have no effect on the flip-flop.

When we again release switch S_1 to its normally closed position, the clock input to the 74LS74 goes to logical 1 level from logical 0 level. It is this action of switch S_1 that transfers the data at the D input of the 74LS74 to the Q and $\overline{Q}$ outputs. This is shown by the timing diagram of Figure 10-11.

The D input and the set input pins 2 and 4 of the 74LS74 flip-flop are connected to +5 volts via the 4.7K-ohm resistor. There is always a logical 1 applied to the set input and a logical 1 applied to the D input of the 74LS74. When data is transferred on the rising edge of the clock input pin, the Q output always goes to logical 1 and the $\overline{Q}$ output always goes to logical 0. This is the condition for forcing the interrupt to the microprocessor under discussion.

The microprocessor must then respond by applying a logical 0 to the clear pin input (1) of the 74LS74 to remove the interrupt request once it has been acknowledged by the CPU. In this way the Q output will go to a logical 0 and the $\overline{Q}$ output will go to a logical 1. It is by applying this clear input to the 74LS74 that the microprocessor effectively resets the interrupt request from the external hardware.

By using the circuit of Figure 10-10 we may indeed apply an interrupt

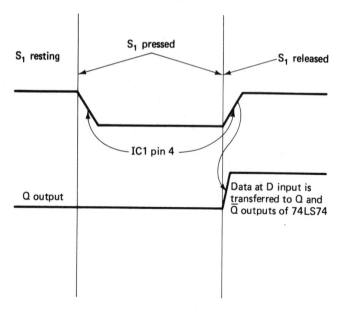

Figure 10-11 Timing diagram showing the relationship between Q output and the switch position. When the switch is released, the Q output is forced to logical 1 and the Q̄ output is forced to logical 0.

input to the microprocessor in a totally asynchronous fashion. The microprocessor can be executing a program and we may push switch S_1 at any time to force an interrupt to occur. When we have discussed the specifics of how interrupts are handled with the microprocessor, we will again refer to Figure 10-10 with more details of why we must use a flip-flop to generate the interrupt request.

10-10 INTERRUPTING THE 8080 MICROPROCESSOR

Figure 10-12 shows how to interface the hardware shown in Figure 10-10 to an 8080 microprocessor. There is only one type of interrupt input for the 8080, maskable interrupt. There is no nonmaskable interrupt. In Figure 10-12 we see that the interface is very simple. The interrupt request line for the 8080 is labeled INT. The INT is pin 14 on the 8080 device. The 8080 has a special output pin labeled INTE (interrupt enabled), pin 16, that will be used to reset the 74LS74. To explain how these two pins interact with the 8080, a general sequence for an interrupt handling with the 8080 will be given. This sequence deals only with the hardware for an interrupt. We discuss the software for interrupts in a later section of this chapter. Let us assume that an interrupt request has been input:

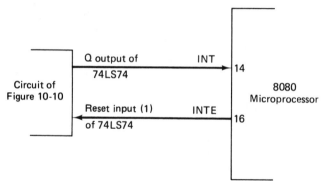

Figure 10-12   Block diagram showing how the circuit of Figure 10-10 interfaces to an 8080 microprocessor.

1. INT pin 14 of the 8080 is forced to logical 1. This indicates that an interrupt is being requested from external hardware.
2. When the 8080 internally acknowledges the receipt of the interrupt request on pin 14, the INTE pin 16 goes to logical 0. This action clears the flip-flop of Figure 10-10 and removes the interrupt request. It is valid to remove the interrupt request because the CPU has acknowledged the interrupt input.
3. The 8080 then asserts the INTA (interrupt acknowledge) control line.
4. External hardware places the correct data on the system data bus to control the execution of the interrupt service routine.

These are the major events that occur for the interrupt interface we have shown. However, there is other hardware that we must discuss for interrupting the 8080. At this point we are discussing only the generation of an interrupt request and the resetting of the interrupt request. When pin 16 of an 8080 is in logical 1 state, the 8080 is in a mode to receive interrupt requests from external hardware. When INTE pin 16 is logical 0, the 8080 is in a mode to ignore all interrupt requests applied to INT pin 14.

There are two software instructions for the 8080 called DI (disable interrupts) and EI (enable interrupts). These two software instructions can be used to manipulate the logical condition of the INTE pin. Further, these instructions will place the 8080 into a mode to ignore interrupt requests. For example, when the 8080 is initially reset, the INTE pin is forced to a logical 0 level. This means the 8080 is in a mode to ignore all interrupt input requests from external hardware. Effectively, the 8080 is in the disable interrupt mode. The only way to force the INTE pin to a logical 1 state is for the 8080 to execute a software instruction EI, or enable interrupt. When the enable interrupt instruction is executed, the INTE pin will go to a logical 1 condition.

With the INTE pin in a logical 1 condition, the clear input to the 74LS74 shown in Figure 10-10 is in logical 1. This allows the interrupt request to be transferred from input D to output Q of the 74LS74. Now suppose that an interrupt has been requested via external hardware. When the interrupt is accepted, the INTE pin is again forced low (to zero) via the hardware internal to the 8080 microprocessor.

What has happened is that the 8080 has again been put into a disable interrupt mode. The only way to reenable the interrupts is for the 8080 to execute an EI software command. So the INTE pin may be forced to a logical 0 state by either the hardware or software for the 8080 microprocessor, but the INTE pin may be set to a logical 1 level only by the execution of software.

What does this particular fact mean to the user of the 8080 when interrupts are designed into the system architecture? It means that the 8080 microprocessor software must contain the proper code to enable interrupts are the correct time. That is, the interrupt is requested by the external hardware and then, during the interrupt routine, the software must again reenable the interrupts. This is logical because during the execution of the interrupt code, we may not want the 8080 to receive further interrupts until a section of program has been completed. We will show detailed software for interrupts later in this discussion. Let's look at Figure 10-13 to examine the general idea of what the software must do.

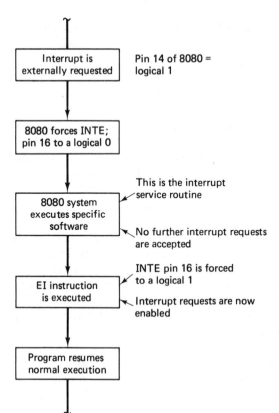

Figure 10-13 Flowchart showing the sequence of events for an 8080 microprocessor to enable interrupts within the service routine.

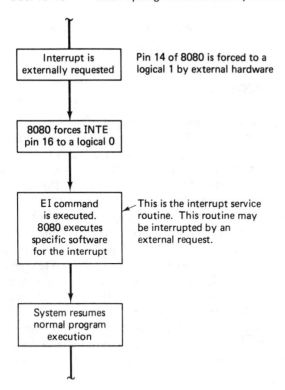

Figure 10-14   Flowchart showing the sequence of events for an 8080 to enable interrupts within the service routine.

Referring to Figure 10-13, the interrupt request is issued via pin 14; then INTE pin 16 goes to logical 0. At this point the system is forced to execute a specific software program in memory. We have not discussed how the system gets to this particular memory location, but we will later. Let us assume now that the system will get to this location in memory. During the execution of this interrupt routine, the INTE pin will be logical 0. When the INTE pin is a logical 0, no interrupt can be accepted by the 8080. After execution of the interrupt routine, the EI instruction is executed, and the INTE pin once again goes to logical 1. The program also resumes normal execution. We have not shown how the program resumes normal execution; this, too, is discussed in detail later.

We can modify Figure 10-13 to appear as shown in Figure 10-14. In Figure 10-14 we see that after the INTE pin is forced to logical 0 via the hardware accepting the interrupt request, the enable interrupt command is executed. When an EI command is executed, the INTE pin goes to logical 1. Now the 8080 is free to accept new interrupt requests.

We see that the system will start to execute the specific software and memory for the interrupt routine requested. But, this routine may again be interrupted from external hardware if another request is generated. It is the function of the software to determine whether the interrupt routine in progress can be further interrupted or whether it should be completed before further interruptions are accepted. This particular aspect of interrupts is common to most microprocessors.

10-11 JAMMING THE RESTART VECTOR

We have now discussed how an interrupt is input to the 8080 and how the 8080 responds to the request with the INTE pin. What we must show next is how the 8080 system will be forced to execute a specific software location within memory when an interrupt occurs. It is a job of the external system hardware to input to the 8080, at a specific time, the address or vector for the memory location to be executed when an interrupt occurs.

A signal on the system-control bus must now be added. Up to this point we have had a system-control bus that comprises four control signals. These are memory write, memory read, I/O write, I/O read. The new signal we must add is an interrupt acknowledge signal. We will label this $\overline{\text{INTA}}$. When the $\overline{\text{INTA}}$ control signal is asserted, logical 0 level, the external hardware of the system must place certain data on the 8080 data bus to handle the interrupt vector. This sequence is shown by the flowchart of Figure 10-15.

In Figure 10-15 we see that the interrupt request will occur, the INTE pin will go to logical 0, and the $\overline{\text{INTA}}$ will go to logical 0. Now certain hardware in the system must be enabled to allow the 8080 to input the interrupt vector. The

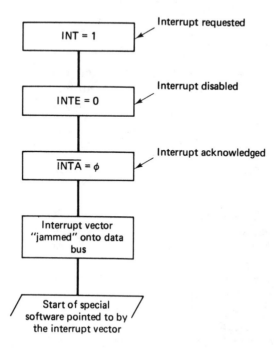

Figure 10-15 Flowchart of the sequence of events for external hardware of the 8080 when an external interrupt is input.

RESTART vector		Data bits								MEMORY address
Name	D_7							D_0		
0	1	1	0	0	0	1	1	1		0_{16}
1	1	1	0	0	1	1	1	1		8_{16}
2	1	1	0	1	0	1	1	1		10_{16}
3	1	1	0	1	1	1	1	1		18_{16}
4	1	1	1	0	0	1	1	1		20_{16}
5	1	1	1	0	1	1	1	1		28_{16}
6	1	1	1	1	0	1	1	1		30_{16}
7	1	1	1	1	1	1	1	1		38_{16}

Figure 10-16   The relationship between the restart vector name, the jammed data bits, and memory address of the restart vector.

8080 has eight interrupt vectors—it can force the system to start executing code from eight specific memory locations. The vectors and the memory addresses to which they correspond are listed in Figure 10-16.

The interrupt vectors for the 8080 are given a special name, *Restart vectors*. Restart is abbreviated RST. The interrupt vectors are further labeled as RST_0–RST_7. RST_0 corresponds to the interrupt vector with address 0. RST_1 corresponds to the interrupt vector with address 8, and so on. If we apply an RST_3 restart vector, we will be forcing the data bits D0–D7 of the data bus to the conditions, or logical states, for the RST_3 vector.

Figure 10-17 shows one way of forcing a restart vector onto the data bus. Forcing a restart vector is also called *jamming* the restart vector on the data bus. We see in Figure 10-17 that this operation is very similar to reading data from memory or reading from I/O. The only difference is that the function of the data input to the 8080 takes on a new meaning.

During an interrupt cycle, the 8080 will accept a restart vector jammed onto the data bus or the op-code for a CALL instruction. If a CALL op-code is jammed onto the system data bus during an interrupt request, the 8080 will assert the INTA control line two more times to read the 16-bit address for the service routine. It is the job of the external hardware to place the CALL op-code and the 16-bit address onto the system data bus during the interrupt cycle. This means that the 8080 will accept either a restart vector 0–7 or CALL instruction. With the CALL instruction the external hardware can force the 8080 to execute an interrupt service routine at any memory location. The restart vectors will force the 8080 to execute a service routine from any of eight predetermined locations.

The hardware implementation of an interrupt is not difficult to accomplish, but it is still one of the five major jobs we listed. That is, the CPU is reading data from the system data bus. The timing of the data onto the data bus is controlled by the microprocessor. The microprocessor frees us from having to consider all the timing constraints for using interrupts. We simply decode the correct data at the correct time and place it on the data bus.

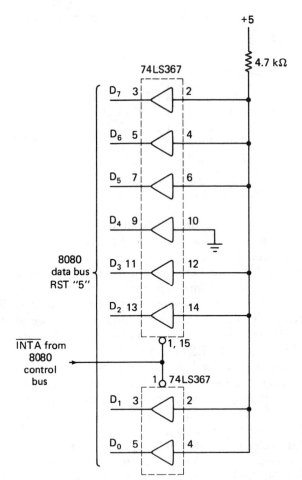

Figure 10-17 Schematic showing one way of jamming a restart vector onto the data bus at the correct time. The restart vector shown is an RST 5.

10-12 INTERRUPT SOFTWARE

We have now discussed how we may input an interrupt request and how we may force, or jam, the 8080 to execute certain codes within the microprocessor when an interrupt occurs. We must next discuss the following:

1. How to set up the system software to handle the interrupts.
2. How to return to the instruction in the program that was being executed prior to an interrupt request.
3. How to write the service routine for an interrupt.

Setting Up the Software to Handle Interrupts

When the external hardware places the restart vector 0–7 on the system data bus as shown in Figure 10-17, the 8080 will go the proper memory address and start executing a program. It is the job of the software to ensure that the program is correct at the location specified. A JMP instruction is usually placed at the restart memory addresses. This is shown in Figure 10-18. The address specified by the jump instruction is the service routine for the particular interrupt.

```
 1                          ;;;;;;;;;;;;;;;;;;;;;;;;;;;;;;;;
 2                          ;
 3                          ; SETUP FOR INTERRUPTS AT
 4                          ; BEGINNING OF ROM SPACE
 5                          ;
 6                          ;;;;;;;;;;;;;;;;;;;;;;;;;;;;;;;;
 7                          ;
 8   0000                             ORG 0000
 9   0000    C3 3B 00                 JMP START      ;RESTART VECTOR
10                          ;
11   0008                             ORG 0008
12   0008    C3 3C 00                 JMP RESTART1
13                          ;
14   0010                             ORG 0010H
15   0010    C3 3D 00                 JMP RESTART2
16                          ;
17   0018                             ORG 0018H
18   0018    C3 3E 00                 JMP RESTART3
19                          ;
20   0020                             ORG 0020H
21   0020    C3 3F 00                 JMP RESTART4
22                          ;
23   0028                             ORG 0028H
24   0028    C3 40 00                 JMP RESTART5
25                          ;
26   0030                             ORG 0030H
27   0030    C3 41 00                 JMP RESTART6
28                          ;
29   0038                             ORG 0038H
30   0038    C3 42 00                 JMP RESTART7
31                          ;
32                          ;
33                          ; DUMMY LOCATIONS FOR THE ASSEMBLER
34                          ;
35   003B    C9            START    RET
36   003C    C9            RESTART1    RET
37   003D    C9            RESTART2    RET
38   003E    C9            RESTART3    RET
39   003F    C9            RESTART4    RET
40   0040    C9            RESTART5    RET
41   0041    C9            RESTART6    RET
42   0042    C9            RESTART7    RET
43   0043                             END
```

Figure 10-18   Setup for interrupts at beginning of ROM space.

Returning to the Main Program after an Interrupt

Before we discuss how to write the interrupt service routines, let's examine how the CPU will return to the main program after the interrupt is over. Recall that the microprocessor may have interrupt requests input at any time in the program. This means that the interrupt may occur at almost any microprocessor address. The question to be answered is, How does the CPU return to the correct memory address after the interrupt is over? To answer this, let's examine the sequence of events that occurs automatically with the 8080 when an interrupt request is honored. This is shown in Figure 10-19.

In Figure 10-19 we see that when the external interrupt request is issued, the INTE pin goes to logical 0 as discussed. The INTA control bus signal goes to logical 0, and the restart vector or CALL instruction is input to the 8080 microprocessor at this time. The 8080 now pushes a return address onto the system stack.

The 8080 waits until the end of an instruction before acknowledging an interrupt input. In this way the program counter will be pointing to the next instruction to be executed if an interrupt did not occur. Therefore, when the return address is pushed onto the stack, it is really the program counter being pushed onto the stack.

This is similar to the sequence of events used when a subroutine is called in a program. When a subroutine is called with the 8080, the 8080 pushes a return address onto the system stack. The return address is the address of the next instruction to be executed after the CALL. When we want to come back to the main program after the subroutine is executed, we simply execute a return

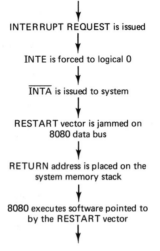

INTERRUPT REQUEST is issued

INTE is forced to logical 0

INTA is issued to system

RESTART vector is jammed on
8080 data bus

RETURN address is placed on the
system memory stack

8080 executes software pointed to
by the RESTART vector

Figure 10-19 Sequence of internal events that the 8080 follows when an external interrupt is requested from external hardware.

(RET) instruction. The execution of an RET pops the return address off the stack so it can be used as the next address for program execution.

This time sequence of events also occurs in the handling of interrupts. The 8080 pushes a return address onto the system stack. When we wish to return to the main program after handling the interrupt, a RET instruction will be executed in the interrupt routine. When the RET instruction is executed, the address that was put onto the stack prior to entering the interrupt routine is popped from the stack. The address popped from the stack now allows the 8080 to resume execution of the program being run prior to the interrupt request. The RET instruction is usually the last instruction of an interrupt service routine.

Let's now examine how the software is written to handle the service routine.

Software for the Interrupt Service Routine

Programs for handling of interrupt requests consist of combinations of instructions to perform the action required of the interrupt. In the example used at the start of this chapter, the service routine could read the clock time and write this time to the CRT memory. If the interrupt indicated that the temperature on some external hardware was excessive, then the service routine would take action to lower the temperature, sound an alarm, or any number of other things. The main point here is that the service routine software is application dependent.

Although the service routine software is application dependent, there are certain guidelines that must be followed when writing the routine. These rules involve saving of registers, reenabling interrupts, and returning from interrupts. The following discussion presents these guidelines for interrupt service routines.

When an interrupt occurs, the 8080 is forced to a certain memory address to execute a specific section of code. You will recall that the interrupt request may occur at any time in the program. This means the 8080 may be just getting ready to make a decision based on the condition of the internal flags. When the interrupt occurs, the 8080 switches to execute the interrupt location of code and in doing so may affect the system flags. We need to save the condition of any registers or flags that may be destroyed during the execution of the interrupt routine. This can be done by pushing the registers onto the stack just after entering the interrupt routine. We can then pop the registers from the stack just before exiting the interrupt routine. This idea is demonstrated by the partial program shown in Figure 10-20.

Recall from earlier chapters that the Z80 had an alternate register set. This register set comes in handy during an interrupt service routine. All we need to do in the service routine is switch to the alternate registers, and the state of the system prior to the interrupt will be intact.

Another guideline for interrupt service routines concerns reenabling of

```
|
PUSH PSW
PUSH H    ⎫
PUSH D    ⎬   Save system status
PUSH B    ⎭
|
⊥
Interrupt software
⫰
POP B
POP D
POP H    ⎫
POP PSW  ⎬   Restore system status
EI
RET
```

Figure 10-20 Partial program showing how the system status is saved and later restored by pushing and popping the system stack.

interrupts. When an interrupt is acknowledged by the 8080, all future interrupts are disabled. They will not be accepted again until the EI instruction is executed. Therefore, if you wish you interrupt service routine to be interrupted by another, higher priority interrupt, then the EI instruction must be executed early in the service routine, usually just after the system status has been saved.

If you do not wish to have your interrupt service routine interrupted, then the EI instruction should be one of the last instructions executed in your service routine, usually just prior to exiting the service routine.

The final guideline for interrupt service routines involves returning to the main program. This action occurs when the CPU executes an RET instruction. Remember that when the RET instruction is executed, the top 2 bytes of the stack are used as the return address. The return address was pushed onto the system stack automatically when the CPU honored the interrupt request. Therefore, you must ensure that the system stack is in the same order it was prior to entry into the service routine. That is, all data that was pushed onto the stack is off. This is exactly the same precaution we discussed when using subroutines in an earlier chapter.

10-13 8080 INTERRUPT SUMMARY

To summarize the interruption of the 8080, we may describe it as follows.

1. The external hardware provides the interrupt request.
2. The 8080 either enables or disables further interrupt requests via the software.
3. The control bus signal $\overline{\text{INTA}}$ is the means by which the interrupt vector is jammed onto the data bus.
4. The interrupt is an address or pointer that indicates to the 8080 where to go to get and execute a particular set of instructions.

5. Prior to the 8080 executing the interrupt code at a particular address, a return address is placed on the system stack similar to the return address used to call a subroutine.

6. To exit an interrupt routine and return to normal programming, a RET instruction is executed.

7. System status should be saved immediately in the interrupt service routine.

8. The EI instruction is used to reenable interrupts. If the service routine may be interrupted, then the EI instruction should occur early in the routine code. If the service routine may not be interrupted, then the EI instruction should occur just prior to the RET instruction.

A typical interrupt service routine located at memory location 5600H is shown in Figure 10-21. This routine simply reads the data at an input port and transfers this data to an output port. Further, this service routine may not be interrupted by another interrupt request. Notice in this figure that the jump instruction to the service routine at location 5600H is executed via a restart vector RST_4.

```
 8   0014                           ORG 0020    ;RESTART VECTOR 4
 9   0014    C3 00 56               JMP RESTART4
10                         ;
11                         ;
12   5600                           ORG 5600H       ;LOCATION OF SERVICE ROUTINE
13                         ;
14   5600    F5            RESTART4 PUSH PSW ;SAVE A REG
15   5601    C5                     PUSH B
16   5602    D5                     PUSH D
17   5603    E5                     PUSH H
18                         ;
19                         ; STATUS IS NOW SAVED
20                         ;
21   5604    DB 43                  IN 43H   ;GET PORT DATA
22   5606    D3 57                  OUT 57H  ;OUTPUT DATA
23                         ;
24                         ; TO RESET INTERRUPT YOU MAY NEED
25                         ; TO OUTPUT DATA TO A SPECIFIC PORT
26                         ;
27   5608    3E 00                  MVI A,00
28   560A    D3 23                  OUT 23H  ;RESET INTERRUPT REQUEST
29                         ;
30                         ; NOW RESTORE SYSTEM STATUS
31                         ;
32   560C    E1                     POP H
33   560D    D1                     POP D
34   560E    C1                     POP B
35   560F    F1                     POP PSW
36                         ;
37   5610    FB                     EI       ;RE-ENABLE INTERUPTS
38   5611    C9                     RET      ;RETURN FROM THIS ROUTINE
39                         ;
40   5612                           END
```

Figure 10-21 Interrupt service routine for restart vector 4. The routine is located at memory address 5600H.

10-14 INTERRUPTING THE 8085 MICROPROCESSOR

Many of the points about interrupts that were given for the 8080 are valid for the 8085 and Z80 microprocessors. However, there are some differences in interrupts between these microprocessors. These differences are discussed in the following sections.

The 8085 may be interrupted in several ways. One way is exactly the same as the 8080 that we described previously. For interrupting the 8085 in the same way as the 8080, pin 10 of the 8085 is used for interrupt request (INTR). The interrupt acknowledge ($\overline{\text{INTA}}$) signal pin 11 of the 8085 is an output that enables the restart vector or CALL op-code to be placed onto the system data bus. This is shown in Figure 10-22. Notice in this figure that when the $\overline{\text{INTA}}$ signal goes low to acknowledge an interrupt, the restart vector 0–7 is jammed onto the data bus, and the 8085 begins executing the interrupt service routine based on the restart vector. The restart vectors are exactly the same address locations as they were for the 8080 previously discussed. These were shown in Figure 10-16.

One difference between the 8085 and the 8080 is that there is no INTE signal available on the 8085. This means that we must provide an output port to reset the interrupt request. In Figure 10-23 we see that after an interrupt request is issued from the D flip-flop circuit of Figure 10-10, the only way to reset the interrupt request is by resetting the D flip-flop. The reset line is generated by the output port shown in Figure 10-23. To remove the interrupt request electrically, the 8085 must write a word to the output port. The writing to the output port must be done early in the interrupt service routine and prior to the interrupts being reenabled, because if the interrupts were enabled before the interrupt request was removed, the 8085 would honor the request, and the service routine would be entered again.

The 8085 also provides an interrupt pin 6 called TRAP. The TRAP input

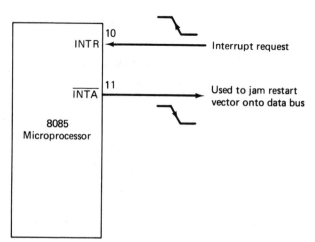

Figure 10-22 Block diagram showing the two physical pins that input and acknowledge the interrupt request of an 8085 microprocessor.

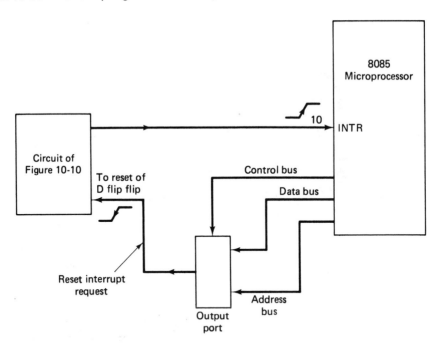

Figure 10-23 Block diagram showing an output port that will be used to reset external interrupt request hardware.

is unmaskable interrupt—that is, a TRAP interrupt cannot be disabled via the software. Recall that the 8080 used only maskable interrupts. The TRAP interrupt is assigned the highest priority of any interrupt. If two or more interrupts are issued at the same instant in time, the TRAP interrupt will take precedence over any of the other interrupt requests. When using a TRAP interrupt, we do not need to jam a restart vector onto the data bus. A restart vector is automatically issued internally on the 8085. The restart address for a TRAP interrupt is 24H. To issue a TRAP interrupt, the hardware shown in Figure 10-24 can be used.

Referring to the figure, note that we have eliminated the D flip-flop from the interrupt request hardware. The TRAP interrupt is both edge and level sensitive. The TRAP interrupt must remain high (logical 1) long enough to be electrically input to the 8085. However, the TRAP input line will not be electrically recognized by the 8085 until the signal goes to a logical 0 and back to a logical 1 again. For this to occur, the switch S_1 of Figure 10-24 must be released and pressed a second time.

There are three other interrupt inputs for the 8085. These interrupts are labeled RST 7.5, RST 6.5, and RST 5.5. The input pins for these are 7, 8, and 9, respectively, of the 8085. These inputs are provided so we do not need to jam a restart address onto the data bus. When these interrupt requests are active, the

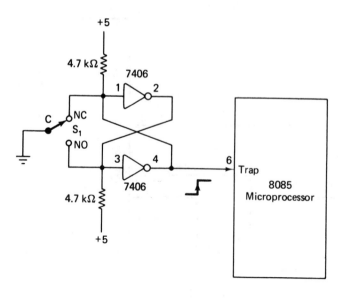

Figure 10-24 Schematic showing how a trap interrupt can be issued for an 8085 microprocessor.

8085 automatically inserts an internal restart vector. The addresses for the restart vectors 5.5, 6.5, and 7.5, are given in Figure 10-25. Notice in this figure that the addresses of the 5.5, 6.5, and 7.5 interrupt vectors are physically between the restart vectors that may be jammed onto the data bus—that is, a restart 5.5 is between restart 5 and restart 6.

On the 8085 microprocessor the restart inputs 5.5 and 6.5 have the same timing as the INTR interrupt input. The RST 7.5 is input differently. Restart vector 7.5 is an active, rising, edge-sensitive interrupt request. Further, only a pulse is required to set the interrupt request on the 8085. This means that an interrupt request can be issued on the RST 7.5 simply by inputting a pulse to the RST 7.5 pin. The pulse input will be remembered until the request is either serviced or reset by the software. This particular feature of the RST 7.5 is shown in the timing diagram of Figure 10-26.

Along with these hardware differences between the 8080 interrupts and the 8085 interrupts, there are software differences also. The main software difference is that the 8085 has an instruction called set interrupt mask (SIM).

INTERRUPT input	RESTART vector address
5.5	$2C_{16}$
6.5	34_{16}
7.5	$3C_{16}$

RESTART vector	MEMORY address
5	28_{16}
6	30_{16}
7	38_{16}

Figure 10-25 The memory addresses of the RST .5 interrupt vectors. Notice that the RST 6.5 is physically between the RST 6.0 and RST 7.0 locations.

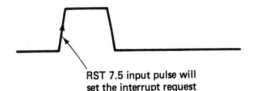

RST 7.5 input pulse will
set the interrupt request

Figure 10-26 Timing diagram showing
the pulse input for an RST 7.5 request.

```
 7   0000   3E 08              MVI  A,08H      ;ENABLE MASK SET
 8   0002   F6 01              ORI  01H        ;DISABLE INT 5.5
 9   0004   30                 SIM             ;SET MASK, RESET 7.5
10                       ;
11   0005   3E 08              MVI  A,08H
12   0007   F6 04              ORI  04H        ;DISABLE INT 7.5
13   0009   30                 SIM             ;SET MASK, RESET 7.5
14                       ;
15   000A   FB                 EI              ;ENABLE INTERRUPTS
16                       ;
17   000B   F3                 DI              ;DISABLE INTERRUPTS
18                       ;
19                       ;
20   000C   3E 08              MVI  A,08H
21   000E   F6 07              ORI  07H
22   0010   30                 SIM             ;NO CHANGE TO INTERRUPT MASK
23                       ;
24                       ;
25   0011                      END
```

Figure 10-27 Program to set the interrupt mask register on the 8085 microprocessor.

The interrupt mask for the 8085 allows the programmer to enable only certain interrupt request lines. That is, RST 5.5, 6.5, and 7.5 can be disabled in addition to the INTR interrupt request. The TRAP interrupt is unmaskable and will always be serviced. See Figure 10-27 for a partial 8085 program that will set different interrupt mask bits.

All the previous information about interrupt service routines that was given for the 8080 is applicable to the 8085. The only new information was the hardware requesting of RST 5.5, 6.5, 7.5 and the setting of the interrupt mask for the maskable interrupts.

10-15 INTERRUPTS FOR THE Z80 MICROPROCESSOR

The Z80 CPU has two interrupt request lines: the *interrupt request* ($\overline{\text{INT}}$) pin 16, and the *nonmaskable interrupt request* ($\overline{\text{NMI}}$) pin 17. $\overline{\text{INT}}$ is a maskable interrupt that can be enabled or disabled via the Z80 software. A nonmaskable interrupt will be accepted by the Z80 CPU at all times. When a nonmaskable interrupt is input, the Z80 automatically starts the interrupt servicing from address 66H. That is, the restart vector for the $\overline{\text{NMI}}$ input is 66H. Further, the $\overline{\text{NMI}}$ interrupt is active on the negative-going edge of the input signal. This is shown by the timing diagram of Figure 10-28.

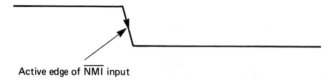

Active edge of $\overline{\text{NMI}}$ input

Figure 10-28 The $\overline{\text{NMI}}$ interrupt input line is active on the negative-going edge of the signal. When this line is asserted it sets an internal latch that remembers the interrupt request.

It makes sense that the $\overline{\text{NMI}}$ input must be edge-sensitive because it can never be disabled with software. Therefore, if it were level sensitive, the system would always be servicing the $\overline{\text{NMI}}$ request. With the input being edge-sensitive, the $\overline{\text{NMI}}$ input must be pulsed again before it will be recognized by the Z80.

Let's now examine the interrupting of the Z80 from the $\overline{\text{INT}}$ pin 16. The Z80 has three separate modes for handling interrupt requests from the $\overline{\text{INT}}$ input pin 16. Recall that the 8080 and 8085 had only one method of handling interrupts. The three modes of interrupts for the Z80 are enabled by the IM0, IM1, and IM2 instructions. When the Z80 is reset, it defaults to interrupt mode 0. Each of these modes of operation is discussed next.

Interrupt Mode 0

Interrupt mode 0 performs in exactly the same manner as the interrupt structure of the 8080. When the interrupt occurs, a restart vector RST_0–RST_7 is jammed onto the data bus via the interrupt acknowledge control line. The $\overline{\text{INTA}}$ control line is generated with a combination of the $\overline{M_1}$ and $\overline{\text{IORQ}}$ control outputs of the Z80. This is shown in Figure 10-29.

The sequence of events that occur during the interrupt mode 0 acknowledge are.

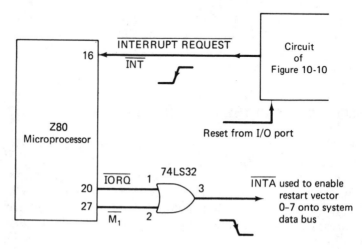

Figure 10-29 Block diagram showing how the external interrupt is input and acknowledged by the Z80 microprocessor.

1. IFF1 and IFF2 (these are internal interrupt flip-flops) are set to a logical 0, thus disabling any further interrupt requests.
2. The $\overline{M_1}$ output line and the $\overline{IORQ}$ output line both go to a logical 0. This situation occurs only on the Z80 during an interrupt and is given the name of *interrupt acknowledge* control line.
3. External hardware decodes the $\overline{M_1}$ and $\overline{IORQ}$ output and enables a single data byte onto the Z80 data bus. The data byte enabled onto the data bus is the opcode for RST_0–RST_7. This byte could also be the op-code for a CALL instruction.
4. The Z80 reads the byte or (bytes if CALL op-code).
5. The program counter is pushed onto the system stack.
6. The Z80 jumps to the memory location specified by the interrupt vector.

This is very similar to the action that occurs during the interrupt acknowledge cycle for the 8080 or 8085.

In order for the Z80 to receive any further interrupt requests from the $\overline{INT}$ input pin 16, the IFF1 flip-flop must be set to a logical 1. This is accomplished when the CPU executes the EI instruction prior to a return from the interrupt service routine.

Finally, the CPU resumes execution of the main program by using the RET or RETI instruction. The RET instruction is identical to the RET instruction for the 8080 or 8085 and pops the top two bytes of data off the stack for use as the return address.

The RETI instruction performs the same operation as the RET instruction. The major difference between these two instructions is that certain peripheral devices, such as the Z80-PIO, are designed to recognize the RETI op-code electrically and remove the interrupt request automatically. Let's now examine the interrupt mode 1 for the Z80 CPU.

Interrupt Mode 1

Interrupt mode 1 is very similar to the 8085 receiving an RST 6.5. When the INT input is active, the Z80 will automatically execute an RST 6.5 and jump to memory location 0038H. No external hardware is required to force this action to occur. This is exactly what occurs when an 8085 receives an RST 6.5 interrupt request. The disadvantage to this type of interrupt mode for the Z80 is that there is only one location to which the interrupt input can jump.

Interrupt Mode 2

The third mode of interrupts for the Z80 is quite powerful. Mode 2 interrupts are set up for the Z80 microprocessor when it executes the IM2 instruction. These interrupts are electrically requested in the same manner as mode 0 and mode 1. Therefore, we will concentrate on how the Z80 responds to a mode 2 interrupt request.

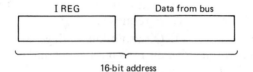

Figure 10-30 Block diagram showing how The I reg and data bus valve form the complete 16-bit address.

When an interrupt occurs, the external hardware must jam the lower 8 bits of a 16-bit memory address onto the data bus. This occurs in a manner similar to the jamming of the restart vector onto the data bus. The upper 8 bits of the memory address are stored in an internal register of the Z80 called the I register.

After the Z80 reads the 8 bits from the data bus, a complete 16-bit address is formed by using both bytes. The most significant byte is read from the I register and the least significant byte is read from the data bus. This is shown in Figure 10-30.

The two bytes of data located at this 16-bit address are another address, which points to the memory location of the interrupt service routine. This allows 256 different interrupt service routines to be accessed directly from the system data bus with a single byte. As an example of how this works, let's examine the following.

Assume that the I register in the Z80 is set to 20H. This is accomplished by the LD I,A instruction. We further assume that when the interrupt occurred, the data jammed onto the bus was equal to 54H. The memory address formed is equal to 2054H.

Let us now assume that the data stored at memory address 2045 and 2055 is equal to

$$2054H \;=\; 48H$$
$$2055H \;=\; 8AH$$

The address of the interrupt service routine would be equal to 8A48H.

This may seem like a lot of work for a single address for an interrupt routine, but it makes the formation of an interrupt vector table quite easy. Further, the interrupt vector table may be located anywhere in system memory. The upper byte of the starting address for the interrupt vector table is located in the I register. The memory address into the table is given by the byte read from the system data bus.

This concludes our discussion of mode 2 interrupts for the Z80 CPU.

10-16 CHAPTER SUMMARY

In this chapter we presented the details of interrupts for the 8080, 8085, and Z80 microprocessors. This started with a general introduction to what an interrupt is. From there we discussed both the maskable INTR and nonmaskable NMI interrupts for the different CPUs. In each case you were presented with a detailed description of how the CPU handles these two types of interrupt requests.

We next discussed the software required for handling an interrupt. This included the setting of jump vectors at the start of a program and guidelines for interrupt service routines.

Using the information presented in this chapter, you will better understand how the CPU handles different interrupt requests from the peripheral system hardware. Further, you will be able to write your own interrupt service routines for various peripheral devices that require them. This knowledge will give you another degree of freedom when designing your own system to meet your specific applications.

REVIEW PROBLEMS

1. How many clock input lines are required for an 8080? What are the voltage levels?
2. When using the 8080 you should also use the _____ clock generator.
3. If you required an 8080 at a clock frequency of 1.6 megahertz, what value of crystal would you need to use on the clock generator chip?
4. What is the purpose of the $\overline{STSTB}$ line output from the 8224 clock-generator chip?
5. If you were using a circuit like the one shown in Figure 10-5 and the clock frequency was equal to 6 MHZ, to what would you set the R_1C_1 product equal?
6. Define the term interrupt.
7. How many type of interrupts are there on (a) the 8080, (b) the 8085, and (c) the Z80?
8. When the 8080 is reset, from what memory address will the first op-code be fetched?
9. What does the term vector mean when relating to interrupts?
10. Interrupt requests to the microprocessor are done in which mode, asyncronous or syncronous?
11. List the events which occur in an 8080 microprocessor when an interrupt has been requested. Assume the EI instruction has been executed.
12. How does the 8080 remove external interrupt requests?
13. Why must the CPU remove the external interrupt requests?
14. If you wish the 8080 to execute the program at RST 4, what 8-bit value is jammed onto the system data bus during an interrupt acknowledge cycle?
15. How can you get the 8080 to execute an interrupt service routine at any location in the system memory? (Hint: Do not use the restart vector.)
16. Draw a schematic that will force the 8080 to execute an RST 2 during an interrupt acknowledge.
17. Write a partial interrupt service routine that saves the system environment.
18. How do you return from an interrupt service routine and keep the system exactly the same as before you were interrupted?

19. How do you prevent your interrupt service routine from being interrupted?

20. How do you get your interrupt service routine to be interrupted by higher priority interrupt requests?

21. What do you gain by using the RST 5.5, 6.5, or 7.5 interrupt inputs for the 8085?

22. If you were only using one interrupt routine in a Z80-based system, which would be the best interrupt mode to choose for system operation? Why?

11

TROUBLESHOOTING USING LOGIC STATE ANALYSIS AND SIGNATURE ANALYSIS

In Chapter 9 we presented a troubleshooting technique called static stimulus testing, or SST. This technique allowed you to troubleshoot your computer system in a static mode. It was mentioned in Chapter 9 that SST could not find dynamic problems in a system. In this chapter you will be presented with two troubleshooting tools that can isolate dynamic faults in a system. These tools are called a *logic state analyzer* and *signature analyzer*. These tools are hardware realizations of the troubleshooting concepts logic state analysis and signature analysis.

We start by discussing logic state analysis, or LSA. From this discussion you will be shown how a logic state analyzer can be used for troubleshooting of dynamic problems. We will discuss examples of using a typical logic state analyzer with a microprocessor system. Then we discuss the signature analyzer. We show how these tools are used to troubleshoot systems that are designed to use them.

The chapter closes with a summary of LSA and signature analysis, pointing out the strengths and weakness of the tools and techniques.

Let's begin with the logic state analyzer. There are several different logic state analyzers commercially available today. These instruments are approximately the same size as an oscilloscope or a desktop computer. Each device implements the function of LSA differently. That is, the hardware of each logic state analyzer is different. Rather than discuss a specific logic state analyzer system to explain LSA, we are going to discuss LSA using a functional system that we have designed ourselves. In using this approach, we present and discuss the major blocks common to most LSA instruments.

Using a logic state analyzer designed with these common hardware blocks, we can then explain how to troubleshoot with this type of instrument. This approach has been chosen because we feel that you should understand the basic concepts of LSA as they apply to a general unit. This approach gives you a firmer understanding of "what should be there." With this common, basic knowledge in hand, you can approach a variety of commercially available logic state analyzers and be on the alert for certain things that are likely to be in a given unit. Learning the details of a specific instrument should be an easy matter after this discussion. Further, the techniques for using an LSA in a troubleshooting mode can certainly be transferred from unit to unit.

11-1 OVERVIEW OF LSA

What is LSA? Let us present a view of LSA that will be of value to anyone working with microprocessors. You can think of LSA as the monitoring of system hardware operation via software execution.

This definition when expanded means that a logic state analyzer is connected physically to the system hardware. However, the stimulus for the hardware to operate is due to the software program. This concept is shown in the block diagram of Figure 11-1. LSA is useful for other types of digital hardware

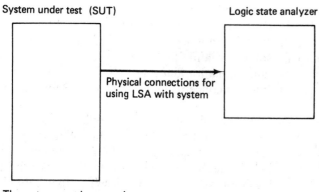

System under test (SUT)   Logic state analyzer

Physical connections for
using LSA with system

The system must be executing
a program for use with a logic
state analyzer

Figure 11-1 The stimulus that enables hardware to operate is controlled by software. The logic state analyzer monitors hardware operations as software is executed.

troubleshooting, not only for a microprocessor-controlled system. Here, however, we will discuss LSA in relation to a microprocessor-controlled system only.

We have already stated that the logic state analyzer is to be connected to the system via the hardware. The hardware response becomes visible on the logic state analyzer display. The particular hardware response in which we are interested is the logic condition (0 or 1) of the device under test. What parts of the system hardware do we wish to monitor? Where does the logic state analyzer connect to the system?

In answer to these questions we should first recall the general block diagram of the microprocessor-controlled system. This general block diagram is shown in Figure 11-2. We see in this figure that there are three major buses in the system. If we can monitor the activity on the system buses in an organized fashion, we will be able to determine exactly what the system hardware is doing.

For example, if we can monitor the address bus, we will be able to determine from where in memory the microprocessor is fetching the instruction. If we can monitor the data bus, then we can determine exactly what data the microprocessor is reading or writing from the system. If we can monitor the control bus, we can determine exactly what type of operation is taking place in the system.

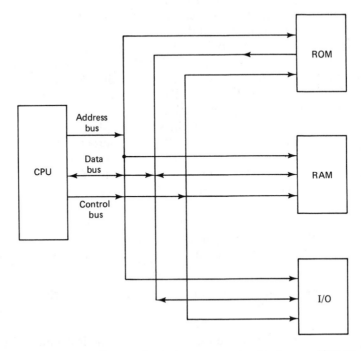

Figure 11-2 Block diagram of typical computer system hardware architecture.

Figure 11-3 The logic state analyzer captures the logical state of the system buses and displays them for the user to view.

We know from our discussion of the system hardware that each of these system buses is changing its logical state very rapidly while the system is executing at normal speed. What the LSA does is to take an electrical "snapshot" of the logic conditions of these buses at a time specified by the user. This feature of being able to monitor the logic state of the system buses while the system is executing at normal system speed is a good one because it is a check of dynamic operation under true operating conditions.

As a first level of troubleshooting we will assume that the LSA is physically connected to the system buses. As the system is executing a program, the LSA will capture the logical state of the system buses and store this in the analyzer's memory. This concept is shown in Figure 11-3. Here we show only one cycle of the system under test being stored in the LSA.

For this single cycle of storage into the LSA we must have a means for storing the logical state of each bit of each system bus. We will store the system conditions for one cycle and then store the conditions for the next system cycle. Ideally, we would like to store the state of each system bus for every cycle that the system under test executes. But this would be impractical to realize with hardware. Therefore, we will assume that the LSA will store a finite number of states of the system under test. The number of states an LSA will store depends on the manufacturer's specifications.

The logical conditions of the system under test are stored while the system is operating at speed. In effect, the microprocessor system is not electrically aware of the logic state analyzer being attached to the system. In other words the logic state analyzer is a passive instrument. The first facts we should keep in mind about the LSA are that it will do the following:

1. Store the logical conditions of the system buses.
2. Store a limited number of clock cycles.
3. Store the logical conditions as the system operates dynamically at its normal rate, or switching speed.
4. Remain transparent to the system under test—that is, it does not load or affect the system.

Storing data in not enough, however. The instrument is useless if the data cannot be viewed. The logic state analyzer must also display the data that was stored. This display can take many different forms, including timing diagrams, binary data, octal data, or hexadecimal data. The medium used for display may also vary from one instrument to another. Some systems use an oscilloscope screen, and others use a video monitor or an ordinary numeric display. The bottom line is that the LSA must somehow display the data taken and stored from the system.

In order for the user to see the stored data on the display, the display cannot be altered rapidly. This means the data on the display must remain visible for a long enough time for the user to interpret it. This fact indicates that the display data is not real-time data. By *real-time,* we mean that the data on the display does not represent data that is present on the system data bus at the time of viewing. This is analogous to a photograph that is taken, processed, and then viewed. We are viewing events that actually took place at an earlier point in time. Moreover, these events are not necessarily taking place at the time we are viewing the picture.

Figure 11-4 shows a flowchart of the general events that occur in logic state analysis. Referring to Figure 11-4 we see that the LSA first inputs the data from the system under test (SUT). This data is stored in real time. Next the LSA processes the real-time data and displays it in a convenient form for later analysis by the user.

At this point, we have introduced the concept of LSA and shown some of the basics. This was intended as an overview. We will build on this general framework in the following sections of this chapter. We will explain why we use the LSA and show exact examples of these devices in use on an 8080 system. We now present some more general information on logic state analysis.

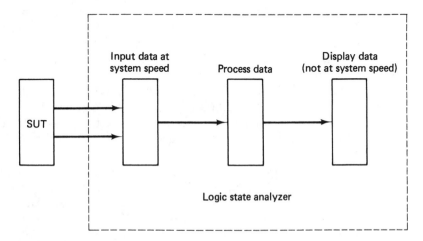

Figure 11-4   Flow of general system events as they occur in a logic state analyzer.

11-2 REASONS FOR USING LOGIC STATE ANALYSIS

We have discussed how the concept and effective use of logic state analysis hinges upon data capture in real time. Why would we be interested in the type of instrument? To answer this, let us present some cases for discussion. Suppose, for example, that we have a malfunctioning system that is not powering up correctly. That is, when the system is first turned on, certain peripheral hardware must have specific data written to it before anything else occurrs in the system. To check that the CPU is writing the correct data to the peripheral hardware, we can connect the LSA to the system buses. This enables us to capture in real time the actual software instructions and data the system under test is executing. The information will be displayed and we can then confirm or disprove our doubts about the proper system operation during power up.

Here is another example where LSA will quickly and easily determine if the system is operating correctly. In examining the system operation we suspect that a particular peripheral device is not being enabled correctly. That is, the system seems to be writing to this port more often than required. By connecting the LSA to the system buses, we can set up the instrument to take data only when the particular I/O device has information written to it. By doing this we can determine if the address decoding is correct for the I/O port and if the CPU is sending the correct data. We could have determined if the address were correct statically, but only the LSA will allow us actually to see the data being written to the device in real time.

A final example where LSA will be of extreme value when troubleshooting concerns memory. This is true of both static and dynamic RAM. During program execution, the CPU will store temporary data into RAM. If one of these RAM locations is defective, then the program will not operate correctly. You can connect the LSA to the system and set up the instrument to take data only when the CPU is reading or writing data to a particular memory address. This will allow you to determine if the system is writing the correct data to memory and reading the correct data from memory at system speed. If the correct data is being written but incorrect data is read, then you know that the system memory is defective.

What we are doing in these cases is determining if the system hardware is operating correctly. The method for making this determination is to have the software stimulate the hardware and monitor the hardware response. Further, we are examining the hardware as it actually performs in the system at system speed.

Logic state analysis is also used for software debugging. If you have written a program and are trying the software out for the first time, using the LSA is helpful. This is because the exact software path is displayed for the user to see. By analyzing the software path shown by the LSA, you can tell if the software has been written correctly.

These are some uses for the logic state analyzer. We are mainly concerned with using the LSA as a hardware troubleshooting tool. These examples were given only to show some simple cases where it may prove useful. However, as we discuss this idea more completely, you will see that there is a wide variety of potential applications for LSA in troubleshooting a microprocesor system.

11-3 MAJOR HARDWARE BLOCKS OF A LOGIC STATE ANALYZER

Let us now get down to some basic considerations concerning the actual hardware of the logic state analyzer. This discussion is undertaken to make you aware of the details. This information should be very helpful when you are using a different logic state analyzer for the first time because it helps to make you aware of what should be there. It then becomes a matter of understanding how a certain function is realized. The concepts presented here are common to most logic state analyzers in industry today.

Major Inputs to the Logic State Analyzer

We have stated that the logic state analyzer monitors the logical state of the three major buses in a microprocessor-controlled system. This fact means that there must be a physical line to the logic state analyzer for each system signal we wish to monitor. To monitor the address lines for the CPU, we would need up to 16 lines for the 8080. The number of lines required to monitor the system address bus depends upon the type of microprocessor used.

If we are to monitor the system data bus, we will need 8 input lines to the logic state analyzer. We also wish to monitor the system control bus. Major signals on this bus are the $\overline{\text{MEMR}}$, $\overline{\text{MEMW}}$, $\overline{\text{IOR}}$, and $\overline{\text{IOW}}$. There are, in addition, different control bus signals that are unique to certain microprocessor-controlled systems. For this discussion we will provide 8 physical input lines for monitoring the system control bus and associated signals.

The inputs to the logic state analyzer will appear as shown in Figure 11-5.

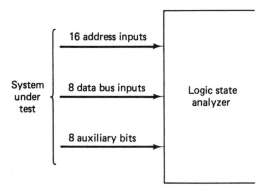

Figure 11-5 Physical inputs to a typical logic state analyzer come from the system under test: address, data, and control buses.

You must keep in mind that the physical inputs we are discussing are found in some form on different logic state analyzers used in industry. The point to be stressed is that the exact number of physical inputs varies from one LSA to another. We are showing a general logic state analyzer so that the many different concepts can be explained fully.

Internal Memory of the Logic State Analyzer

The next major hardware block we will discuss is the internal memory of the logic state analyzer. We recall that the function of the internal memory is to capture the logical state of CPU and system signals in real time. Each line or physical input to the logic state analyzer must have a location for captured data to be stored in internal memory.

In our logic state analyzer we have provided 32 physical inputs. This means our internal memory must be organized in $n \times 32$, where n is the number of different address locations and 32 represents the number of parallel data inputs to the memory. The value of n determines how many consecutive *snapshots* of system data we can store in real time. That is, if the number n were equal to 256, we could store up to 256 unique states of the system under test.

If the logic state analyzer is to capture data at the speed of the system under test, then the internal memory must be capable of storing the data as fast as the system executes. This will require some type of high-speed memory devices usually contained in the ECL (emitter coupled logic) family of logic. When using a logic state analyzer, you must know the maximum speed at which the instrument will capture data. If your system changes states more rapidly than this, the logic state analyzer will not be useful for your particular system.

You can see that the size of the internal memory determines how much data we can store. For example, the bit storage required for the 256×32 memory we are discussing is $256 \times 32 = 8192$ bits of data. We mention this fact about internal memory because we must be aware of how much data can be captured in real time by the logic state analyzer before it reaches the memory storage limits.

Writing Data to the Internal Memory

We have now discussed the physical inputs to the system and show how large a memory is required. We next discuss the concept of writing data data into the internal memory. First of all, we know that the internal memory must have a write pulse applied in order to store data. This means that the logic state analyzer must have some physical means of inputting a write pulse to its internal memory.

The write-enable pulse for the logic state analyzer has a number of functions. Its first function is to write the data into internal memory. Remember that the internal memory must have the data present before a write-enable

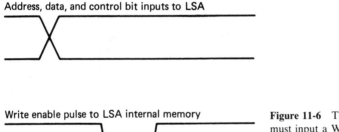

Address, data, and control bit inputs to LSA

Write enable pulse to LSA internal memory

Figure 11-6 The system under test must input a WE pulse to the logic state analyzer to qualify the data input to the logic state analyzer's internal memory.

pulse is applied. This means that the write-enable pulse for the logic state analyzer must not occur until the logical information levels on the address bus, data bus, and control bus are all stable. This is shown in Figure 11-6.

In effect, we see that the write-enable pulse for the logic state analyzer is the *qualifier* for the data input. That is, we are qualifying the events we wish to examine with the write-enable input. This is the second job of the write-enable pulse. Finally, write-enable pulse works in conjunction with some other logic signals to trigger the logic state analyzer into the data capture mode.

Triggering the Logic State Analyzer

By *triggering* the logic state analyzer we mean starting the sequence of events by which the device begins to input data. We have discussed the fact that the logic state analyzer can record only a finite number of system states in real time. In our logic state analyzer we can record up to 256. After the 256th state is recorded, the 257th state is lost because we have exhausted the memory-storage locations. All states past the 256th are lost also. This means we must be selective about what states of the system we save or record into the internal memory. If your system is executing thousands of instructions each second, it will not require very much time to fill up the recording space within the internal memory of the logic state analyzer. So we do not want to record everything. We may want to wait until a certain address in memory is accessed before we start recording.

For example, if our program starts at address locations 0000, we may wish to wait until the system reaches address 00F3 before we start recording. Most logic state analyzers provide a means to select a predetermined address locations for the system to compare to. When this address is obtained, the logic state analyzer is set into the recording mode. The predetermined address is sometimes referred to as the *trigger word*. The write-enable pulse works in conjunction with the trigger word to start the logic state analyzer into action.

Figure 11-7 shows a block diagram of a logic state analyzer with all of the features we have discussed thus far.

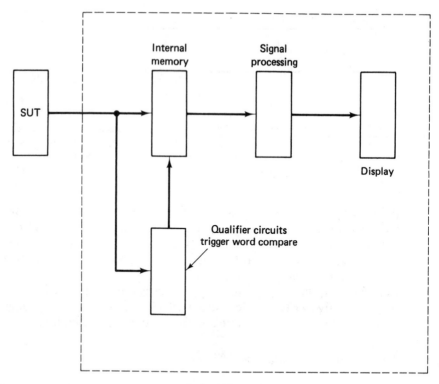

Figure 11-7   Block diagram showing a general logic state analyzer.

Delayed Trigger of the Logic State Analyzer

In addition to a trigger word to a trigger the logic state analyzer, some units provide the option of delaying the start of the recording until a specified length of time after the trigger word has been input. Suppose we use a 25 for the delayed trigger; this means that 25 write enables after the trigger word is input, the logic state analyzer begins to record data.

Output Display for the Logic State Analyzer

We now discuss, in general terms, how the logic state analyzer can display the stored data. Again, we will present a variety of ways in which the data may be displayed. The display modes shown are common to a majority of logic state analyzers used in industry today.

Let us examine a block diagram of the data storage and data retrieval mechanism that could be employed in a logic state analyzer. This block diagram is shown in Figure 11-8. As we examine Figure 11-8, we see that the data is stored into the internal memory via path A. The data is read and displayed via path B. Display of the data, we recall, is in a nonreal-time mode.

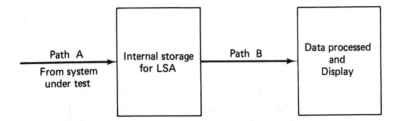

Figure 11-8 Block diagram showing the path for data storage and display in a logic state analyzer.

After we retrieve the data from memory, the question is how we wish to display it. That is, how should the data appear to the user? Let us assume that we will be displaying the data on a video monitor, so the display data will show up in some form on the face of a cathode-ray tube. We want to output the data we have stored in a way that is easily understood by the user.

One form that the output data may take is shown in Figure 11-9a. Here, we simply show the logical state of all bits in all buses. That is, the logical state

```
          Address Output                        Data

msb                            lsb    msb                 lsb

0 0 0 0  0 0 0 1  0 0 0 0  0 0 0 0    1 1 0 1  1 0 1 1
0 0 0 0  0 0 0 1  0 0 0 0  0 0 0 1    0 1 1 0  0 0 1 0
0 0 0 0  0 0 0 1  0 0 0 0  0 0 1 0    1 1 0 1  0 0 1 1
0 0 0 0  0 0 0 1  0 0 0 0  0 0 1 1    0 1 1 0  0 0 0 0
0 0 0 0  0 0 0 1  0 0 0 0  0 1 0 0    1 1 0 0  0 0 1 1
0 0 0 0  0 0 0 1  0 0 0 0  0 1 0 1    0 0 0 0  0 0 0 0
0 0 0 0  0 0 0 1  0 0 0 0  0 1 1 0    0 0 0 0  0 0 0 1
0 0 0 0  0 0 0 1  0 0 0 0  0 0 0 0    1 1 0 1  1 0 1 1
```

(a)

```
          Address                        Data

          0100                           DB
          0101                           62
          0102                           D3
          0103                           60
          0104                           C3
          0105                           00
          0106                           01
          0100                           DB
```

(b)

Figure 11-9 (a) Display format where each bit of the internal memory is written as a logical 1 or logical 0; (b) more compact display format where each group of 4 bits is converted to a hexadecimal display number.

of each bus line is displayed to the viewer by the logic state analyzer writing a 1 or a 0 to the CRT. Notice also that each line of the display represents a different point in time. The first, or top, line represents the logical state that each CPU bus line was in at the time the trigger word was accepted by the logic state analyzer.

Note, too, in Figure 11-9a that the 1s and 0s displayed are grouped so that they represent the system buses that were monitored. This is one form of logic state analyzer display that some instruments use.

One disadvantage of the display format of Figure 11-9a is that it is not compact. That is, exactly the same information could occupy less display space if it were presented in a different form. A possibility for this form would be to represent that 1s and 0s in a different number system, such as a hexadecimal or octal format. This is illustrated in Figure 11-9b. Each line of display has the same meaning as before. It is simply in a different form.

One other possible output display is in the form of a timing diagram, as shown in Figure 11-10. Each signal displayed will have the logical 1 and 0 show up as a level of the signal in the diagram. The tic marks on the bottom of the x axis show the times when the system was strobing data into the internal memory. This is also the minimum resolution of the waveform. There may have been a "glitch" that occurred between write enables to the internal memory. The timing diagram does not show these glitches on the screen. Further, it would be confusing to display all the input signals as timing diagrams. Therefore, logic state analyzers

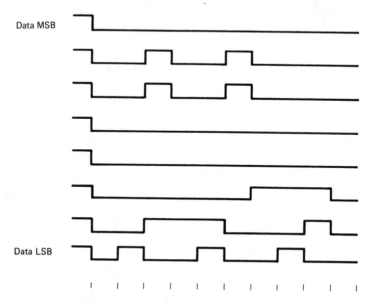

Figure 11-10 Another form the output of a logical state analyzer can take is timing diagrams. This shows each bit of the data field as a separately timed signal.

will usually allow you to select which signals you desire for output display. However, this flexibility varies from machine to machine.

If you are going to use a logic state analyzer extensively, then the type of output display may be a feature that should be investigated further. It is always nice to have the logic state analyzer output data in a manner that is comfortable for the user.

Summary of Logic State Analyzer Hardware

In summary of the logic state analyzer hardware, we refer back to the block diagram of Figure 11-7. These major hardware blocks are common to many instruments. Figure 11-7 also shows the inputs, which are the physical paths on which data are entered into the logic state analyzer. Next, you notice the internal memory of the analyzer. The physical size of this memory determines how much data the logic state analyzer is able to record at any given time. Another major input for the logic state analyzer is the trigger word and the write pulse, or qualifier, for the internal memory. Finally, we see the display section of the logic state analyzer. It is the function of the display section to format the data in the internal memory in a form easy for the user to interpret.

11-4 APPLYING THE LOGIC STATE ANALYZER

We next present various applications for the logic state analyzer. These applications are designed to make you aware of different types of uses for the logic state analyzer. The applications given are by no means the only ones. The intent is to show a variety of applications that can lead you into thinking of different possible applications of the logic state analyzer.

In our first example, we will use the logic state analyzer to determine if our program flow is correct. That is, we want to determine if we are reading the correct data from memory and if the microprocessor is executing the code correctly. This application will use an 8080 microprocessor executing the program shown in Figure 11-11. Here we have a simple program that will input the data from the input port and output this data to an output port. The program then jumps to location 100H again.

To determine if the system is executing the program correctly, we must monitor the address bus. The logical condition of this bus will show how the system is executing the program. The question is, What should we expect the

```
              ORG 100H
       START  IN 62H
              OUT 60H
              JMP START
              END         Figure 11-11
```

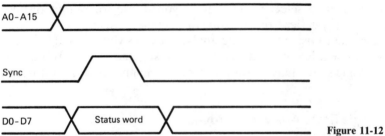

A0–A15

Sync

D0–D7   Status word

Figure 11-12

logic state analyzer to display? To answer this we must examine the program in detail. We will set the trigger word to be address 100H. The write-enable pulse to the logic state analyzer will be the sync output of the 8080.

The reason we chose the sync output of the 8080 is that the pulse is valid each time a new address is presented to the system address bus. This allows us to strobe in the addresses as they change while the system is executing. The timing for this operation is shown in Figure 11-12.

We now have a means to monitor the events of the address bus as they occur within the system. When the 8080 generates address 0100, the logic state analyzer starts to capture data. The logical conditions for the system address bus are shown in Figure 11-13a. The information shown in Figure 11-13 is exactly the same data that the logic state analyzer should store.

Let us examine Figure 11-13 and determine if the addresses shown are correct. The first line of Figure 11-13 shows the address bus equal to 0100H.

Address Output				Data	
msb			lsb	msb	lsb
0 0 0 0	0 0 0 1	0 0 0 0	0 0 0 0	1 0 1 0	0 0 1 0
0 0 0 0	0 0 0 1	0 0 0 0	0 0 0 1	1 0 0 0	0 0 1 0
0 1 1 0	0 0 1 0	0 1 1 0	0 0 1 0	0 1 0 0	0 0 1 0
0 0 0 0	0 0 0 1	0 0 0 0	0 0 1 0	1 0 1 0	0 0 1 0
0 0 0 0	0 0 0 1	0 0 0 0	0 0 1 1	1 0 0 0	0 0 1 0
0 1 1 0	0 0 0 0	0 1 1 0	0 0 0 0	0 0 0 1	0 0 0 0
0 0 0 0	0 0 0 1	0 0 0 0	0 1 0 0	1 0 1 0	0 0 1 0
0 0 0 0	0 0 0 1	0 0 0 0	0 1 0 1	1 0 0 0	0 0 1 0
0 0 0 0	0 0 0 1	0 0 0 0	0 1 1 0	1 0 0 0	0 0 1 0
0 0 0 0	0 0 0 1	0 0 0 0	0 0 0 0	1 0 1 0	0 0 1 0

Figure 11-13 LSA output showing the address and data (status word) that are strobed by the sync output from the 8080.

This is the address for the op-code for the instruction IN 62H. The data displayed is not the data which is being read from memory. Rather, this data is the status word being output by the 8080. The status word is the data that is strobed because the sync output line is used as the write input to the logic state analyzer.

Next, the address is equal to 0101H. The 8080 is reading the address of the input port from memory. All the data for the instruction has been read into the CPU, and the instruction will now be executed. The third line of Figure 11-13 shows the address 6262H. This is the input port address being output on the system address bus. All addresses shown in the first three lines are for the execution of the input instruction.

The fourth line of Figure 11-13 shows the address 102H. This is the address of the next instruction in the program, namely, the OUT 60H instruction. In the fifth line of Figure 11-13 the address is equal to 103H. The CPU is reading the output port address from memory at this time. With the data from addresses 102H and 103H in memory, the CPU will now execute the instruction. This is shown by the sixth line of Figure 11-13, which shows an address of 6060H.

The next address of Figure 11-13 is equal to 104H. This is the address of the JMP 0100H op-code. The CPU will next output addresses 105H and 106H to obtain the jump target address value for the jump. Address 0106H is the final address output for a single pass through the program. The next address will be 0100H, which is the start of the program loop.

If the LSA displays the addresses as described, then you know that the system is executing the program correctly. If at any time the address output does not agree with what you know should be there, then you must investigate the problem further. For example, if in line 3 of the LSA display, the address is 0081H instead of 0062H, you know that the information being read from memory is incorrect. You could set up the LSA differently to record the data being read from memory instead of the addresses output by the CPU. How to accomplish this will be shown in a later example.

11-5 EXAMINING DATA WRITTEN TO MEMORY

In Section 11-4 we showed how the logical state analyzer can be used to monitor program flow by recording the logic conditions of the system address bus as the CPU is executing a program. We now present an application in which we examine data being written to memory. This type of application is useful if you suspect that an incorrect data byte or sequence of data bytes is being transferred to memory from the CPU.

To approach this problem, we must review the electrical events in the system that occur during a memory-write operation. In brief these are the events:

1. The address is placed on the address bus.
2. The data is placed on the data bus.
3. The memory-write enable is asserted.

Therefore, if we are to examine the data that is being written into the memory, we must know two facts:

1. To what address location is the data being written?
2. When (in time) is the data valid on the data bus?

We need to know the exact address location to examine because the program may be writing data to many different locations. We want to examine only what is written to a specific location. This means that we must know the precise system address where the data in which we are interested is supposed to be written.

The second fact for concern is when, exactly, is the data valid on the system buses? We need to know this because the LSA requires a strobe for the data at the correct instant in time. To capture the data in a memory-write operation, we use the memory-write enable signal of the system. In this example we trigger off the WE input to the memory that is enabled. The physical origin of the strobe input to the logic state analyzer varies from system to system and operation to operation. It is up to you as a hardware troubleshooter to decide which signals in the system will be the best to use for your purpose. A good understanding of how the CPU communicates within a system environment will aid in making this decision. The information given in the previous chapters on memory and I/O are very good sources for achieving a good understanding of how the hardware of a system operates.

Let us assume that the 8080 is executing the program shown in Figure 11-14. In this program we see that the 8080 will write to memory location 00F2H and output to ports 60H and 61H. The data written will be 55 to memory location F2H.

Figure 11-15 shows the LSA output for examining data written to memory location 00F2H. We have set the address to trigger on the value 00F2H and the $\overline{\text{MEMW}}$ control line to trigger the logic state analyzer. Note that this data was taken on a system which used static RAM and a $\overline{\text{MEMW}}$ control line. Each line

```
 7   0100                         ORG  100H
 8   0100   3E AA       BEGIN     MVI  A,0AAH
 9   0102   D3 60                 OUT  60H      ;A TO PORT 60H
10   0104   2F                    CMA           ;COMPLEMENT A
11   0105   D3 61                 OUT  61H      ;A TO PORT 61H
12   0107   32 F2 00              STA  00F2H    ;A TO ADDRESS 00F2H
13   010A   C3 00 01              JMP  BEGIN
14   010D                         END
```

Figure 11-14

Address Output				Data	
msb			lsb	msb	lsb
0 0 0 0	0 0 0 0	1 1 1 1	0 0 1 0	0 1 0 1	0 1 0 1
0 0 0 0	0 0 0 0	1 1 1 1	0 0 1 0	0 1 0 1	0 1 0 1
0 0 0 0	0 0 0 0	1 1 1 1	0 0 1 0	0 1 0 1	0 1 0 1
0 0 0 0	0 0 0 0	1 1 1 1	0 0 1 0	0 1 0 1	0 1 0 1
0 0 0 0	0 0 0 0	1 1 1 1	0 0 1 0	0 1 0 1	0 1 0 1
0 0 0 0	0 0 0 0	1 1 1 1	0 0 1 0	0 1 0 1	0 1 0 1
0 0 0 0	0 0 0 0	1 1 1 1	0 0 1 0	0 1 0 1	0 1 0 1
0 0 0 0	0 0 0 0	1 1 1 1	0 0 1 0	0 1 0 1	0 1 0 1

Figure 11-15   LSA output showing the address and data that are strobed by the MEMW output from the 8080 system. The program being executed is shown in Figure 10-15.

of Figure 11-15 shows the address and the corresponding data. Using this type of setup will enable you to see exactly what data the CPU is writing to any memory location. Further, this data is obtained at system speed.

If the data is not correct, then you know that either the program flow is in error or the system hardware is defective. If the system once worked and is now malfunctioning, the hardware is most likely suspect.

11-6 EXAMINING DATA WRITTEN TO I/O

Let us now use the logic state analyzer to examine data written to an output port. To accomplish this, we must know the address of the output port. We must also know when data will be valid on the data bus. These facts are the same as the facts needed to capture data for the memory-write operation just discussed.

To get started, in the examination of data written to an output port, the logic state analyzer will need to trigger on the port address. Next, the strobe input signal input to the logic state analyzer will be the $\overline{\text{IOW}}$ signal. In the system we are using for this example, we use the port-write strobe signal. If you are unfamiliar with the port-write strobe, please refer to Chapter 8.

The program we use as an example is the same one as for the memory-write example, only the addresses set up on the logic state analyzer are different. This program was shown in Figure 11-14. In this figure we see that the 8080 will write data AAH to output port 60H and write data 55H to output to port 61H. Therefore, we trigger the logic state analyzer at the first port address 60H. The strobe input is the $\overline{\text{IOW}}$, as previously stated. Figure 11-16 shows the logic state analyzer output for recording data in this mode.

Address Output				Data	
msb			lsb	msb	lsb
0 1 1 0	0 0 0 0	0 1 1 0	0 0 0 0	1 0 1 0	1 0 1 0
0 1 1 0	0 0 0 1	0 1 1 0	0 0 0 1	0 1 0 1	0 1 0 1
0 1 1 0	0 0 0 0	0 1 1 0	0 0 0 0	1 0 1 0	1 0 1 0
0 1 1 0	0 0 0 1	0 1 1 0	0 0 0 1	0 1 0 1	0 1 0 1
0 1 1 0	0 0 0 0	0 1 1 0	0 0 0 0	1 0 1 0	1 0 1 0
0 1 1 0	0 0 0 1	0 1 1 0	0 0 0 1	0 1 0 1	0 1 0 1
0 1 1 0	0 0 0 0	0 1 1 0	0 0 0 0	1 0 1 0	1 0 1 0
0 1 1 0	0 0 0 1	0 1 1 0	0 0 0 1	0 1 0 1	0 1 0 1

Figure 11-16   LSA output showing the address and data which is strobed by the IOW output form the 8080. The program being run is shown in Figure 11-15.

Notice in Figure 11-16 that each line shows the address of the port and the data being written to the port. By using the LSA in this manner you can determine if the system is sending the correct data to any output port in the system. If the data is not correct then you must investigate further.

No Display

One point to mention at this time about using a logic state analyzer is this. If the instrument never receives a trigger signal, then the recording of data will not take place. When this occurs the display will remain blank. Some logic state analyzers will display the message "waiting for trigger." If this occurs, then you can be certain that something is wrong with the system.

The problem could be that the system address is never equal to the port address. This fact could indicate that the CPU is reading the incorrect port address from memory. Another possibility that would cause a no-trigger condition is if the $\overline{\text{IOW}}$ were never asserted.

When using a logic state analyzer and the display remains blank when you know something should be displayed, investigate the system hardware to ensure that the conditions for which the instrument is waiting will occur.

11-7 EXAMINING DATA READ FROM MEMORY

We next discuss how you can use the logic state analyzer to examine data that is being read from system memory. To start the discussion, let us review the sequence of events for the 8080 when reading data from memory. It should be stated that you must be familiar with the CPU for your system in order to know to which control lines to connect the logic state analyzer.

The steps for reading data from memory are as follows:

1. The address for memory is placed on the address bus.
2. The data bus is prepared to read data.
3. The memory-read signal ($\overline{\text{MEMR}}$) is asserted.

You can use these facts to set up the logic state analyzer to monitor the system buses during the execution of a memory-read operation. As an example, we will use the logic state analyzer to monitor data read from memory while executing the program shown in Figure 11-17.

In Figure 11-17, notice that there is no "read data from memory" instruction, such as

```
MOV A,address
```

But consider what happens when the CPU is executing the program. The CPU is always reading data from memory as it executes the program. The CPU must always fetch an instruction from memory. So, even though we have not specifically instructed the microprocessor system to read data from memory, the program will inherently do it.

The data that being read is the instruction data. This data may be from the object code shown in Figure 11-17. We set the trigger word of the logic state analyzer to 0100H. This is the memory location of the first instruction in memory. The strobe input to the logic state analyzer is the $\overline{\text{MEMR}}$ signal. We assume our system has a $\overline{\text{MEMR}}$ control line. In our system we will use the signal in the system that enables the data onto the system data bus from the ROM or RAM to clock the logic state analyzer. This signal must be active at the correct time. You coulld physically obtain this signal directly from the ROM chip enable if you were monitoring ROM.

The results of this type of logic state analyzer setup will give a display similar to Figure 11-18. This shows the addresses, starting with the trigger address 0100H. Adjacent to the address is the data that was present on the data bus at the time the $\overline{\text{MEMR}}$ control signal line was active.

You should be careful which edge of the $\overline{\text{MEMR}}$ control signal is used to strobe the data. Many logic state analyzers have the capability of strobing or capturing data on either the negative-going or the positive-going edge of the

```
 7    0100                          ORG 0100H
 8    0100   DB 62        AGAIN     IN 62H       ;READ DATA FROM PORT 62H
 9    0102   D3 60                  OUT 60H      ;OUTPUT DATA TO PORT 60H
10    0104   C3 00 01               JMP AGAIN    ;LOOP BACK
11    0107                          END
```

Figure 11-17

Address Output				Data	
msb			lsb	msb	lsb
0 0 0 0	0 0 0 1	0 0 0 0	0 0 0 0	1 1 0 1	1 0 1 1
0 0 0 0	0 0 0 1	0 0 0 0	0 0 0 1	0 1 1 0	0 0 1 0
0 0 0 0	0 0 0 1	0 0 0 0	0 0 1 0	1 1 0 1	0 0 1 1
0 0 0 0	0 0 0 1	0 0 0 0	0 0 1 1	0 1 1 0	0 0 0 0
0 0 0 0	0 0 0 1	0 0 0 0	0 1 0 0	1 1 0 0	0 0 1 1
0 0 0 0	0 0 0 1	0 0 0 0	0 1 0 1	0 0 0 0	0 0 0 0
0 0 0 0	0 0 0 1	0 0 0 0	0 1 1 0	0 0 0 0	0 0 0 1
0 0 0 0	0 0 0 1	0 0 0 0	0 0 0 0	1 1 0 1	1 0 1 1

Figure 11-18 LSA output showing the address and data that are strobed by the MEMR output from the 8080. Notice that the data bits are equal to the object code shown in Figure 11-18.

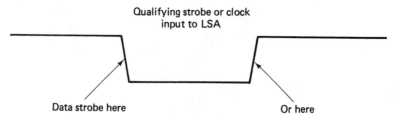

Figure 11-19 The user may select either the positive-going or the negative-going edge of the qualifier strobe as the active edge.

strobe signal. This concept is shown in Figure 11-19. In the 8080 system you should use the positive-going edge of the $\overline{\text{MEMR}}$ signal. The reason you do not want to use the negative-going edge is explained as follows: The data on the address bus will be stable; hence this bus presents no problems. However, the data on the data bus will not reflect the logical state of what is in the memory until a finite time after the $\overline{\text{MEMR}}$ signal goes to a logical 0. This is shown in Figure 11-20. Further, the data on the data bus stays valid until a finite time

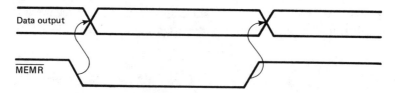

Figure 11-20 The data on the data bus will not be stable on the negative-going edge of the $\overline{\text{MEMR}}$ signal. Therefore, the user should specify the positive-going edge as the active one.

after the $\overline{\text{MEMR}}$ signal has gone away, due to the delays in the chips used in the system. For these reasons, you should use the positive-going edge of the $\overline{\text{MEMR}}$ for strobing data into the logic state analyzer. This type of reasoning must be applied to all strobe inputs to the logic state analyzer, not just the memory-read strobe.

This point about strobe edges is important when using a logic state analyzer. You should recognize when data is valid in the system and understand the details of how a particular logic state analyzer strobes data. First, can the edges be used by the logic state analyzer? Can the active edge be changed? We call your attention to these questions here so that you can ask them and find answers when using a specific logic state analyzer.

11-8 LSA SUMMARY

In the first part of this chapter we have presented the basics of LSA. The discussion has been designed to provide you with a solid understanding of what to look for. When using a logic state analyzer for the first time, knowing what to expect helps one to master the exact details quickly; of course, these details change from one logic state analyzer to another.

Also shown in this unit were various common applications for the logic state analyzer. These applications are straightforward, with many similarities in actual practice. Most LSAs are capable of these types of applications. One should understand how a general logic state analyzer performs these functions.

A system troubleshooter who has access to a logic state analyzer can quickly expand this simple set of applications to include many others. The basic operation steps for each application given permits complete understanding of how the hardware executes the major jobs of the CPU.

Let's leave the logic state analyzer and examine a different troubleshooting tool called the signature analyzer.

11-9 INTRODUCTION TO SIGNATURE ANALYSIS

In this part of the chapter we present *signature analysis,* another hardware troubleshooting technique. After defining signature analysis, we examine the essential components of a signature analysis troubleshooting tool. We do not discuss in detail any particular signature analyzer used in industry today. The objective of this presentation is to give you some insight into the general techniques and use of signature analysis rather than to provide a detailed understanding of the hardware of any one instrument that is commercially available. These insights will allow you to approach and use different signature analyzers available today and to point out what to look for when using signature analyzers. Further, this discussion gives enough information to enable you to apply signature analysis to a microprocessor-based system.

Signature analysis may be defined as a technique for compressing a large bit stream of digital data into a compact 3- or 4-digit signature. The previous sentence may sound ambigous and nebulous. So, as we proceed, we will examine what lies behind the definition, and the concept will become clear. At the end of this chapter we will again give this definition. At that point, you may accept the definition as commonplace, perhaps even trivial. The point is, you must have a feeling for the problem to be solved before you can truly appreciate how easy signature analysis is to understand.

11-10 THE PROBLEM

To illustrate the problem that is solved by signature analysis, let's use an example: Suppose that a computer system is not functioning correctly. We know this because an output device is not responding correctly.

We will assume that this output device is connected to the system by a single 8-bit port. This port comprises an 8-bit latch with an LED connected to each output line of the latch, as shown in Figure 11-21. When the system is functioning correctly, the display on the LEDs shows the binary number 11110101, or hexadecimal number F5. However, we have examined the display and find that it is showing 00000011. How can we determine where the problem is? There are many ways of doing this, but since we are discussing signature analysis, let us discuss how it might be applied to solve such a problem. We start by analyzing the general nature of the problem confronting us. One possibility for determining the hardware component that has failed is to monitor different outputs in the system and verify if they are correct.

If the outputs of one hardware block are valid and the outputs of the following hardware block are not valid, then we know the general location of the hardware fault. This is illustrated in Figure 11-22. For example, if we monitor the outputs of the system under discussion that are the inputs to the output port, we will gain meaningful information about the system. If these logic levels are valid, then we know that the problem in the system occurs after the output port. This is shown in Figure 11-23. These last few statements sound very simple. DO NOT BELIEVE IT! Monitoring the various outputs in a microprocessor system while the system is running is a difficult task to realize. Some of the many reasons why this is true are as follows:

1. Many outputs in a microprocessor controlled system are capable of being in the tristate mode. This means that you must determine the validity of a logical 1 or a logical 0 and the off state. The validity includes the voltage levels as well as the time dependence. A typical output of this nature is shown in the oscilloscope wave form display of Figure 11-24.

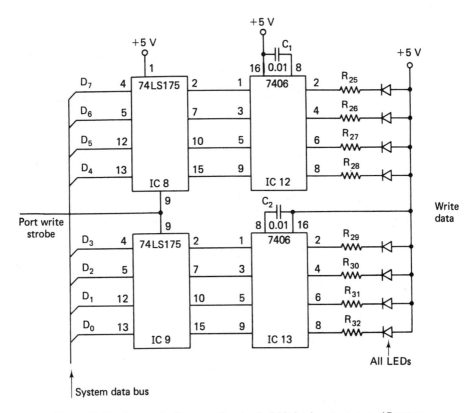

Figure 11-21   Schematic diagram of a simple 8-bit latch output port. (Courtesy
Creative Microprocessor Systems)

2. The output wave forms in a microprocessor system are usually nonperiodic. This fact makes observation with an oscilloscope extremely difficult.
   The waveform does not repeat at a high frequency rate, such as a square
   wave or clock. The question is, How long must you examine a waveform
   that is nonperiodic to determine if it is valid? The concept is shown in the
   timing diagrams of Figure 11-25a and b.

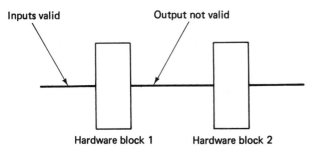

Figure 11-22   When inputs to hardware
block 1 are valid but outputs are not
valid, the problem could be either defective outputs in hardware block 1 or defective inputs to hardware block 2. Regardless of which is the real fault, the
problem has been narrowed down to a
limited number of possibilities.

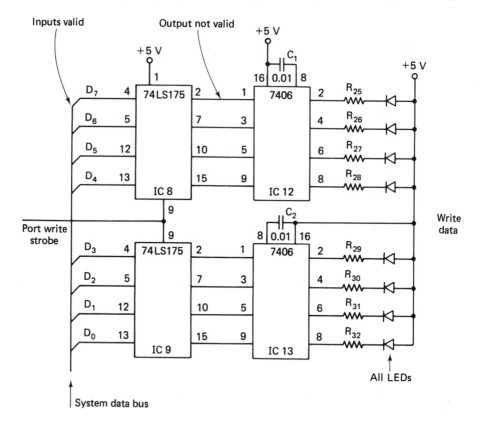

Figure 11-23   (Courtesy Creative Microprocessor Systems)

3. Many outputs have narrow output pulses that occur at very long intervals. For example, an ouput pulse may be used to strobe data into an output port. This strobing of data may occur only once or twice in an entire program. The pulse may be only 300 nanoseconds and occur two or three times in 1 second. This is shown in the timing diagram of Figure 11-26. To capture and analyze this pulse with confidence is a real challenge.

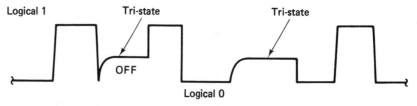

Figure 11-24   Typical output waveform of a microprocessor system signal. Notice the tristate of the waveform.

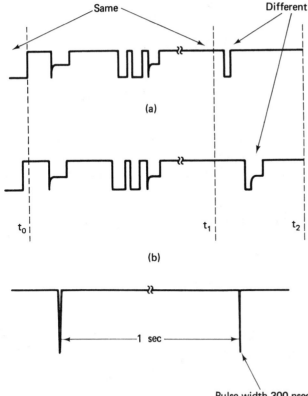

(a)

(b)

t_0 t_1 t_2

Figure 11-25   (a) Timing diagram of a system signal; (b) if the sample time is t_1, the waveforms are equal; if the sample time is t_2, then the waveforms are different.

Pulse width 300 nsec

Figure 11-26   Timing diagram showing a very low duty cycle strobe signal. This type of signal is extremely difficult to monitor with an oscilloscope.

These are only a few of the reasons that monitoring ouputs in a microprocessor system is a difficult job with an oscilloscope. (If you have ever tried to monitor such outputs with an oscilloscope, you may have a few reasons of your own to add to this list.)

If we examine the difficulties in monitoring signals that have the characteristics we have been discussing, one common fact stands out. The reason for the monitoring difficulty is that we are trying to monitor these signals *visually*. This is the whole idea behind the oscilloscope. It will let you visually monitor the signal action in the system. But when the signals to be observed are brief, fleeting, nonperiodic pulses that must be sorted out from the other pulses in the system, the difficulties of capture, observation, and analysis often become severe.

11-11 THE SOLUTION

If we pause for a moment to think about it, why do we really need to see what the output waveforms are? The real-world objective of visual monitoring of these signals is to enable the user to make a go–no go decision from what is

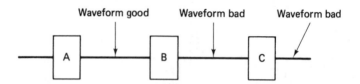

Figure 11-27 Block diagram showing how one can work backward through a system to find the first location where the waveform is valid.

seen. That is, if the waveform is valid we simply acknowledge the fact. But if the waveform is not correct, we must move further back along the logic path to check the output waveform that is the input to the block we just checked. This is shown graphically in Figure 11-27.

When we analyze this type of troubleshooting where we check outputs and then either move forward or backward in the logic path based on the validity of the waveform that we observed, we may state the following as a basic trouble-shooter's maxim:

> We do not care what the actual timing or voltages of the waveforms are; we simply care if the timing and voltages are what they are supposed to be.

This may sound like double talk. But it is true. If we had some way other than our oscilloscope of determining if the waveforms we were monitoring were valid, we would not care what the actual waveforms were. Of course you may ask the question, How will we know if the waveform is valid or not this way? The answer to this very basic question will be delayed until later in this presentation; what we wish you to accept at this point is the discussion is this:

> If we have some way of determining if a waveform is valid on a yes (valid) or no (not valid) basis, then we do not need to monitor the waveform visually with an oscilloscope. The signature analyzer is an instrument that gives us a means to do this.

11-12 COMPRESSING A BIT STREAM

Now that we have an idea of what we would like the signature analyzer to do, let us proceed with the discussion. We will introduce some new concepts that are realized in signature analysis. The first new concept we will discuss is that of compressing a bit stream.

This concept has two major points that must be explained if our discussion is to be useful:

1. The bit stream: What is it?
2. Compressing: How is it defined?

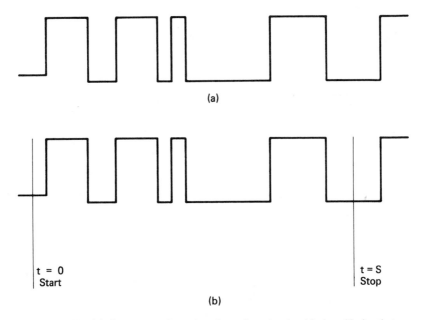

(a)

t = 0
Start

t = S
Stop

(b)

Figure 11-28 (a) Output waveform changing voltage levels with time. Notice that the waveform is nonperiodic; (b) waveform with time restrictions. We observe the waveform between $t = 0$ and $t = S$, and we do not care about the waveform before $t = 0$ or after $t = S$.

Let us first discuss what we mean by a bit stream. If we consider a digital output in a system when the system is running, the output within the system will probably be changing with time. This is shown in Figure 11-28a. If we put some restrictions on Figure 11-28a, this output can be defined even further.

The first restriction is that the time dependence of this output must be relative to some starting point in time. This means that we must provide a time equal to 0. Next, we let this time dependence go from time = 0 to some ending time. We call this time S (for stop). Now we have an output that must have a certain shape or waveform from time = 0 to time = S. This is shown in Figure 11-28b.

Next, we divide the time interval shown in Figure 11-28b from time = 0 to time = S into smaller, equal time units. This is shown in Figure 11-29. Now we have a waveform that can be divided into still smaller units of time between starting time 0 and ending time S. To relate this to already familiar equipment, we might think of Figure 11-29 as a display on an oscilloscope and time = S as the right side of the screen on the oscilloscope. The smaller units of time are analogous to the time grid on the screen of the oscilloscope.

Now we further divide the time span by a *qualifying* strobe. This means that within the small time unit, the qualifying strobe will be present at the same time. This is shown in Figure 11-29.

The purpose of the qualifying strobe is the following. The instant the

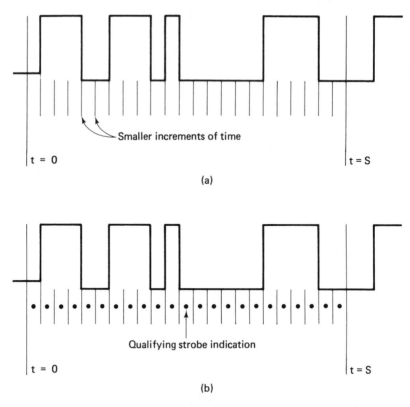

Figure 11-29 Waveform of Figure 12-8 divided into small time increments between $t = 0$ and $t = S$; (b) waveform of Figure 12-8 with a qualifying strobe between each small time increment.

qualifying strobe goes to a logical 1 level, the logical level of the waveform is sampled. This means that if the waveform were a logical 1 state at the moment of sampling, then a logical 1 would be stored in the signature analyzer as the value of the waveform for that single strobe. Figure 11-30a shows the resulting logical 1s and 0s that would be stored by sampling a waveform.

The resulting series of logical 1s and 0s shown in Figure 11-30 may be referred to as a *bit stream*. We see that a bit stream is simply a group of time-dependent logical 1s and 0s. If we were saving this bit stream as the bits were occurring, the flow of the bits into some storage medium would be as shown in Figure 11-30.

In this figure we see that the stream originates with the bit generated at time = 0. The stream ends with the bit generated at time = S. This is consistent because the bits generated at time = 0 occur first chronologically. Therefore, they must be the first ones that flow into storage.

From the previous discussion of a bit stream, it is clear that the stream may

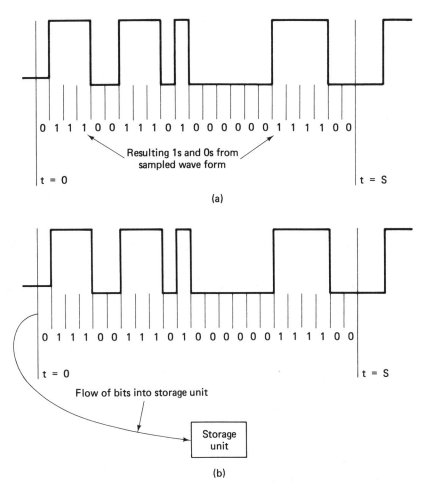

Figure 11-30   (a) Resulting 1s and 0s obtained at the time of the qualifying strobe; (b) the bit stream would be input to the storage unit from bit at $t = 0$ to bit at $t = S$.

be anywhere from a few bits to many millions of bits in length. One way to examine the bit stream would be one bit at a time—that is, to store the logical 0 or logical 1 value of each bit in the stream and then compare these stored values against known good values. Each bit in the stream should have a certain predetermined value. This 0 or 1 value is what we check for when we measure logical voltage levels with a DC voltmeter when SST is used. Here, we store all the predetermined values in memory and compare each bit electrically as it is generated in the stream, as shown in Figure 11-31.

If we examine this technique, a theoretical solution to our initial problem appears. We could store and check all bits of the program to be executed, but this is an impractical solution. To store each bit of the stream in a memory

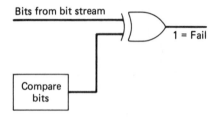

Figure 11-31 One way to compare a bit stream; each bit of the stream is compared against a reference stream stored in memory.

might quite easily run the memory size into megabits. Let's take an example to show exactly what we mean. We will store each bit of a bit stream. The waveform to be stored will be measured from $t = 0$ to $t = S$, with the time interval equal to 1 second. We will strobe the wave form every 100 nanoseconds and store the resulting logical 1s and 0s. This means we will have

$$\frac{1 \text{ second}}{100 \text{ nanoseconds}} = .01 \times 10^9 \quad \text{bits}$$

in the stream, which equals 10^7, or 10,000,000, bits for this one waveform!

In a normal system we would have many waveforms to examine, say 200. The total storage for this type of testing would be;

$$200 \times 10,000,000 = 2,000,000,000 \quad \text{bits}$$

Although memory technology has progressed greatly, total storage would still require a significant amount of memory. So, for practical reasons, this is not a valid solution to the problem.

A Practical Solution

Another technique—elegant in its simplicity—is used instead of the preceding technique to accommodate the storage or representation of a bit stream. The technique has many options, so keep in mind that what we are discussing in general terms is the basis for all.

A shift register can be used to store the bit stream. The shift register need only be 16 bits wide! This is quite amazing. We will store a bit stream, or its representation, in a 16-bit shift register. As you might have guessed, the shift register is realized in very special way. A block diagram of this shift register is shown in Figure 11-32.

In this figure, we see that the 16 bits of the shift register are the circuit outputs. The clock for the register is the strobe we previously discussed. The truly different thing about this shift register is the use of feedback. Different outputs of the shift register are combined logically with the logical state of the bit stream input. Also, the outputs are combined together logically to form the inputs to additional stages of the shift register.

We are purposely avoiding the topic of how the outputs are combined logically. The mathematics is rigorous. The logical combining of the shift regis-

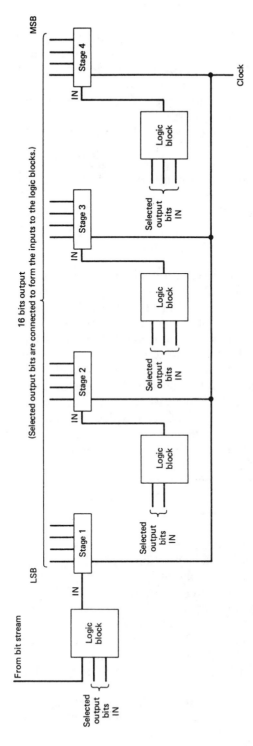

Figure 11-32 Block diagram of a special 16-bit shift register used for storing the bit stream.

ter outputs is calculated so that there is little chance of two different bit streams generating the same 16-bit output of the shift register. The bit stream can be many millions of bits long; yet the shift register can still generate a 16-bit word that is a unique representation of that single bit stream. Another bit stream may differ in only one bit, and yet the 16-bit word representation is vastly different from the previous 16-bit word.

Using the shift register with the special feedback incorporated, it is now possible to represent a very long (megabit) bit stream with only 16 bits. It is to this megabit number—reduced to one unique 16-bit word—that we refer when we speak of compressing the bit stream. In truth, we are making the data field smaller.

11-13 SIGNATURE ANALYSIS

The 16-bit word representation of the bit stream is often referred to as a *signature*. What we are doing is analyzing the signatures obtained from the bit stream and determining if they are correct. If a signature is correct, we know the bit stream was exactly what it was supposed to be on every bit position. When this is the case, we can move on in the troubleshooting process and check more signatures. If the signature is not correct, we can check the signatures at different points in the logical path. We have realized our goal of monitoring a digital waveform without using an oscilloscope, or viewing the waveform.

In effect, signature analysis has made the checking of a dynamic system exactly like checking a static one. We place the measurement probe on a point in the system and read the face of the instrument to check the resulting signature. This is very similar to that we do when we measure a voltage. We place our voltage probe on a point in the system and read the meter to see if the reading is what it is suppose to be.

Realities of Signature Analysis

Although we have presented the essential concepts of signature analysis, some necessary additional details must be covered. To do this, let us examine Figure 11-33, a functional block diagram of a machine to accomplish signature analysis.

Referring to Figure 11-33, we see the block labeled *input level detect*. This section of the signature analyzer must make the decision if the input voltage is a valid logical 1 or a valid logical 0. On many signature analyzers these logic thresholds are adjustable. In others the logic threshold may be set to accommodate a certain family of logic, such as TTL.

Next, look in Figure 11-33 at the three input lines labeled *strobe input*, *time* = 0, and *time* = S. These inputs are provided to the signature analyzer from the system under test. The hardware of the signature analyzer must be electrically informed about when to start and stop sampling the wave form.

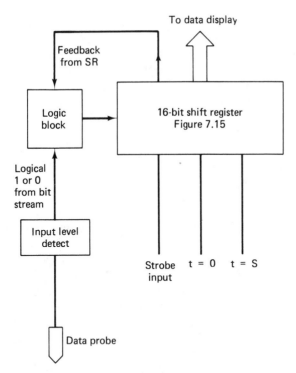

Figure 11-33 Block diagram of a machine to perform signature analysis.

Sometimes the two signals time = 0 and time = S are combined as a single signal. The sample time would be as shown in Figure 11-34.

The system under test will usually supply the strobe, start, and stop pulses for the signature analyzer. Thus at a minimum, the section of the system under test must be working before the signature analyzer can be used as a trouble-shooting tool. Many times in a microprocessor system the start, stop, and strobe inputs must be designed into the hardware when the system is being built. We mention these facts so that one is aware of them when contemplating the use of signature analysis.

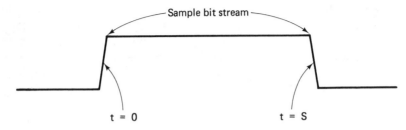

Figure 11-34 Start and stop time can be combined in a single enable strobe input to the signature analyzer.

Some signature analyzers have made provisions for systems that were not originally designed for its use. That is, the signature analyzer will provide the start, stop, and strobe inputs. However, not all signature analyzers will do this. You must be aware of the capabilities and limitations of any unit you intend to use. If provisions for using the signature analyzer on your system were not designed into the system, then you must make them before you can use the signature analyzer.

11-14 USING THE SIGNATURE ANALYZER

Initial Signatures

Throughout this discussion we have stated that we get a signature and compare it with what should be there. The question is, How do we first determine what should be there? This is usually done in the following way. A properly working system is first obtained. The signature analyzer is then attached to the working system. The signature analyzer probe is connected to the output, or node, in the system where the signature is wanted. Then, the resulting signature of the properly operating system becomes the reference signature from then on.

Note that all signatures must be obtained prior to working on a defective system. This is another point about signature analysis of which you should be aware when contemplating its use in troubleshooting. Further, if the signatures are not known, it is up to the troubleshooter to obtain them. This can, at times, constitute a problem of rather large proportions.

Using the Signature Analyzer

In systems that have been designed to use the signature analyzer, the reference signatures are usually written on the schematic, a shown in Figure 11-35. These signatures and their placements on the schematic are very similar to the voltage wave forms that are drawn on linear circuits.

Let us now return to our initial problem with the malfunctioning output port and see how a signature analyzer would help in troubleshooting. We connect our signature analyzer to the system. The system under test must provide the strobe, the qualifying pulse, and the start and stop inputs to the signature analyzer. (We assume that our system has been designed to accommodate the use of a signature analyzer.) We now apply the dynamic probe from our signature analyzer to the inputs of our malfunctioning output port.

At each different signal line, we verify that the signature obtained is equal to the reference signature. If a particular signal line has an incorrect signature, then we know that there is a problem in logic path. If all inputs to the output port have correct signatures, then we may safely assume that the problem is on the "other," or output, side of the hardware block. That is, if inputs are correct,

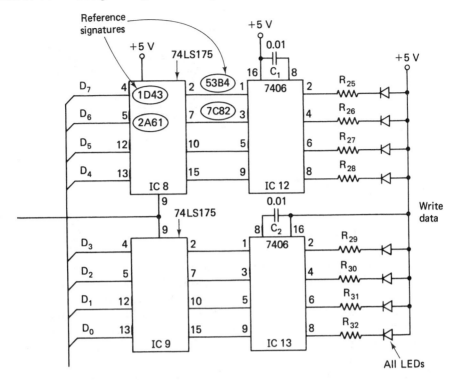

Figure 11-35 Schematic diagram showing how reference signatures are printed on a circuit for use with a signature analyzer. (Courtesy Creative Microprocessor Systems)

then the problem must be caused by some hardware not responding properly to these correct inputs.

We can see that by using signature analysis the visual examination of all digital signals is not necessary. Furthermore, signature analysis is very similar to making voltage measurements with a voltmeter. The dynamic probe is designed to be used very much like a static voltage probe on a voltmeter.

These facts are high points of signature analysis. The troubleshooting technique can be very useful to personnel with limited microprocessor background. The systems designed for use with signature analyzers have checkout procedures that require one to measure different signatures within the system. The basic premise is that the troubleshooter does not have to know what the system does or how it does it in order to repair it. The troubleshooter simply checks system outputs at various points and verifies their validity or non-validity, as the case may be.

An area in which signature analysis is being used quite extensively is the checkout of computer boards prior to being loaded into a system. These boards sometime have as many as 100 integrated circuits, (ICs), on them. Here is the

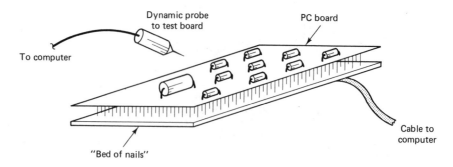

Figure 11-36   Signature analysis can be used to checkout printed circuit boards.

way signature analysis is used to check these boards. The boards are placed, one at a time, on a test system. This is shown in Figure 11-36.

The computer that runs the test system has the signature analyzer built into it. This computer also has the reference signatures stored in its internal memory. After the operator has connected the board to the test system, the computer will print a message on the system CRT that tells the operator where to place the probe on the board. Sometimes the CRT will show a graphic representation of the circuit board and blink at the spot on the board where the probe is to be placed.

Once the probe is in place, the operator presses a *foot switch*, which informs the computer that the probe is ready. The computer then runs the test and reads the signature obtained. If the signature is correct, the computer will visually indicate the next test point. If the signature is not correct, the computer will indicate the next test point, which will lead to finding the malfunction. The computer actually leads the operator through the board and, based on the signatures obtained, prints out the defective IC or pronounces the board working.

The beauty of this setup is that the boards may be checked quickly, and a highly skilled technician or engineer is not required to operate the test. Finding the problem on a large board is difficult. But the difficulty has been mastered by the person who programmed the tester. Once this has been done the operator follows the computer instructions which were written by the person who instructed the computer on the proper procedures for testing the board.

11-15 SUMMARY OF SIGNATURE ANALYSIS

In summary, signature analysis is another technique that you should understand. However, its effectiveness is very dependent on the user, as is the case with most troubleshooting techniques. It is the intent of this text to provide some insight into the strengths and limitations of the various troubleshooting techniques available for microprocessor systems. Signature analysis is an effective technique if the conditions for its effective use can be established and used properly. An effective microprocessor troubleshooter should be capable of knowing and using many different types of electronic testing equipment.

The main point of signature analysis to be stressed is a knowledge of how it works. It is also important to know what hardware in the system must be working in order to use signature analysis. The troubleshooter must also be aware of any limitations of signature analysis or special precautions to observed in making a particular check. This is true of any troubleshooting technique. Like signature analysis, every troubleshooting technique is built upon certain basic premises and has its own unique compliment of strengths and weaknesses as well as its own unique demands upon the knowledge of the troubleshooter.

REVIEW PROBLEMS

1. Write a definition of logic state analysis.
2. How does the logic state analyzer connect to the microprocessor system?
3. For the 8080 CPU how many logic analyzer inputs are required to monitor the address and data bus?
4. Explain the statement, the logic state analyzer is a passive instrument.
5. Draw a block diagram of the major parts of a logic state analyzer.
6. Why must we use a trigger input for the logic state analyzer?
7. What is the purpose of a delayed trigger on a logic state analyzer?
8. Does the logic state analyzer display data in real time? Explain your answer.
9. If you wished to use the logic state analyzer for monitoring data read from an input port, what signal would you use as a clock input?
10. How could you use the logic state analyzer to determine if the software was executing correctly?
11. What is menat by the term compressing a bit stream?
12. What is a signature?
13. What is the function of the start and stop inputs on the signature analyzer?
14. How are initial signatures obtained?
15. Why does a system need to be designed to use signature analysis as a troubleshooting technique?
16. What are the functions of the signature analyzer probe?
17. You have a system that uses signature analysis. The hardware has been modified since the initial signatures were obtained. Can you think of a problem that may occur now in using signature analysis with this modified system?

12

USING THE 8255 PIA WITH THE 8080, 8085 AND Z80

The first large scale integration (LSI) pheripheral device we will discuss in this text is the 8255 programmable interface adapter. We will examine it to see the following:

1. How it operates.
2. How it connects electrically to the CPU busses.
3. How it can be programmed to operate as a versatile I/O chip.

12-1 OVERVIEW OF THE 8255

The 8255 is a 40-pin dual-in-line package (DIP) LSI device, designed to perform a variety of interface functions in a microprocessor system environment. This device was first manufactured by Intel Corporation for use with the 8080 microprocessor. There are several manufacturers of the 8255 at the present time, and it is used in many 8- and 16-bit microprocessor-based systems.

Let's begin with an explanation of how the 8255 functions prior to its connection to the CPU buses and prior to programming. Once you are familiar with the 8255 at this basic level, you can easily explore its uses with your own and other system applications.

Figure 12-1 shows a block diagram of the 8255. Let's examine the function of each block. In this figure, we can see that there are four blocks that can connect external devices via physical lines from the PIO. These lines are labeled PA0–PA7, PB0–PB7, and PC0–PC7. The groups of signals are logically divided into three different I/O ports, labeled port A (PA), port B (PB), and port C (PC). Each port has two separate I/O blocks associated with it, and each block is connected to the 8255 internal data bus. It is via this internal data bus that information is exchanged within the 8255.

In Figure 12-1 there are two blocks, labeled group A control and group B control, that define how the three I/O ports operate. (There are several different operation modes for the 8255, and these modes must be defined by the CPU writing programming or control words to the device.) Notice that group C of the 8255 consists of two 4-bit ports. One of the groups is associated with group A and the other with group B device signals. (*Note:* The reason for this division will become clear later on. For now it is only necessary to recognize the fact.)

The final logic blocks shown in Figure 12-1 are labeled data bus buffer and read/write control logic. These blocks provide the electrical interface between the microprocessor and the 8255. The data bus buffer block buffers the data input and output lines to and from the CPU data bus. The read/write control logically routes the data to and from the correct internal registers with the right timing. The internal path being enabled depends on the type of operation being performed by the CPU; that is, it depends on whether the operation is an I/O read or an I/O write.

12-2 PINOUT DESCRIPTION OF THE 8255

In this section we examine each pin of the 8255 and discuss its function. This information will be useful later on, when we discuss how each pin connects with the microprocessor system busses. A pinout of the 8255 appears in Figure 12-1. Let's now examine each pin.

D0–D7.   These are the data input and output lines for the device. All information written to and read from the 8255 by the CPU occurs via these eight data lines.

CS (Chip-select input).   Whenever this input line is a logical 0, the microprocessor can electrically read and write to the 8255.

8255A/8255A-5
PROGRAMMABLE PERIPHERAL INTERFACE

- ■ MCS-85™ Compatible 8255A-5
- ■ 24 Programmable I/O Pins
- ■ Completely TTL Compatible
- ■ Fully Compatible with Intel Microprocessor Families
- ■ Improved Timing Characteristics

- ■ Direct Bit Set/Reset Capability Easing Control Application Interface
- ■ Reduces System Package Count
- ■ Improved DC Driving Capability
- ■ Available in EXPRESS
  - — Standard Temperature Range
  - — Extended Temperature Range
- ■ 40 Pin DIP Package or 44 Lead PLCC

  (See Intel Packaging: Order Number: 231369)

The Intel 8255A is a general purpose programmable I/O device designed for use with Intel microprocessors. It has 24 I/O pins which may be individually programmed in 2 groups of 12 and used in 3 major modes of operation. In the first mode (MODE 0), each group of 12 I/O pins may be programmed in sets of 4 to be input or output. In MODE 1, the second mode, each group may be programmed to have 8 lines of input or output. Of the remaining 4 pins, 3 are used for handshaking and interrupt control signals. The third mode of operation (MODE 2) is a bidirectional bus mode which uses 8 lines for a bidirectional bus, and 5 lines, borrowing one from the other group, for handshaking.

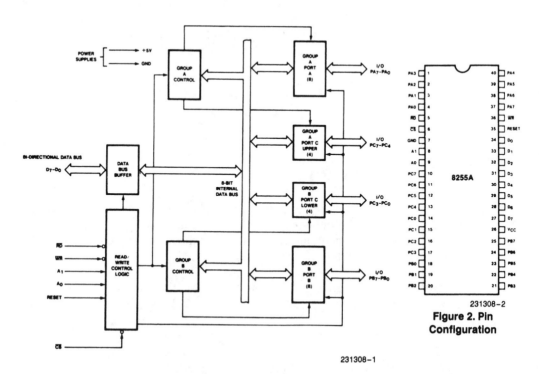

231308-1

Figure 12-1 A block diagram and device pinout of the 8255 PIA. (Courtesy of Intel Corp.)

RD̄ (Read input).   Whenver this input line is a logical 0 and the C̄S̄ input is a logical 0, the 8255 data outputs are enabled onto the system data bus.

W̄R̄ (Write input).   When this input line AND the C̄S̄ input line are both logical 0, the data inputs to the 8255 are written into the selected register.

A0–A1. (Address inputs).   The logical combination of these two input lines determines which internal register of the 8255 data is written to or read from.

Reset.   Whenever this input line is a logical 1, the 8255 is placed in its reset state. All peripheral ports are set up as input ports.

PA0–PA7.   These signal lines are used as an 8-bit I/O port that can be connected to peripheral devices.

PB0–PB7.   These signal lines are used as an 8-bit I/O port that can be connected to peripheral devices.

PC0–PC7.   These signal lines are used as an 8-bit I/O port that can be connected to peripheral devices. Also, this group of signals may be divided into two 4-bit groups and used to control the PA0–PA7 lines and the PB0–PB7 lines in certain 8255 operating modes.

Now that we are familiar with the 8255 device pins, let's use this information to connect the chip to the microprocessor system buses.

12-3 CONNECTING THE 8255 TO THE CPU

Let's now use the 8255 as an output port in a microprocessor system which is 8080-, 8085-, or Z80-based. Further, the system will have the three buses realized as discussed in Chapter 5. We will connect the 8255 to the system buses and assume that all communication will occur as I/O operations. The 8255 has two address inputs, labeled A0 and A1; this implies that the 8255 contains four unique I/O addresses. In normal use, the A0 and A1 input lines will be connected to the A0 and A1 address lines of the microprocessor system address bus. See Figure 12-2.

The C̄S̄ input line on the 8255 is used to map the device logically into the system I/O space. For example, let's suppose we assign the I/O addresses as 60H, 61H, 62H, or 63H. The schematic for realizing this decoding is shown in Figure 12-2.

In Figure 12-2 we see the decoding required to map the 8255 into an 8-bit microprocessor system. The address lines A4–A7 are used to select the chip. When these four lines are 0110 (6XH), the 8255 is selected. Address lines A0 and A1 will determine with which port of the 8255 communication will occur.

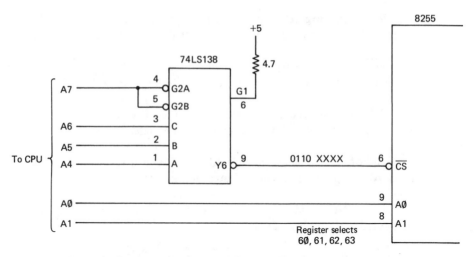

Figure 12-2 Decoding scheme for selecting the 8255. The 8255 will be selected whenever address lines A7-A4 equals 6.

In Figure 12-2 we see that address lines A2 and A3 are not used. This is true only for this example. These two lines could have been used along with A4 and A7 to map the device into the I/O space. By not using input address lines A2 and A3, we have reduced the complexity of the system hardware. However, in doing this we have given up some of the available system I/O space.

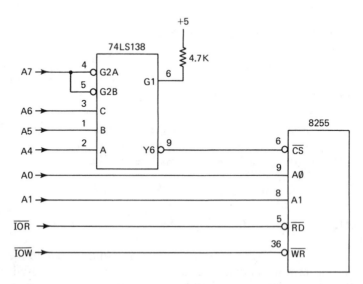

Figure 12-3 Partial schematic showing the addition of the $\overline{\text{IOR}}$ and $\overline{\text{IOW}}$ control line connected to the $\overline{\text{RD}}$ and $\overline{\text{WR}}$ inputs of the 8225.

If system I/O space is not critical in your application, then the decoding technique given in Figure 12-2 will do the job. If your application needs all available I/O space, then it will be necessary to decode address lines A2–A7 to generate the chip-select input line to the 8255 device.

The next thing we do is to connect the $\overline{RD}$ and the $\overline{WR}$ input lines to the $\overline{IOR}$ and $\overline{IOW}$ signals of the CPU system. Figure 12-3 shows the $\overline{IOR}$ and $\overline{IOW}$ control bus signals connected to the 8255.

The final control signal that connects to the 8255 is the RESET input. An important point to note about the reset input is that it is active in the logical 1 state. Reset input to the 8080 is active logical 1. For the 8085 and Z80 the reset is active logical 0. Therefore, if the 8255 is to be used with 8080, the reset input may be connected directly to the 808 reset input. If the 8255 is to be used with the 8085 or the Z80, the reset from the CPU must be inverted and then connected to the 8255 reset input.

Finally, the data lines D7–D0 of the 8255 may be connected directly to the D7–D0 system data bus lines of the CPU system. Figure 12-4 shows the complete connection of the 8255 to the three bus architecture.

12-4 THE 8255 READ AND WRITE REGISTERS

Now that we have the 8255 connected to the CPU buses, let's learn how to program the device so that it will operate in any manner you desire. We start by discussing the four internal registers (i.e., the four read and write registers) contained in the 8255. In our decoding example, the addresses of these registers were 60H, 61H, 62H, 63H. The basic register definitions are as follows:

$\overline{RD}$	$\overline{WR}$	A1	A0	REGISTER NAME
1	0	0	0	write port A data
0	1	0	0	read port A data
1	0	0	1	write port B data
0	1	0	1	read port B data
1	0	1	0	write port C data
0	1	1	0	read port C data
1	0	1	1	write control word date
0	1	1	1	illegal read register

(DEVICE PINS)

The function of registers 0–2 is defined by the word written to the control register 3. Figure 12-5 shows the bit definition of the control register. Let's now examine some of the important operating modes of the 8255. We start by discussing the three operating modes for the 8255 device and learn how it is used in each mode. As each mode is shown, examples of 8080 programs will be given to aid in understanding.

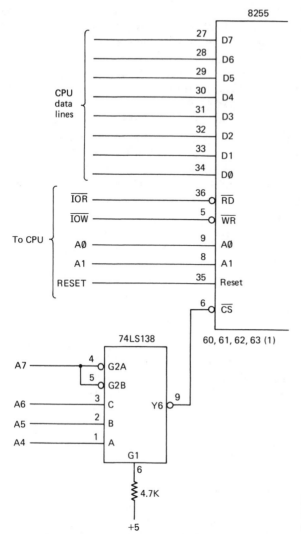

Figure 12-4 Complete schematic showing the connection of the 8255 to the CPU system bus lines.

12-5 MODE 0—BASIC REGISTER I/O

To set up mode 0 for basic register I/O, the programmer must first write a control word to the control register. This word will define how the registers are to be used in the 8255 device. The bits of the control register used for programming the standard I/O function should appear like this:

D8	D7	D6	D5	D4	D3	D2	D0
1	0	0	0	0	0	0	0

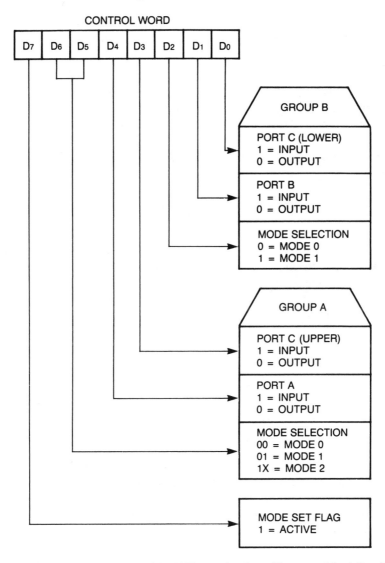

Figure 12-5   Bit definitions of the 8255 control register. (Courtesy of Intel Corp.)

If we refer to Figure 12-5, we can see the following:

Bit D7 defines this word as a control word to set the mode.

Bits D6 and D5 define the mode of operation for the 8255 port A. This is mode 0.

Bit D4 = 0 indicates that port A is an output.

Bit D3 = 0 sets the upper 4 bits of port C as outputs.

Bit D2 sets the mode for port B. A logical 0 sets mode B for port B as 0.
Bit D1 = 0 sets port B up as an output port.
Bit D0 = 0 sets the lower 4 bits of port C as an output port.

This control word written to the 8255 with an OUT instruction to port 63H defines all three ports—A, B, and C—as output ports. For the 8255 this gives 24 (three 8-bit ports) unique output lines that can be used to interface to external devices. The 8080 instructions for setting up the 8255 in mode 0 are as follows (we assume the 8255 device is connected to the CPU buses as shown in Figure 12-4):

```
MVI  A,80H      set the control word
OUT  63H        output control word to 8255
```

Once the 8255 has been programmed using these two instructions, we can write data to any output port using OUT instructions. For example, let's suppose that we wish to write 23H to port A outputs, 41H to port B, and 73H to port C outputs. To accomplish this, we would use an instruction sequence similar to the following:

```
MVI  A,23H      set up port A data
OUT  60H        output data to 8255
MVI  A,41H      set up port B data
OUT  61H        output data to 8255
MVI  A,73H      set up port C data
OUT  62H        output data to 8255
```

After these instructions have been executed, the output ports A, B, and C of the 8255 will be programmed to the specified data values. This is a convenient way for three individual output ports to be contained on a single chip.

It is possible to program the ports for a combination of input and output. For example, ports A and C could be programmed as output ports, and port B could be programmed as an input port. Referring to Figure 12-5, the control word to set up the 8255 in this configuration would be:

D7	D6	D5	D4	D3	D2	D1	D0
1	0	0	0	0	0	1	0

After this control word is programmed into register 3 of the 8255, the device will be logically set up, as shown in Figure 12-6. We could use an IN instruction to read the data input from port B:

```
IN  61H      read data from port B
```

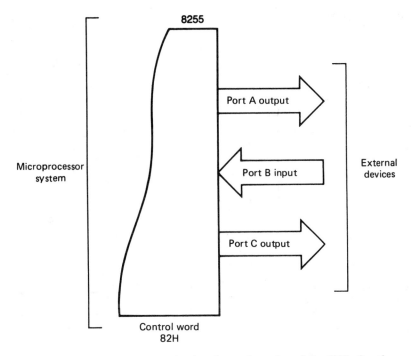

Figure 12-6 A block diagram showing the configuration of the 8255 after the control word has been written to it.

Once the 8255 has been set up in the correct configuration, it becomes an easy matter to read or write data at any device port. There are several different combinations for setting up the 8255 in mode 0 basic I/O. Figure 12-7 shows all these combinations as well as the appropriate control word for programming the 8255.

12-6 A MODE 0 EXAMPLE

As an example of using the 8080 CPU with the 8255 in mode 0, let's examine a simple keyboard interface. We will examine both the hardware and software for the keyboard. The keyboard is organized as a 4-by-4 matrix of SPST (single-pole–single-throw) switches. We will use the 8255 to interface the keyboard switches to the CPU. The hardware schematic for this example appears in Figure 12-8. Here is an overview of how the circuit and program operate.

Port B outputs are connected to the keyboard columns. Port A inputs are connected to the keyboard rows. All the row inputs are pulled up to +5 volts via the 10K-ohm pull-up resistors. The column lines are forced to a logical 0, one at a time. After each column line has been set to a logical 0, the input port

MODE CONFIGURATIONS

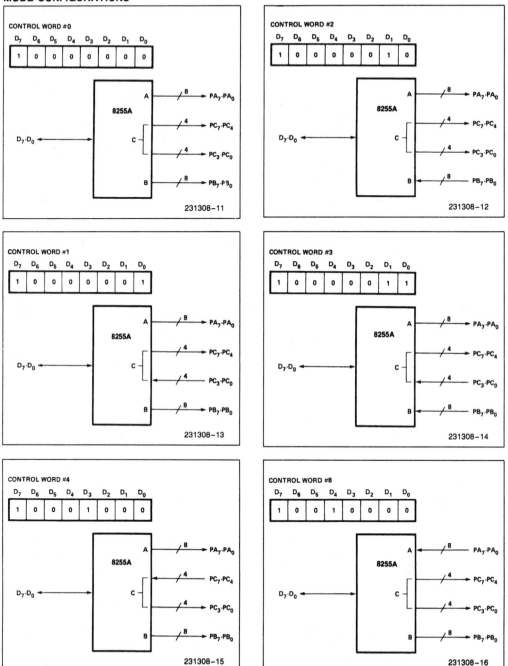

Figure 12-7   These are all of the possible configurations that 8255 can be used in with mode 0. (Courtesy of Intel Corp.)

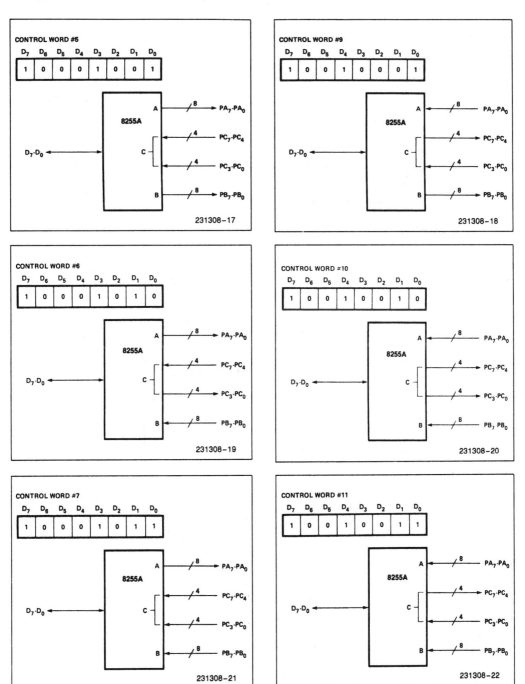

Figure 12-7 (*continued*)

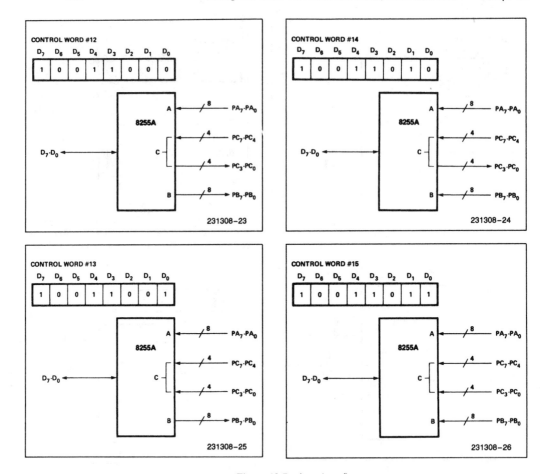

Figure 12-7 *(continued)*

A is read. If any bit read from the input port is a logical 0, then a key has been pressed. We can determine which key has been pressed by determining the bit that is a logical 0 and the column line that is active. Figure 12-9 shows a program that operates in the manner just described.

Although the previous example was for a simple 4-by-4 matrix, we can easily expand it to fit a matrix of any size. Using an 8255 is a simple way to provide the necessary hardware for the interface.

12-7 OPERATING MODE 1 FOR THE 8255

Another mode of operation for the 8255 is that of an I/O device in a handshake environment. Ports A and B are the 8-bit data ports, whereas upper bits of port C are used as handshake lines for port A and the lower 4 bits of port C are used as handshake lines for port B.

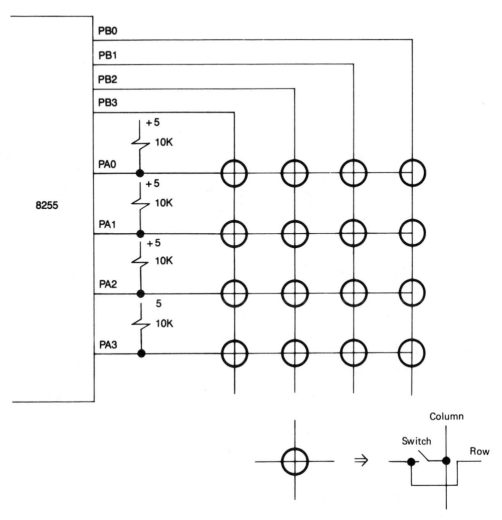

Figure 12-8 A schematic diagram showing the interconnections for a simple 4-by-4 keyboard array interfaced to an 8255.

The main idea of an I/O transfer using a handshake is this: The peripheral device electrically informs the 8255 that data is either available to be sent or ready to be taken. The handshake lines are used to provide this information. (See Figure 12-10.)

In the diagram labeled (a) in Figure 12-10, data is being sent to the 8255 from the peripheral device. Before the peripheral device writes data to the 8255, however, it must check the input buffer full flag. If this flag is true, then the data in the 8255 buffer has not be read by the CPU. That is, data has been sent to the 8255 from the external device, but the CPU has not read the data. If the flag is false, then the external device will write data to the 8255 and the

```
 7                          ;;;;;;;;;;;;;;;;;;;;;;;;;;;;;;;;
 8                          ;
 9                          ; Port B is the output port
10                          ; Port A,C are the input ports
11                          ;
12                          ;;;;;;;;;;;;;;;;;;;;;;;;;;;;;;;;
13                          ;
14        0060     PORTA    EQU  60H
15        0061     PORTB    EQU  61H
16        0062     PORTC    EQU  62H
17        0063     CONTP    EQU  63H ;CONTROL PORT
18        0099     CONWD    EQU  99H ;CONTROL WORD
19                          ;
20                          ;
21  0000  3E 99                    MVI  A,CONWD    ;CON WORD TO A
22  0002  D3 63                    OUT  CONTP      ;OUT TO CONTROL PORT
23                          ;
24  0004  06 FE     COL      MVI  B,0FEH     ;COL 0 ACTIVE
25  0006  78        COL1     MOV  A,B
26  0007  D3 61              OUT  PORTB      ;ASSERT COL LINE
27  0009  DB 60              IN   PORTA      ;READ ROWS
28  000B  2F                 CMA             ;INVERT DATA
29  000C  E6 0F              ANI  0FH        ;MASK OFF UP NIBBLE
30  000E  C2 1C 00           JNZ  KEYIN      ;IF NZ, KEY PRESSED
31  0011  78                 MOV  A,B
32  0012  07                 RLC             ;ACTIVE NEXT COL
33  0013  47                 MOV  B,A        ;SAVE VALUE
34  0014  FE EF              CPI  0EFH       ;LAST COL??
35  0016  CA 04 00           JZ   COL        ;YES START OVER
36  0019  C3 06 00           JMP  COL1       ;NEXT COL OUTPUT
37                          ;
38                          ; HANDLE A KEY PRESS
39                          ;
40  001C  4F        KEYIN    MOV  C,A        ;ROW POS IN C
41  001D  78                 MOV  A,B        ;COL POS IN A
42  001E  2F                 CMA             ;INVERT COL POS
43  001F  E6 0F              ANI  0FH        ;MASK OF UP NIBBLE
44  0021  C9                 RET             ;RETURN TO MAIN
45                          ;
46                          ; RETURN FROM SUBROUTINE
47                          ;
48                          ; RETURN WITH:
49                          ;
50                          ; A = COLUMN POSITION, 1 IN ACTIVE COL
51                          ; B = ROW POSITION, 1 IN ACTIVE ROW BIT
52                          ;
53  0022                     END
```

Figure 12-9 Program for deleting a key press on the hardware shown in Figure 12-8.

input buffer full flag will go true. When the CPU reads the data, the input buffer full flag will go false.

In the diagram labeled (b) in Figure 12-10, the 8255 will soon be sending data to the peripheral device. Before the 8255 can send the data, however, it must first check to see if the output buffer full flag is set. This flag is a signal to the external device that the 8255 has data ready for the external device to take. When the external device strobes the data out of the 8255, the output buffer full flag goes false, thus indicating that the external device has taken the data. At

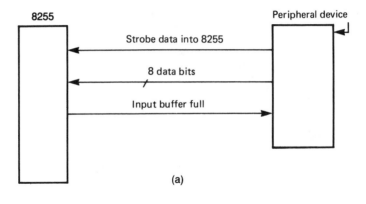

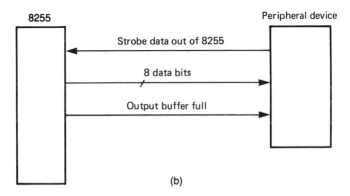

Figure 12-10   A block diagram showing how handshake lines can be used between the 8255 and an external peripheral device.

this point, the CPU can send more data to the output buffer for the external device to take.

The technique of handshaking data from one device to the next is very useful in applications where the external device is slower than the system microprocessor. With this technique, the microprocessor can put data in the output buffer and then perform other tasks. When the slow external device has taken the data, the microprocessor can send more data. Let's now go over the details of how the 8255 can be made to work in the handshake environment.

Let's assume that the 8255 is sending data to the external device from port A and receiving the data at port B. This order could be reversed in that port A could be receiving the data and port B could be sending. Another alternative is that both ports could be receiving or both sending. The main idea is that both port A and port B can be used to transfer data in the handshake mode as inputs or outputs. Furthermore, each port is independent of the other. You need only modify the control word to set up the 8255 for the proper register operation.

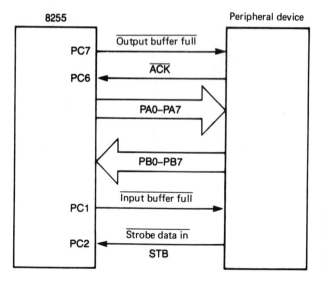

Figure 12-11 An 8255 configuration that will output data from port A lines to the peripheral device and input data to port B lines from the peripheral device.

The block diagram displayed in Figure 12-11 shows the configuration we desire for this application. To place the 8255 in this configuration, the correct control word for this mode must be written to the device prior to its use. The control word for this configuration would be 10100110, or A6H. This word is derived from the proper control bits shown in Figure 12-5 being set.

In Figure 12-11, data output is on the 8255 pins PA0–PA7; the output buffer full is on PC7; the acknowledge from the external device is on PC6; data input is on PB0–PB7; the input buffer full is on PC1; and strobe data into 8255 is on PC2. Figure 12-12 shows the port C pin definitions for mode 1 input and output.

Now that we have defined the hardware, let's examine the software that allows the 8255 to operate. For this discussion, we will assume that there will be no interrupts. The CPU will poll the status lines of the 8255 to check whether data has been received from the input port or taken from the output port.

	Out	In
PC0	$INTR_B$	$INTR_B$
PC1	IBF_B	$\overline{OBF}_B$
PC2	$\overline{STB}_B$	$\overline{ACK}_B$
PC3	$INTR_A$	$INTR_A$
PC4	$\overline{STB}_A$	I/O
PC5	IBF_A	I/O
PC6	I/O	$\overline{ACK}_A$
PC7	I/O	$\overline{OBF}_A$

Figure 12-12 Port C pin definitions for mode 1 input and output.

Further, we will assume that the external device is electrically capable of taking the data and sending it to the 8255. It is important to note that the 8255 provides only one-half of the solution. The other half is that the external device must be made to communicate electrically in a manner consistent with the 8255. Figures 12-13a and b shows the timing for a typical transfer with the 8255 in the configuration required.

To send data to the external device, the 8255 must first examine the output buffer full line, PC7, for a logical 1. This is done by reading the port C data using an input instruction. If port C, bit D7, is a logical 1, then the external

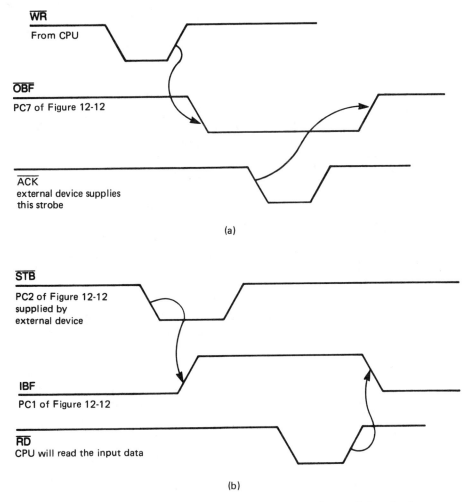

(a)

(b)

Figure 12-13   A timing diagram for a typical transfer on the port B lines for the 8255 configurations shown in Figure 12-11.

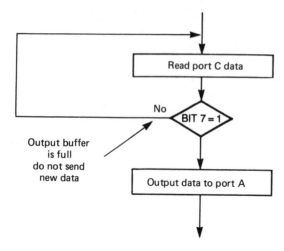

Figure 12-14 A flowchart showing the sequence of events required to transfer data to the external peripheral device using the port A lines and upper port A lines of the 8255.

device has taken the data that was present in the output buffer. If bit D7 of port C is a logical 0, the CPU should not send any more data. When bit D7 or port C gives to a logical 1, the CPU can again write data into the port A data register. At this time the output buffer line port C, bit D6, goes to a logical 0. This indicates to the external device that data is present and can be taken. The external device responds to this line going true by strobing the $\overline{ACK}$ input line, which is port C, bit D7, as described previously. Figure 12-14 shows a flowchart for the complete output operation in the handshake mode. The 8080 mnemonics to realize this flowchart appear in Figure 12-15.

Let's now learn how the CPU reads data from the external device via the 8255. The first operation to occur is that the program checks the input buffer full line, port C, bit D1. When this line is a logical 1, it indicates that the external device has strobed data into the 8255 via the $\overline{STB}$ input line, port C, bit D2. At this time the CPU reads the port B data register, using the IN instruction. At this point, the input buffer line, port C, bit D1, goes to a logical 0, thus indicating to the external device that another byte may be strobed into the 8255.

```
 9              24E8           ODATA  EQU 24E8H  ;ADDRESS OF DATA IN MEM
10                            ;
11   0000   DB 62             BACK   IN  62H    ;READ PORT C DATA
12   0002   E6 80                    ANI 80H    ;MASK OFF LOWER 7 BITS
13   0004   CA 00 00                 JZ  BACK   ;BIT NOT SET, IN AGAIN
14                            ;
15                            ; BIT WAS SET
16                            ;
17   0007   21 E8 24                 LXI H,ODATA ;MEM ADD IN HL
18   000A   7E                       MOV A,M    ;INPUT DATA
19   000B   D3 60                    OUT 60H    ;OUTPUT DATA
20   000D   C9                       RET        ;RETURN TO MAIN
21   000E                            END
```

Figure 12-15 Program to realize flowchart of Figure 12-14.

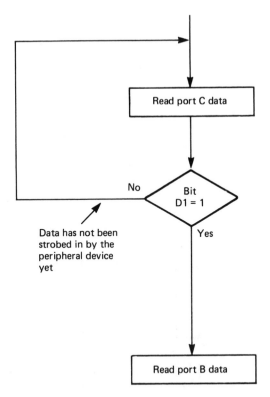

Figure 12-16   A flowchart showing the sequence of events required to transfer data into the 8255 from the external device via the port B and lower port C lines.

```
 9                            ;
10   0000   DB 62      START  IN  62H  ;READ PORT C DATA
11   0002   E6 02             ANI 02H  ;TEST BI D1
12   0004   CA 00 00          JZ START ;NOT SET, READ AGAIN
13   0007   C9                RET      ;RETURN TO MAIN
14   0008                     END
```

Figure 12-17   Program to realize the flowchart of Figure 12-16.

Figure 12-16 shows a flowchart for this type of transfer. The mnemonics to realize the flowchart appear in Figure 12-17.

12-8 OPERATING MODE 2 FOR THE 8255

Another mode of operation for the 8255 is labeled mode 2. In this mode the device can use port A as a bidirectional data port. That is, the eight lines of port A can transmit data to and from the peripheral equipment. When the 8255 is programmed in this mode, port A has a block diagram like the one shown in Figure 12-18.

Figure 12-18 shows an output and input latch. The output latch holds the data written to the port by the CPU and waits for the external device enable to

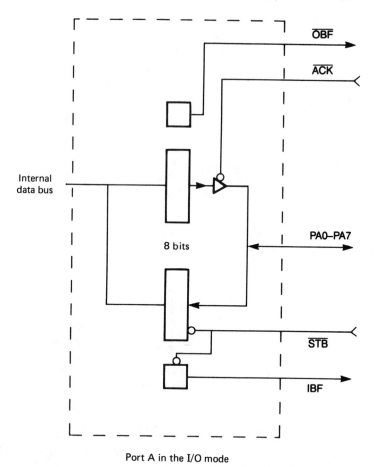

Port A in the I/O mode

Figure 12-18   A block diagram designed to help you understand the 8255 device
when it is programmed into the mode 2 configuration.

transfer the data onto the port A output lines. The input latch stores the data
sent to the port A input lines by the external device.

Let's now examine the block diagram in Figure 12-18 and see how data is
transmitted to the external device. The first event to occur is that the CPU
writes data into the output latch of port A. When this occurs, the output line
labeled $\overline{\text{OBF}}$ (for output buffer full) is set true. This signal electrically informs
the external hardware that data is available from port A. The $\overline{\text{OBF}}$ signal also
indicates to the CPU that the output buffer is full and has not been read by the
external device.

Next, the external device sends an $\overline{\text{ACK}}$ input to the 8255. This input
enables the output latch data onto the port A data lines. It also resets the $\overline{\text{OBF}}$
line, thus indicating to the CPU that the output data in port A has been read.
At this time the CPU can send more data to output port A.

Port C line	Definition
PC0	I/O
PC1	I/O
PC2	I/O
PC3	INTR$_A$
PC4	$\overline{STB}_A$
PC5	IBF$_A$
PC6	$\overline{ACK}_A$
PC7	$\overline{OBF}_A$

Figure 12-19   Port C pin definitions with the 8255 in the mode 2 configuration.

Before data can be received by the 8255 from the external device, the external device must first examine the logical state of the IBF (input buffer full) output line. If this line is a logical 1, then the input buffer is full and the input data has not yet been read by the microprocessor. Let's assume that the IFB line is a logical 0, indicating that there is no data in the buffer. The external device places the data on the port data lines and asserts the $\overline{STB}$ input to the 8255. This action will strobe the data into the internal latch of port A and set the IBF line true. The CPU will monitor the logical level of the IBF line by reading the port C data. When the IBF line is true, data is available to be read. When the CPU reads the data, the IBF line goes to a logical 0.

At this time the external I/O device may send more data to the 8255 in the manner just described. To allow the CPU to examine the status of the external interface lines, port C is used to reflect the logical state of these lines. Figure 12-19 shows the bit definitions for port C when the 8255 is programmed for mode 2 operation.

12-9 CHAPTER SUMMARY

In this chapter we have seen how the 8255 device is used with a microprocessor. We have examined the organization of the device and demonstrated one way that it can be connected to the CPU system buses. Finally, we discussed the software necessary to operate the 8255.

The 8255 I/O device is an extremely versatile interface chip. It can simplify the hardware required to realize some system applications.

REVIEW PROBLEMS

1. How many I/O lines does the 8255 have?
2. What is the purpose of the $\overline{CS}$ input to the 8255?
3. Describe the conditions of the 8255 registers after a reset has been applied.

4. How many internal ports are on the 8255?

5. Draw a schematic diagram showing the minimum 8255 connection to the 8080 CPU. Assume the 8255 is mapped into the I/O space 50–53H.

6. Write a partial program that will set the 8255 up for mode 0 I/O. In your program use the variables for port A, port B, port C, and control port for the register addresses. Set up each of the ports as outputs.

7. Change your program in Problem 6 to set port A as an input, port B as an output, and port C as an output.

8. Write a program that reads data from port A and writes data to port B. The program continually loops doing this function.

9. Using the circuit shown in Figure 12-8, write a program that will set all column lines to a logical 0 and then detect which row line is active when a key is pushed. Do not scan the keyboard. Be sure the initialize the 8255 prior to using the device.

13

SERIAL COMMUNICATION (RS-232)

In this chapter we discuss the basic concepts of serial communication. During our discussion, we will examine an LSI device, called the 8251, which is used for the hardware implementation of a serial communication scheme.

By the end of this chapter you should have a firm, basic understanding of how serial communication can be realized in a microprocessor-based system.

13-1 OVERVIEW OF SERIAL COMMUNICATION

Serial communication is the transmission of data in a bit stream, one bit at a time. All transmission occurs in a time sequential fashion.

Parallel communication, on the other hand, is the opposite of serial communication—all bits of the data transfer are received or transmitted at the same time. A good example of parallel communication is an I/O read or write operation of the CPU, in which all 8 data bits are transmitted (written) or

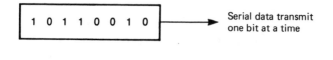

(a)

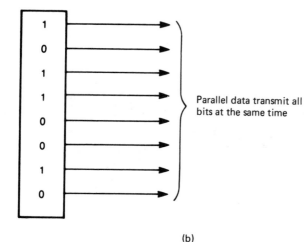

Figure 13-1   (a) A serial data bit stream to be transmitted. The complete word is transmitted one bit at a time; (b) a parallel data byte to be transmitted. Each bit is transmitted in a parallel fashion.

(b)

received (read) at the same time. In fact, all the data in the microprocessor communications that we have discussed so far have been transmitted and received in a parallel mode.

To further illustrate what serial communication is and how it differs from parallel communication, let's consider this example. We wish to send 8 bits of data from one piece of hardware in a microprocessor system to another. We plan to send the data in two ways: parallel and serial.

Figure 13-1a shows how data appears when transmitted in a serial fashion; Figure 13-1b shows how it appears when transmitted in a parallel fashion. Notice that the parallel transmission requires 8 separate lines for the communication—1 for each parallel bit to be transferred. In the serial transmission, only one physical line is required—the 8 bits of data are sent over the single wire, one bit at a time.

13-2 SERIAL TIMING

Let us now extend the simple serial example given in the previous section to illustrate some additional concepts. One of the critical points of serial communication is the frequency of the transmitted data bit stream. This frequency is called the *baud rate*. Baud rate is defined as the transmitted bits per second over a single line.

Typical baud rates are 110, 150, 300, 1200, 2400, 4800, and 9600. Let's

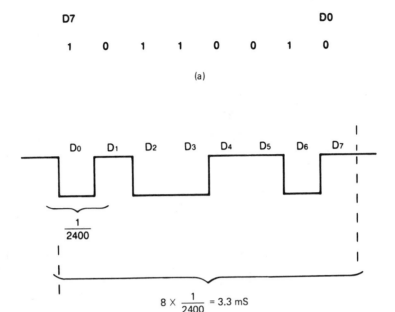

Figure 13-2   (a) Eight bits of data to be transmitted in a serial fashion; (b) waveform that would be generated if the byte were transmitted in a serial fashion; that is,
bit D0, through D1, D2, and so on.

suppose that we wish to transmit 8 bits of data at 2400 baud. The data to be
transmitted is shown in Figure 13-2a. Figure 13-2b shows the wave form that
would appear on the oscilloscope as the data is being transmitted.

   Figure 13-2 also shows that the width in seconds of a single transmitted bit
is equal to 1/baud rate. For this example the width of a single bit is equal to 1/
2400 = .000416 second, or 416 microseconds. Knowing this, we can calculate
the amount of time required to transmit the entire 8 bits. This time is equal to 8
× 416 microseconds = 3328 microseconds. The transfer time for the same 8 bits
would normally be less than 1 microsecond for a parallel transfer.

13-3 CONVERTING PARALLEL DATA TO SERIAL DATA

A major part of serial communication is the conversion of parallel data to a
serial bit system. Serial communication must take the parallel data from the
microprocessor and transform it into a bit stream. The steps in the conversion
process are as follows:

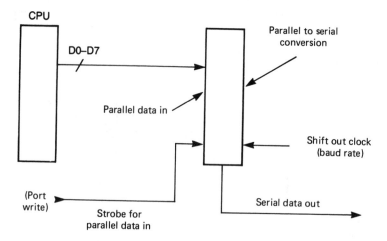

Figure 13-3 A block diagram showing the process of converting a parallel byte into a serial bit stream to be transmitted. The microprocessor will write the data to be transmitted into parallel-to-serial shift register. The data will then be shifted out at the correct rate.

1. Store the parallel 8-bit word in a shift register.
2. Shift the 8 bits out of the register one bit at a time at the correct baud rate.

These two steps are shown in the block diagram in Figure 13-3. We see that the data to be transmitted is first generated by the microprocessor and then stored in the parallel load, 8-bit shift register. The data is then shifted out D0 first and D7 last.

13-4 START BIT

So far in our discussion of serial transmission we have explained baud rate and parallel-to-serial conversion. The data that is transmitted during a serial transmission must be capable of being received and electrically interpreted. To make this possible, another bit, called the *start bit,* is automatically added to the data bit stream.

The function of the start bit is to let the receiving hardware know electrically when a new data stream is starting and thus allow it to synchronize its clock to the input bit stream. Each data stream represents a single data character. You may want to think of each data stream as a byte of parallel data. Of course, the data does not have to be 8 bits in length, but thinking of it in this way makes it easier to understand if you are new to the subject of serial transmission.

When the data transmission output line is not sending information, it is in

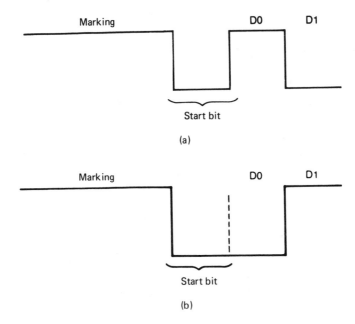

Figure 13-4   (a) A timing diagram showing how the start bit would appear if the first transmitted bit were a logical 1; (b) a timing diagram showing how the start bit would appear if the first transmitted bit were logical 0.

a state called *marking*. This is the idle state of a serial transmission line. Let us assume that the marking state of a transmission line is a logical 1. The start bit that is added to the data bit stream is of the opposite logical level to that of the marking logical level. In this case, the start bit is a logical 0.

The start bit is really 1 bit added to the front end of the data bit stream. Whatever the baud rate of the bit stream is, the start bit will be equal to 1 bit width. This is shown in Figure 13-4. The receiving hardware will detect this start bit and enable the receiving section to take the new data.

13-5 PARITY BIT

Another transmitted bit that can be added to a serial bit stream is the *parity bit*. This bit is inserted by the transmitter and used by the receiver. Here is an explanation of what the parity bit does.

When a word is set up to be transmitted, it contains a certain number of logical 1s. The number of logical 1s may be either odd or even. For example, the 8-bit word 54H has three 1s in it; the word 55H has four. The receiving hardware is set up to receive the data and to encounter either an even or odd number of 1s.

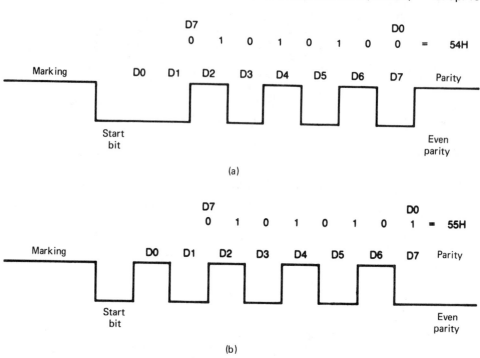

Figure 13-5 (a) A timing diagram showing an even parity bit for the data 54H. The parity bit is a logical 1 in this case; (b) a timing diagram showing an even parity for the data 55H. Notice that the parity bit is a logical 0 for this byte.

Suppose that the receiving hardware is set up to expect an even number of 1s in every bit stream. The number 55H would be all right, but the number 54H would not. Therefore, the transmitter would insert an additional 1 into the bit stream with the 54H as the data is being sent. This would add an additional bit to the data stream. This bit would either be a logical 1 or a logical 0, whichever makes the total number of bits in the new length bit stream an even number. Figure 13-5 shows the parity bit for the transmission of 54H and 55H. The parity bit is used by the receiving hardware to detect error in the data transmission.

13-6 STOP BIT

The last bit added to the bit stream by the transmitting hardware is the *stop bit*. The receiving hardware expects to detect a stop bit at the end of the bit stream. There are different numbers of stop bits. There are 1, 1½, or 2. Figure 13-6 shows a complete 8-bit data word with start bits, data, parity, and stop bits added. Thus a complete bit stream would consist of 12 bits instead of the 8 with

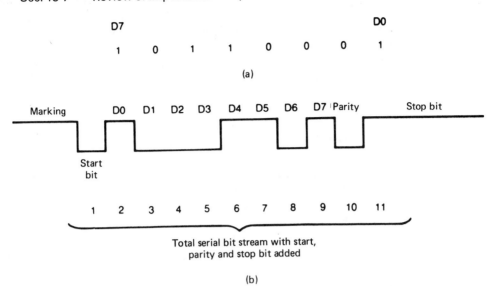

Figure 13-6 (a) Data to be transmitted in a serial fashion; (b) waveform generated by the complete transmission bit stream with start, stop, and parity bits added.

which we started. At 2400 baud, the absolute transmission time for the data shown in Figure 13-6 is equal to 12 × 416 microseconds = 4.99 milliseconds.

13-7 REVIEW OF IMPORTANT CONCEPTS OF SERIAL COMMUNICATION

In review, here is a list of the important points of serial communication:

1. Serial data is transmitted one bit at time, the LSB first.
2. Data is transmitted at a fixed rate called the baud rate. The baud rate is equal to the number of bits per second transmitted. For example, 1200 baud transmits data with a clock frequency of 1200 hertz.
3. Parallel data in a microprocessor system must first be converted to serial data to be communicated with in a serial fashion.
4. Marking is a term used for the logical state of the serial transmission line when no data is present.
5. The transmitter of the serial data will add 1 data bit, called a start bit, at the front of the data. This bit will be of the opposite logical level to the marking state of the serial transmission line.
6. The transmitter may add a single bit, called a parity bit, to the end of the data stream. The parity bit can be inserted to generate even or odd parity.

Parity is used by the receiver at a first-level error check on the transmitted data.

7. Transmitting hardware will add stop bits to the data stream after the parity bit. The stop bit will be 1, 1½, or 2 bits wide. Stop bits are the same logical level as the marking level of the serial transmission line.

13-8 OVERVIEW OF THE 8251

The concept of serial data transmission in a microprocessor system is often used. One of the most common uses is in computer-to-terminal interfacing. Since this type of data transmission is so common, special LSI devices called USARTS (Universal Synchronous Asynchronous Receiver Transmitters) have been designed to accommodate it. The 8251 is such a device. Figure 13-7 shows how the 8251 fits into the hardware architecture of a microprocessor system.

We can see in Figure 13-7 that the 8251 is treated exactly like the 8255 I/O device that we discussed previously. The CPU communicates through the I/O system. The software instructions set up or program the device to operate in a certain way. In this chapter, we will show you how you can connect the 8251 to a microprocessor system to achieve reliable communication. First, however, let's discuss the important software concepts involved in using the 8251. By the end of this discussion you should know how a typical serial I/O device operates.

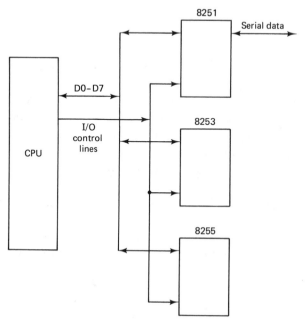

Figure 13-7 Block diagram showing how the 8251 USART will fit into a standard microprocessor system architecture.

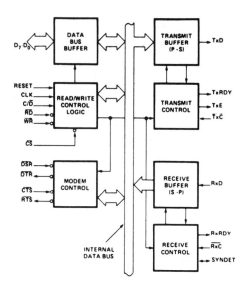

Figure 13-8 A block diagram of the internal architecture of the 8251 Universal Synchronous Asynchronous Receiver Transmitter (USART). (Courtesy of Intel Corp.)

Figure 13-8 shows a block diagram of the 8251. Let's discuss the function of each block in a serial communication scheme, starting with the *data bus buffer*. This block is the physical connection between the 8251 and the microprocessor system data bus.

Next is the *read/write control logic*. This block ensures that the data of the 8251 is read or written to the correct internal location with the correct timing.

The block labeled *modem control* is used to simplify the interface between the 8251 and a modem. A modem is a device that is used to enable serial transmission over a telephone line. Figure 13-9 shows how a modem fits into a serial communication system over such a line.

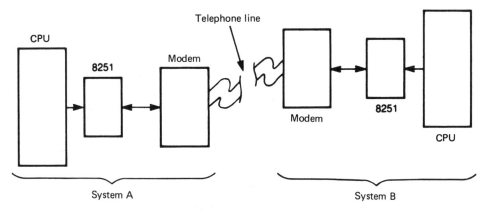

Figure 13-9 This block diagram shows how a modem fits into the process of transmitting information from one place to another over a telephone line.

Referring to Figure 13-8, we can see that the two blocks labeled *transmit buffer* (P-S) and *transmit control* operate together to output the serial data stream. (*Note:* P-S stands for parallel-to-serial conversion.) Data is output from the device on the line labeled TxD. Tx stands for transmit, and D is for data. The transmit-control block electrically monitors the status of the transmit buffer to determine if the buffer is empty, that is, if it is ready for another character to be transmitted.

The last two blocks shown in Figure 13-8 are the *receiver buffer* (S-P) and the *receive control.* A receiver buffer inputs a bit stream from another serial device and converts it into a parallel word, which can then be read by the microprocessor using an input instruction. The receiving control block monitors the status of the receiver buffer to determine electrically when the buffer is full.

13-9 PINOUT OF THE 8251

Figure 13-10 shows the pinout of the 8251 device. Let's now discuss the function of each pin. This will help you understand how the 8251 can be connected physically to the microprocessor system buses.

D0–D7.   These are the eight data lines that connect to the microprocessor system data bus. All information and programming words are input and output from the 8251 via these lines.

Reset.   This input is active logical 1. The 8251 is set to an idle state when the RESET is asserted. It then remains in this state until the appropriate programming wrods are sent to it via the data bus.

CLK (clock).   This input to the 8251 is used to synchronize internal data transfers. This is not the baud rate clock input. This clock must be at least 30 times as fast as the baud rate of the receiver or transmitter.

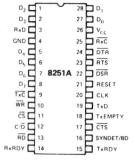

Figure 13-10   Pinout of 8251 device. (Courtesy of Intel Corp.)

$\overline{\text{WR}}$ **(Write data, active logical 0).**   When this input line is a logical 0, the data on the system data bus is written to the 8251 internal registers.

$\overline{\text{RD}}$ **(Read data, active logical 0).**   When this input line is a logical 0, the internal data from the 8251 is placed on the system data bus.

C/$\overline{\text{D}}$ (Control/data input lines).   This line is used to define the internal registers of the 8251. When this line is a logical 1, data will be written or read from the control register. When it is a logical 0, data will be written or read from the data register of the device.

$\overline{\text{CS}}$ **(Chip-select active logical 0).**   A logical 0 on this input line enables data to be written or read from the device.

Modem Control

The following four pins are used to simplify the interface to a modem:

$\overline{\text{DSR}}$	Data set ready
$\overline{\text{DTR}}$	Data terminal ready
$\overline{\text{CTS}}$	Clear to send
$\overline{\text{RTS}}$	Ready to send

Let's examine them in detail.

$\overline{\text{DSR.}}$ This input can be tested by the CPU. It is normally used to test modem conditions, such as data set ready.

$\overline{\text{DTR.}}$ This output can be set low by sending a certain bit pattern to the 8251. This output is normally used for data terminal ready.

$\overline{\text{CTS.}}$ A logical 0 on this input enables the 8251 to transmit serial data if internal conditions on the device permit. The CTS line may also be used as a hardware handshake line.

$\overline{\text{RTS.}}$ This output bit may be set low by writing the correct word to the 8251.

TxD (Transmit data output).   This is the output line on which data will be transmitted.

RxD (Receive data input).   This is the input line on which data will be received when it goes to the 8251 from the serial bit stream.

TxC (Transmit baud rate clock input).

RxC (Receive baud rate clock input).

TxRDY.   This output signals the CPU that the 8251 is ready to accept another character to be transmitted. This line can be used as an interrupt for the CPU.

TxEMPTY.   When the device has no characters to transmit, this line will go to a logical 1. It will be set automatically to a logical 0 when the CPU writes a character to the transmitter buffer.

13-10 CONNECTING THE 8251 TO THE CPU BUSES

Figure 13-11 shows the complete interface schematic diagram for connecting the 8251 USART to the 3-bus architecture system as described in Chapter 5. Let's review some of the important points about Figure 13-11. To start, we see that address lines A7–A1 are decoded to allow the chip-select input pin 11 of the 8251 to respond to an I/O select code of 7CH or 7DH. The decoding is accomplished via the 74LS04 and 74LS30.

Address line A0 is connected to the C/$\overline{D}$ (command/data) input pin 12 of the 8251 device. This address line will logically inform the USART whether the information on the system data bus is programming or transmitting data for the device. Port 7CH is the data port, whereas port 7DH is the command port. Notice that the C/$\overline{D}$ input line is a different logical level for each of these two input port codes.

We next see in Figure 13-11 that system-control signals $\overline{IOR}$ and $\overline{IOW}$ are connected directly do the $\overline{RD}$ and $\overline{WR}$ input pins of the 8251. Whenever the CPU is communicating with an I/O device, one of these two lines will be active. If the CPU is communicating with the 8251, the $\overline{CS}$ input pin 12 will be active with either the $\overline{RD}$ or $\overline{WR}$ input.

The reset input to the 8251 is asserted in the logical 1 state. This is the same active state as the 8080 CPU. Therefore, the reset input to the 8251 can be connected directly to the reset input of the 8080 CPU.

Pin 20 of the 8251 is labeled *clock*. This input line is not the baud rate clock but is a clock that will synchronize all data transfers between the CPU and the 8251. The input frequency of this clock line must be at least 30 times as fast as the baud rate clock input on the TxC and/or RxC clock lines.

We notice that the TxC and RxC clock input lines are connected together in Figure 13-11. This allows the 8251 to transmit and receive data at exactly the same baud rate. However, you do not have to do this. The 8251 is designed to allow the user to transmit and receive data at two completely different baud rates.

The final section of Figure 13-11 is the data buffer, 74LS245, which is controlled by the direction line input pin 1. When the $\overline{CS}$ input is a logical 0 and the $\overline{RD}$ input line is a logical 0, the direction input pin 1 of the 74LS245 is a

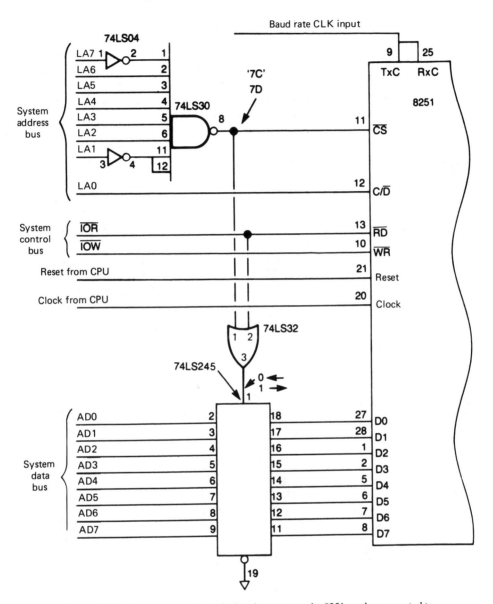

Figure 13-11 A complete schematic showing one way the 8251 can be connected to the system buses. Notice the use of the data buffer between the CPU data bus and the 8251 data input pins.

logical 0. This allows the data from the 8251 to be placed on the system data bus. At all other times the 74LS245 will be inputting data from the system data bus to the 8251.

As can be seen from the schematic in Figure 13-11, connecting the 8251 to the system buses is very similar to connecting most other LSI devices to the 8080, 8085, and Z80. Once the connection is made, the task becomes one of controlling the 8251 with the correct software commands.

13-11 THE SERIAL CONNECTION

Let's now connect the 8251 to the serial transmission and serial reception lines. There are various serial transmission standards available. For this example, we will choose one that is in wide general use today: the RS-232 electrical standard for serial transmission.

The RS-232 uses a high-voltage-level pulse instead of the normal TTL levels. The voltage levels on the RS-232 transmission line are approximately ±12 V. Figure 13-12 shows a block diagram of a typical interface to the RS-232 bus. This figure shows that the TTL level from the 8251 must be converted to the RS-232 levels prior to transmission on the line.

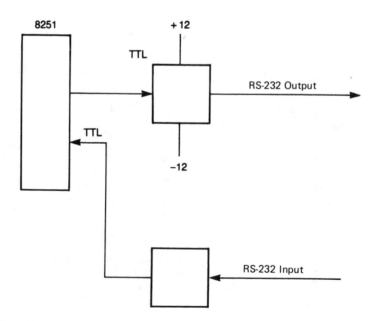

Figure 13-12   A block diagram showing how the TTL voltage levels of the 8251 USART are translated into RS-232 voltage levels. Also shown is the translator from RS-232 voltage levels into TTL levels for the USART.

At the receiving end of the RS-232 lines, the logical levels must be reconverted into a TTL level for input to the serial receiver. The conversion of TTL voltage levels to RS-232 voltage levels is so common that special integrated circuits have been designed to perform this function. Examples of these special integrated circuits are the MC1488 and MC1489. These devices are shown in partial data sheet form in Figures 13-13 and 13-14.

Figure 13-15 shows how the MC 1488 and the MC1489 are used in a typical system performing RS-232 serial communication. Notice that to accomplish complete communication between one serial device and another requires only three wires, Tx (transmit), Rx (receive) and GND.

The preceding discussions of the system hardware have presented the main points of serial transmission that you should keep in mind when considering the use of serial communication. In a system that uses serial communication, the preceding hardware blocks must exist in some form. Once you know what should be there, then finding and analyzing the circuits is a much easier task.

13-12 PROGRAMMING THE 8251

Now that we have connected the 8251 to the CPU buses, let's focus on the software necessary to make the device operate in an asyncronous serial communication environment. In a general sense, the software informs the 8251 of the parameters for the transmission. These include baud rate, number of data bits, number of stop bits, and parity information.

Once the device has been set up, the actual transmission of the data takes place. The microprocessor first determines if the transmit buffer is empty. If it is, then a character is loaded into the buffer and sent out via the serial line. This action takes place for all characters transmitted by the CPU.

For the CPU to receive a character, it must first examine a status register to determine if the receiver will read the character. It must then wait for the status to indicate when another character has been received.

The preceeding description of the reception and transmission of serial characters is only a simplified version of what actually occurs. We will now explain how we can accomplish this operation using the 8251.

Prior to transmission, a set of control words must be loaded into the 8251. To accomplish this, the CPU must write to the command register immediately after the system reset has been asserted. The 8251 will electrically assume that the first I/O write to the command register is the mode instruction and that a command word will follow immediately. There may be one or more bytes in the command word. The number of bytes depends on the byte sent for the mode instruction.

Figure 13-16 is a diagram showing bit definitions for the the mode words.

 **MOTOROLA**

QUAD LINE DRIVER

The MC1488 is a monolithic quad line driver designed to interface data terminal equipment with data communications equipment in conformance with the specifications of EIA Standard No. RS-232C.

Features:

- Current Limited Output
  ±10 mA typ
- Power-Off Source Impedance
  300 Ohms min
- Simple Slew Rate Control with External Capacitor
- Flexible Operating Supply Range
- Compatible with All Motorola MDTL and MTTL Logic Families

QUAD MDTL LINE DRIVER
RS-232C
SILICON MONOLITHIC
INTEGRATED CIRCUIT

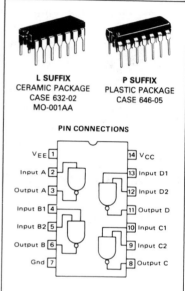

L SUFFIX
CERAMIC PACKAGE
CASE 632-02
MO-001AA

P SUFFIX
PLASTIC PACKAGE
CASE 646-05

PIN CONNECTIONS

V_{EE} [1]	[14] V_{CC}
Input A [2]	[13] Input D1
Output A [3]	[12] Input D2
Input B1 [4]	[11] Output D
Input B2 [5]	[10] Input C1
Output B [6]	[9] Input C2
Gnd [7]	[8] Output C

TYPICAL APPLICATION

LINE DRIVER MC1488   INTERCONNECTING CABLE   LINE RECEIVER MC1489

MDTL LOGIC INPUT — INTERCONNECTING CABLE — MDTL LOGIC OUTPUT

CIRCUIT SCHEMATIC
(1/4 OF CIRCUIT SHOWN)

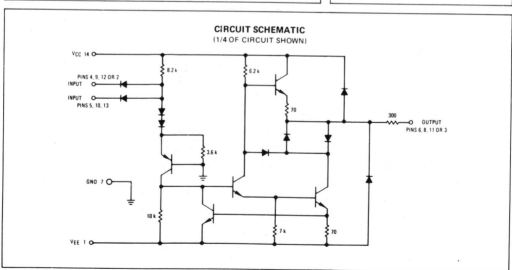

Figure 13-13 A partial data sheet of the Motorola MC1488 voltage translator from TTL to RS-232 levels. (Courtesy of Motorola, Inc.)

MOTOROLA

MC1489
MC1489A

QUAD LINE RECEIVERS

The MC1489 monolithic quad line receivers are designed to interface data terminal equipment with data communications equipment in conformance with the specifications of EIA Standard No. RS-232C.

- Input Resistance — 3.0 k to 7.0 kilohms
- Input Signal Range — ± 30 Volts
- Input Threshold Hysteresis Built In
- Response Control
  a) Logic Threshold Shifting
  b) Input Noise Filtering

**QUAD MDTL
LINE RECEIVERS
RS-232C**

**SILICON MONOLITHIC
INTEGRATED CIRCUIT**

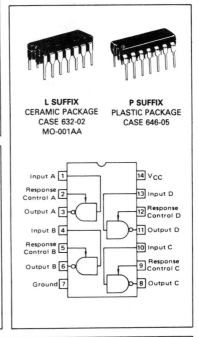

L SUFFIX
CERAMIC PACKAGE
CASE 632-02
MO-001AA

P SUFFIX
PLASTIC PACKAGE
CASE 646-05

Input A 1	14 V_{CC}
Response Control A 2	13 Input D
Output A 3	12 Response Control D
Input B 4	11 Output D
Response Control B 5	10 Input C
Output B 6	9 Response Control C
Ground 7	8 Output C

TYPICAL APPLICATION

LINE DRIVER
MC1488

INTERCONNECTING
CABLE

LINE RECEIVER
MC1489

MDTL LOGIC INPUT — INTERCONNECTING CABLE — MDTL LOGIC OUTPUT

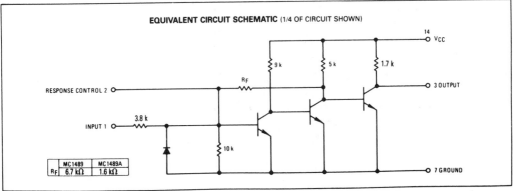

EQUIVALENT CIRCUIT SCHEMATIC (1/4 OF CIRCUIT SHOWN)

14 V_{CC}

9 k   5 k   1.7 k

RESPONSE CONTROL 2   R_F   3 OUTPUT

INPUT 1   3.8 k

10 k

R_F	MC1489	MC1489A
	6.7 kΩ	1.6 kΩ

7 GROUND

Figure 13-14   A partial data sheet of the Motorola RS-232 to TTL voltage converter. (Courtesy of Motorola, Inc.)

359

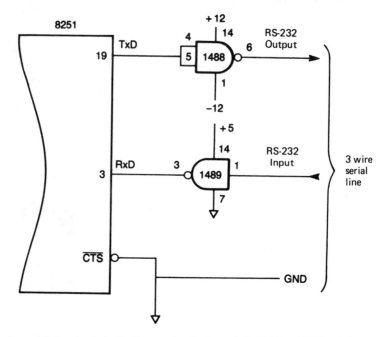

Figure 13-15   A schematic diagram showing how the MC1488 and MC1489 devices can be used in a serial communication scheme.

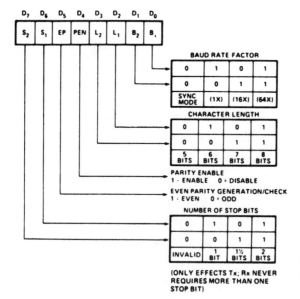

Figure 13-16   A diagram of the mode word definitions for the 8251 device. (Courtesy of Intel Corp.)

Let's use this figure to set up a mode word for a typical transmission. In this transmission we will define our serial line to be the following:

1. The baud rate equals 2400. Our baud rate generator will have a frequency equal to 16 times the baud rate. This allows for slight variations in the clock rates between the receiver and transmitter. D1 and D0 of the mode word will be set to 1 and 0, respectively.
2. There will be 8 data bits transmitted for each character.
3. The parity bit will be set for even parity.
4. The transmitter will insert 2 stop bits per character.

With the preceding definition, the mode word written to the 8251 will be:

$$1\ 1\ 1\ 1\ 1\ 1\ 1\ 0$$

Let's break this word down and examine it closely:

Bits D7, D6 = 11	This sets up 2 stop bits.
Bits D5, D4 = 11	This sets even parity and enable parity.
Bits D3, D2 = 11	These bits set character length to 8 bits.
Bits D1, D0 = 10	The baud rate factor is set to x16.

The 8080 instructions to write the mode word are

```
MVI A,0FEH
OUT 7DH          port number=7DH
```

The next output word written to the 8251 is the command instruction. Bit definitions for this byte are shown in Figure 13-17. Based on the bit definitions given in this figure, the byte written will be:

$$0\ 0\ 0\ 1\ 0\ 1\ 0\ 1$$

This byte enables the receiver and transmitter and resets the error registers. At this point the 8251 is ready for transmission or reception of characters.

A final register that we will now discuss is the *status register*. This register is read only. It is used so that the microprocessor can monitor the status of the serial transmission and check for any errors in transmission that the 8251 can detect. The bit definitions for the status register are shown in Figure 13-18. It is possible to read the status register with a simple input read operation, thus ensuring that the C/$\overline{D}$ input to the 8251 is a logical 1.

13-13 FRAMING ERROR

The first type of error we will examine is the *framing error*. A framing error occurs when a character is received in the serial input buffer and no stop bits are detected in the stream. This type of error usually results in faulty data being read.

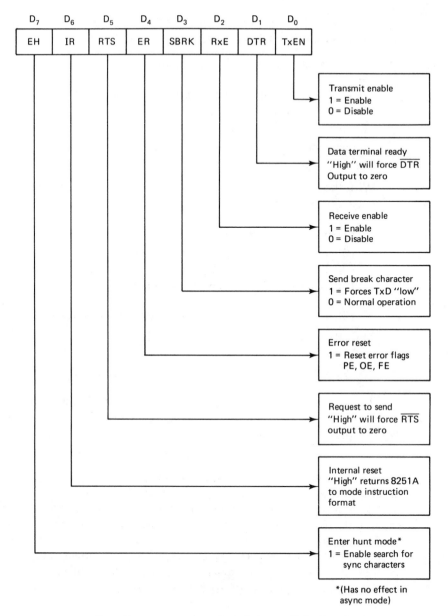

Figure 13-17 Bit definition of the command instruction word format for the 8251 device.

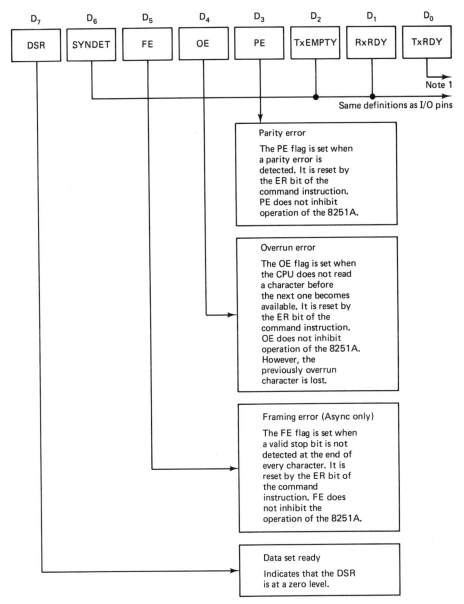

Figure 13-18 Bit definitions of the status words for the 8251 device. The microprocessor can be informed electrically of the 8251 status by examining different bits of the status word. The status word is obtained from the 8251 by using an input instruction from the CPU.

13-14 OVERRUN ERROR

An *overrun error* occurs when a character is received in the serial input buffer and the previous character has not yet been read by the CPU. This results in the previous character being lost and the new character being written over it.

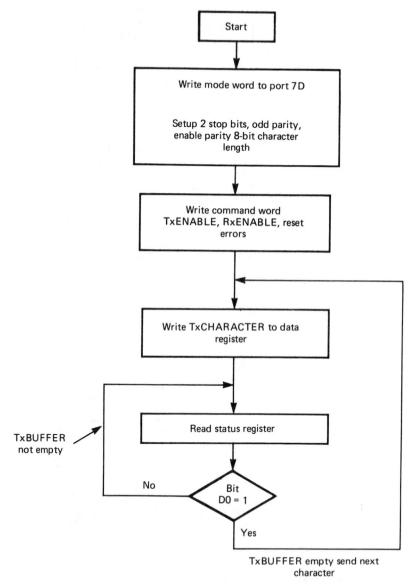

Figure 13-19   A flowchart for the sequence of events required to transmit a single character repeatedly to another serial receiving device.

```
10   0000   3E FE                         MVI  A,0FEH    ;MODE WORD IN A
11   0002   D3 7D                         OUT  7DH       ;OUTPUT TO 8251
12                                  ;
13                                  ; 2 STOP BITS, EVEN PARITY, ENABLE PARITY
14                                  ; 8 BITS/CHAR, X16 MULTIPLIER
15                                  ;
16   0004   3E 15                         MVI  A,15H     ;COMMAND WORD IN A
17   0006   D3 7D                         OUT  7DH       ;OUTPUT TO 8251
18                                  ;
19                                  ; TX ENABLE, RX ENABLE, RESET ERRORS
20                                  ;
21   0008   3E 49           LOOP1   MVI  A,49H     ;ASCII "I"
22   000A   D3 7C                   OUT  7CH       ;OUTPUT TO TX
23   000C   DB 7D           LOOP2   IN   7DH       ;READ STATUS REG
24   000E   E6 01                   ANI  01H       ;TEST BIT D0=1
25   0010   CA 0C 00                JZ   LOOP2     ;IF 0, KEEP POLLING
26   0013   C3 08 00                JMP  LOOP1     ;BUFFER EMPTY, NEXT CHAR
27   0016                           END
```

Figure 13-20   Program to realize the flowchart of Figure 13-19.

13-15 8251 APPLICATION EXAMPLE

The following is a simple 8080 program for using the 8251. This program repeatedly sends the same character to the terminal. This program is not very useful except that it shows some important points about polled transmission. Figure 13-19 presents a flowchart for this program. The 8080 mnemonics to realize this flowchart are shown in the program in Figure 13-20.

13-16 AN EXPANDED APPLICATION FOR THE 8251

Let's now examine a program that allows the 8251 to echo a character sent to it by the terminal. This program shows how the 8251 operates in the transmission and reception of characters. Figure 13-21 shows a block diagram of the hardware interconnections. The flowchart for this program is shown in Figure 13-22.

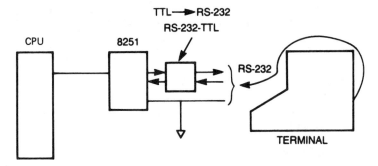

Figure 13-21   A block diagram showing how a USART fits into the communication scheme between a microprocessor system and a terminal.

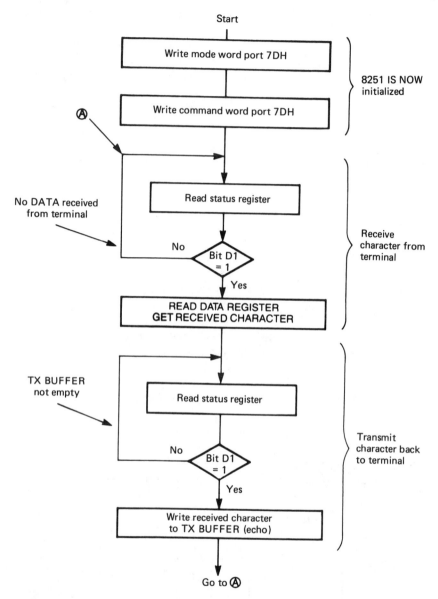

Figure 13-22 A flowchart showing the sequence of events required to echo a character sent to the microprocessor system from the terminal and back to the terminal.

```
10   0000   3E FE                    MVI A,0FEH    ;MODE WORD IN A
11   0002   D3 7D                    OUT 7DH       ;OUTPUT TO 8251
12                           ;
13                           ; 2 STOP BITS, EVEN PARITY, ENABLE PARITY
14                           ; 8 BITS/CHAR, X16 MULTIPLIER
15                           ;
16   0004   3E 15                    MVI A,15H     ;COMMAND WORD IN A
17   0006   D3 7D                    OUT 7DH       ;OUTPUT TO 8251
18                           ;
19                           ; TX ENABLE, RX ENABLE, RESET ERRORS
20                           ; THE 8251 DEVICE IS NOW ENABLED
21                           ; FIRST THE DEVICE WILL WAIT FOR A CHARACTER
22                           ; TO BE SENT TO IT FROM THE TERMINAL
23                           ;
24   0008   DB 7D            LOOP1   IN 7DH        ;READ STATUS REG
25   000A   E6 02                    ANI 02H       ;TEST BIT D1=1
26   000C   CA 08 00                 JZ LOOP1      ;IF 0, KEEP POLLING
27                           ;
28                           ; WE HAVE A CHARACTER IN THE RECEIVE BUFFER
29                           ; OF THE 8251
30                           ;
31   000F   DB 7C                    IN 7CH        ;READ THE CHARACTER
32   0011   4F                       MOV C,A       ;CHAR IN C
33                           ;
34                           ; WE WILL NOT CHECK THE DATA FOR ERRORS
35                           ;
36                           ; ECHO THE DATA BACK TO THE TERMINAL
37                           ;
38   0012   DB 7D            LOOP2   IN 7DH        ;READ THE STATUS REGISTER
39   0014   E6 01                    ANI 01H       ;TEST D0=1
40   0016   CA 12 00                 JZ LOOP2      ;IF ZERO, NOT READY
41                           ;
42                           ; XMIT IS READY TO OUTPUT ANOTHER CHARACTER
43                           ;
44   0019   79                       MOV A,C       ;GET CHARACTER
45   001A   D3 7C                    OUT 7CH       ;SEND TO 8251
46   001C   C3 08 00                 JMP LOOP1     ;WAIT FOR NEXT CHAR
47   001F                            END
```

Figure 13-23   Program to realize flowchart of Figure 13-22.

The 8080/8085 mnemonics for realizing the flowchart of Figure 13-22 are given in Figure 13-23.

13-17 CHAPTER SUMMARY

In this chapter we have examined the basics of serial communication. We have covered several important topics, including baud rate, start bits, stop bits, and parity bits. We have also discussed the meaning of two basic errors sometimes encountered in serial communications: framing and overrun errors. In the second part of this chapter we have examined a real serial communication device: the 8251. We have selected this device because it is simple to use and understand and it is in wide general use in industry today. In addition, it solves the problem of serial communication simply and elegantly.

REVIEW PROBLEMS

1. Write a brief definition of serial communication.
2. In general, which type of communication is faster, serial or parallel? Why?
3. What is baud rate?
4. What is the shortest time a signal could be true if you were transmitting at 9600 baud?
5. What is the function of the start bit?
6. Define the term marking.
7. What is the function of the parity bit?
8. What is the function of the stop bit?
9. What do the letters USART stand for?
10. Draw a block diagram of how a modem fits into the serial transmission scheme.
11. Why do you need to use devices like the MC1488 and MC1489 when using an RS-232 serial bus?
12. What is a framing error?
13. What is an overrun error?
14. Expand the program shown in Figure 13-23 to keep inputting characters from the terminal and saving them in sequential fashion in memory. When a CR (0DH) character is received, you stop storing the characters. This type of activity is what an operating system does when getting commands from the keyboard of the computer.
15. Write a program that will output an entire ASCII message from the system to the terminal. Assume the message starts at memory location 5000H and is terminated by the ASCII character $, which equals 24H. An example of a message to transfer would be the word BYE. In memory this message would appear as

```
5000H = 42H  B
5001H = 59h  Y
5002H = 45H  E
5003H = 24H  $  message terminator
```

You would continue to transfer the data starting at memory location 5000H until you encountered the 24H. You can make the transfer message as long as you choose; just be certain to terminate the message with a 24H.

14

USING THE 8253 PROGRAMMABLE INTERVAL TIMER

This chapter will show you how to use the 8253 programmable timer chip with the 8080, 8085, or Z80 microprocessor. We will begin by examining and describing this timer chip. Next, this device will be interfaced to the microprocessor. Finally, a software program will be written that shows how to use the 8253 in a variety of applications.

14-1 BLOCK DIAGRAM OF THE 8253 PROGRAMMABLE TIMER

Let's begin by examining the block diagram in Figure 14-1 and defining the internal registers and operating modes of this device. The diagram shows that the timer has three independent, programmable counters—and they are all identical. In this chapter we will learn how to apply and use each counter.

Also shown in Figure 14-1 is the *data bus buffer*. This block contains the logic that buffers the data bus to and from the microprocessor to the 8253 internal

8253/8253-5
PROGRAMMABLE INTERVAL TIMER

- **MCS-85™ Compatible 8253-5**
- **3 Independent 16-Bit Counters**
- **DC to 2.6 MHz**
- **Programmable Counter Modes**

- **Count Binary or BCD**
- **Single +5V Supply**
- **Available in EXPRESS**
  - **— Standard Temperature Range**
  - **— Extended Temperature Range**

The Intel® 8253 is a programmable counter/timer device designed for use as an Intel microcomputer peripheral. It uses NMOS technology with a single +5V supply and is packaged in a 24-pin plastic DIP.

It is organized as 3 independent 16-bit counters, each with a count rate of up to 2.6 MHz. All modes of operation are software programmable.

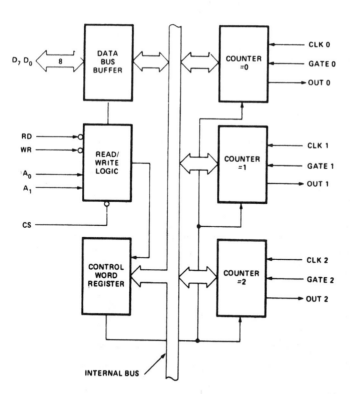

Figure 14-1 A block diagram of a 8253 programmable interval timer. (Courtesy of Intel Corp.)

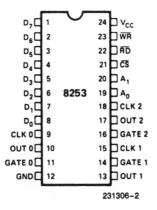

Figure 14-2   Pinout of the 8253 programmable interval timer. (Courtesy of Intel Corp.)

231306-2

registers. In addition, there is the block labeled *read/write logic,* which controls the reading and writing of the counter registers. The final block—the *control word register*—contains the programmed information that is sent to the device from the system microprocessor. In effect this register defines how the chip logically operates. Figure 14-2 shows a pin configuration for the 8253 device.

14-2 THE THREE COUNTER LINES: CLOCK, GATE, AND OUT

Each counter in the block diagram in Figure 14-1 has three logical lines connected to it. Two of these lines, *clock* and *gate,* are inputs. The third, labeled *out,* is an output. The function of these lines changes and depends on how the device is initialized or programmed. Here is a general definition of the lines.

Clock.   This input is the clock input for the counter. The counter is 16 bits. The maximum clock frequency is 1/380 nanoseconds, or 2.6 megahertz. The minimum clock frequency is DC, or static operation.

Gate.   The input can act as a gate for the clock input line or it can act as a start pulse, depending on the programmed mode of the counter.

Out.   This single output line is the signal that is the final programmed output of the device. Actual operation of the out line depends on how the device has been programmed.

14-3 8253 INTERNAL REGISTERS

A list of the internal registers of the 8253 device appears in Figure 14-3. Let's first examine and discuss the *mode word register.* This register defines the overall operation of the device. Since each of the three counters is fully indepen-

	$\overline{RD}$	$\overline{WR}$	A0	A1	
Counter 0	1	0	0	0	Load counter 0
	0	1	0	0	Read counter 0
Counter 1	1	0	0	1	Load counter 1
	0	1	0	1	Read counter 1
Counter 2	1	0	1	0	Load counter 2
	0	1	1	0	Read counter 2
Mode word or	1	0	1	1	Write mode word
control word	0	1	1	1	No operation

Figure 14-3   A list of the internal 8253 registers that will program the internal counters of the device.

dent, each one can be programmed by the microprocessor outputting the correct data to the mode word register. We will show how this can be accomplished as our discussion proceeds. Let's first define the four internal registers shown in Figure 14-3.

Control word register.   This register is addressed when A0 and A1 inputs are logical 1s. The data in this register control the operating mode and the selection of either binary or BCD (binary-coded decimal) counting format. This register can only be written to.

Counter 0, 1, 2.   Each counter is identical, and each consists of a 16-bit, presettable down counter. Each is fully independent and can be made to count in BCD (Binary Coded Decimal) or binary. Contents of the counters can be easily read by the microprocessor. When the counter is read, the data within the counter is not disturbed. This allows the system to monitor the counter's value at any time, without disrupting the overall function of the device.

14-4 CONNECTING THE 8253 TO THE MICROPROCESSOR BUSES

Before we program the 8253, let's learn how to connect it with the microprocessor system address, data, and control buses. The 8253 may be thought of as four separate I/O ports. The main selection of the I/O ports is accomplished with the

$\overline{\text{CS}}$ input. When this input is a logical 0, the 8253 is selected for communication with the CPU.

Address input lines A0 and A1 on the 8253 determine with which I/O ports communication occurs during the input or output cycle. This I/O architecture can be referred to as device port I/O. The device can be thought of as the 8253, and the ports associated with the device are the internal registers.

Decoding of the 8253 $\overline{\text{CS}}$ input is accomplished with address lines A2–A7 for the 8080 CPU. Let's assume that we have selected the I/O space 30H, 31H, 32H, or 33H. The hardware to perform this decoding is shown in Figure 14-4.

The $\overline{\text{RD}}$ and $\overline{\text{WR}}$ inputs to the 8253 are connected directly to the $\overline{\text{IOR}}$ and $\overline{\text{IOW}}$ system control lines. This setup is identical to the one for the 8255 device that was presented in Chapter 12. Figure 14-5 shows the connection to the $\overline{\text{RD}}$ and $\overline{\text{WR}}$ input lines for the 3-bus system. We assume the $\overline{\text{IOR}}$ and $\overline{\text{IOW}}$ control lines are generated the same way as shown in Chapter 5.

Data lines on the 8253 can be connected directly to the data bus lines in the system data bus. Figure 14-6 shows a complete schematic for connecting the 8253 to the 3-bus system architecture. Figure 14-7 shows the 3-bus system with a data buffer for the 8253. The data bus buffer used is the 74LS245. In Figure 14-7, whenever the 8253 is selected and the $\overline{\text{RD}}$ input is a logical 0, the 74LS245 is set to place the data from the 8253 onto the CPU data bus. At all other times,

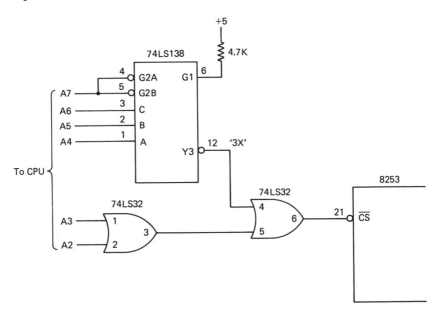

Figure 14-4   Port-select decoding of ports 30H, 31H, 32H, and 33h for use in a microprocessor system.

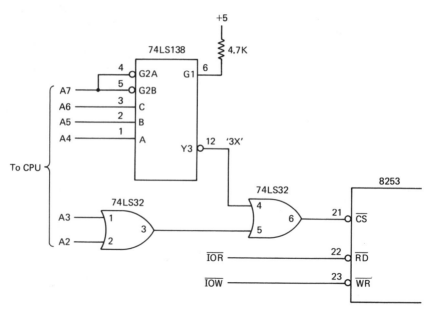

Figure 14-5 Port-select decoding of ports 30H, 31H, 32H, and 33h for used in a microprocessor system with $\overline{IOR}$ and $\overline{IOW}$ connected.

the 74LS245 places the data from the microprocessor system data bus onto the 8253 data pins.

Using the circuit connections shown in Figure 14-6 and 14-7, it is possible to communicate reliably between the CPU and the 8253 device. With the hardware connection complete, the task now becomes one of programming the device so that it will perform in a way that your system application demands.

14-5 PROGRAMMING THE DEVICE (CONTROL WORD FORMAT)

All the operating modes for the counters are selected by writing bytes to the control register. Figure 14-8 shows the control word format. The address for the control word is A0 = 1 and A1 = 1.

In Figure 14-8, bits D7 and D6 are labeled SC1 and SC0. These bits select the counter to be programmed. Before any counter can be programmed, it is necessary to define, using the control bits D7 and D6, which counter is being set up. It should be noted that once a counter is set up, it will remain that way until it is changed by another control word. The bits D7 and D6 are defined as follows:

D7	D6	COUNTER SELECT
0	0	0
0	1	1
1	0	2
1	1	illegal value

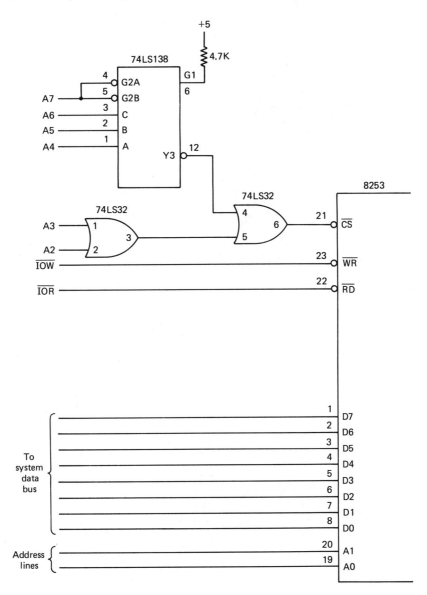

Figure 14-6   Schematic diagram of the hardware connection between the CPU and the 8253.

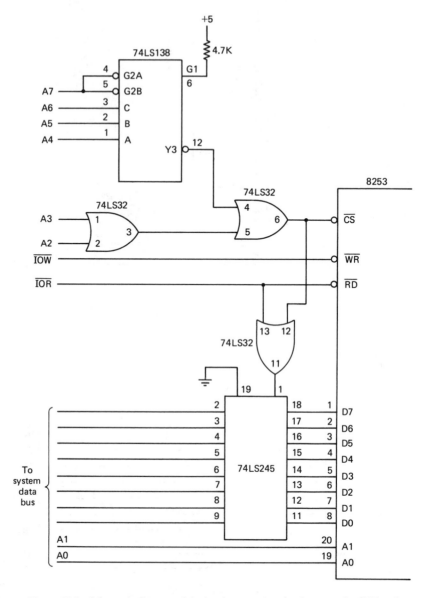

Figure 14-7 Schematic diagram of the hardware connection between the CPU and the 8253 with data buffers used. The data buffer is the 74LS245.

Control byte D7–D0

D7	D6	D5	D4	D3	D2	D1	D0
SC1	SC0	RL1	RL0	M2	M1	M0	BCD

Figure 14-8 Bit definition of the counter control register.

Bits D5 and D4 of the control word in Figure 14-8 are defined as the read/load mode for the register that is selected by bits D7 and D6. Bits D5 and D4 define how the particular counter is to have data read from or written to it by the microprocessor. These bits are defined as follows:

D5	D4	R/L DEFINITION
0	0	Counter value is latched; which means that the selected counter has its contents transferred into a temporary latch; which can then be read by the CPU
0	1	Read/load least significant byte only
1	0	Read/load most significant byte only
1	1	Read/load least significant byte first; then most significant byte.

What do these bits tell the device to do? The first value, 00H, is the *counter latch mode.* It is useful for taking a counter reading during the counter operation. When this mode is specified, the counter value is latched into an internal register at the time of the I/O write operation to the control register. When a read of the counter occurs, it is this latched value that is read.

If the latch mode is not used, then it is possible that the data read back may be in the process of changing while the read is occurring. (See Figure 14-9.) This could result in erroneous data being input by the CPU. To read the counter value while the counter is still in the process of counting, we must first issue a latch control word and then issue another control word that indicates the order of the bytes to be read.

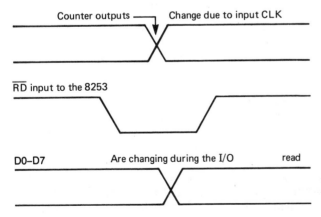

Figure 14-9   A timing diagram showing how the counter outputs may be changing while the microprocessor is electrically reading the 8253 device. This can be avoided by latching the counter output count prior to reading the counter value.

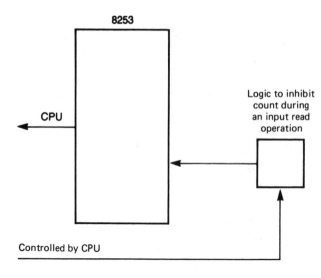

Figure 14-10 A block diagram showing how the counter output can be made stable during a CPU read operation by externally disabling the clock (count) input line.

An alternative method of obtaining a stable or valid count from the timer chip is externally to inhibit counting while the register is being read. This technique is shown in Figure 14-10. Each technique has certain disadvantages. The latching method may give the microprocessor a reading that is "old" by several cycles, depending on the speed of the count and which byte of the counter is being read. The external inhibiting function requires additional hardware. In addition, it may change the overall system operation. It is up to you to determine the best way to operate the device in a specific application.

The next three bits of the control word in Figure 14-8 are D3, D2, and D1. These bits determine the basic mode of operation for the counter. We describe the mode and then follow with examples showing how to use the counter in each of the five modes. Here are the mode descriptions:

```
D3  D2  D1                   MODE VALUE

 0   0   0       mode 0: interrupt on terminal
                 count
 0   0   1       mode 1: programmable one-shot
 x   1   0       mode 2: rate generator
 x   1   1       mode 3: square wave generator
 1   0   0       mode 4: software triggered
                 strobe
 1   0   1       mode 5: hardware triggered
                 strobe
```

The final bit of the control register D0 determines how the register will count—that is, if it will count in BCD or binary. If D0 is a logical 1, the count will be in BCD; if it is a logical 0, the count will be in binary. The maximum values for the count in each count mode are 2^{16} in binary and 10^4 in BCD.

14-6 EXAMPLE OF MODE 0: INTERRUPT ON TERMINAL COUNT

Let's now learn how to use mode 0 with the 8253. We describe the function of mode 0 and define each important pin. Finally, we write a software program using 8080 mnemonics to set up the 8253 for use in mode 0.

Mode 0 allows the 8253 to do the following. When the internal count is equal to 0, the OUT pin of the counter will be set to a logical 1. When the counter is loaded with a nonzero value, the OUT pin is set to zero. The counter will be programmed to an initial value and then count down at a rate equal to the input clock frequency. When the count is equal to 0000, the OUT pin will be a logical 1. In this application the OUT pin could be connected to interrupt the microprocessor. The output will stay a logical 1 until the counter is reloaded with the same count or a new count or until a mode word is written to the device.

Once the counter starts counting down, the gate input can disable the internal counting when the gate is a logical 0. To show how the mode operates, let's look at an example. In this example we will be counting events with counter number 0. The count will be 125 in BCD. When the count is 125, the 8080 will be interrupted. The interrupt service routine is at location 1400H. A block diagram of this example is shown in Figure 14-11.

To set the counter, we must program the control word. Based on the bit definitions of the control given in Figure 14-9, we will choose a control word equal to 00110001. Here is why:

Bits D7 and D6	= 00	This defines counter 0.
Bits D5 and D4	= 11	This sets up the register to load the least significant byte first and the most significant byte next.
Control bits D3, D2, and D1	= 000	This defines the mode as 0.
Finally, D0	= 1	This indicates that the count mode will be BCD. The total control word written to port 33H is then 00110001.

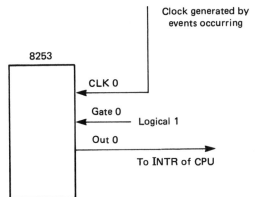

Clock generated by events occurring

8253

CLK 0

Gate 0

Logical 1

Out 0

To INTR of CPU

Figure 14-11 A block diagram of the system in the example. The gate input is enabled, and the counter 0 clock is input to the device. When the count reaches 0, the output will interrupt the CPU.

```
10                              ; WE ASSUME THAT INTERRUPTS ARE DISABLED
11                              ;
12   0000    3E 31                      MVI  A,31H     ;SET UP CONTROL WORD
13   0002    D3 33                      OUT  33H       ;OUTPUT TO 8253
14   0004    3E 25                      MVI  A,25H
15   0006    D3 30                      OUT  30H       ;OUTPUT LSB OF CTR 0
16   0008    3E 01                      MVI  A,01
17   000A    D3 30                      OUT  30H       ;OUTPUT MSB OF CTR 0
18                              ;
19                              ; COUNT IS IN BCD
20                              ;
21                              ; COUNTING STARTS WHEN MSB IS OUTPUT
22                              ;
23   000C    FB                         EI             ;TURN ON INTERRUPTS
24   000D                               END
```

Figure 14-12   Program to initialize the 8253 for counting in BCD with counter 0.

Next, we must write the start count into the counter register. Since we are counting in BCD, the least significant byte will be 25 hexadecimal and the most significant byte will be 01. These two bytes are written to port 30H. Figure 14-12 shows a partial 8080 program that will initialize the 8253 device.

As soon as the second data byte is loaded into register 0, the counting starts. When 125 count pulses are input, the counter register will equal 0 and the OUT pin will become a logical 1, thus interrupting the 8080. The interrupt service routine will take the appropriate action for the application. Part of the interrupt service routine reinitializes the 8253 device. Figure 14-13 shows an example of an interrupt service routine with the 8253 operating in mode 0.

```
10                              ; WE ASSUME THAT THE SERVICE ROUTINE
11                              ; IS AT LOCATION 1400H.  THE COUNT HAS
12                              ; REACHED 125.  AT THIS TIME WE WILL
13                              ; RELOAD THE COUNTER 0.
14                              ;
15   1400                               ORG  1400H     ;ORIGIN OF ROUTINE
16   1400    3E 25                      MVI  A,25H
17   1402    D3 30                      OUT  30H       ;RELOAD LSB
18   1404    3E 01                      MVI  A,01H
19   1406    D3 30                      OUT  30H       ;RELOAD MSB
20                              ;
21   1408    FB                         EI
22   1409    C9                         RET            ;RETURN FROM INTERRUPT
23                              ;
24                              ; PART OF THE INTERRUPT ROUTINE
25                              ; WOULD BE TO PERFORM SOME ACTION
26                              ; REQUIRED WHEN THE COUNTER COUNTED DOWN.
27                              ; THE PART SHOWN HERE WOULD COME AT THE
28                              ; END OF THE SERVICE ROUTINE.
29   140A                               END
```

Figure 14-13   Interrupt service routine for Figure 14-12.

14-7 MODE 1: PROGRAMMABLE ONE-SHOT

In mode 1, the 8253 can be made to give an output pulse that is an integer number of clock pulses. The one-shot is triggered on the rising edge of the gate input. If the trigger occurs during the pulse output, the device will again be retriggered. This is shown in Figure 14-14.

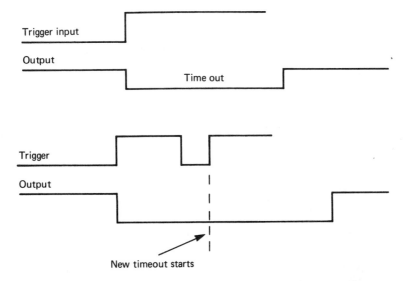

New timeout starts

Figure 14-14 A timing diagram showing how the 8253 is used as a programmable one-shot. When the trigger input goes to a logical 1, the output will go to a logical 0 for the programmed time of the counter. If the trigger is again applied before the counter times out, the time-out will be retriggered.

Let's assume that the input clock frequency is 1 megahertz. We will program the 8253 for an output pulse of exactly 75 microseconds. The 8080 program to perform this setup is shown in Figure 14-15. In this figure we have assumed that counter 1 is being used as the programmable one-shot.

```
 9  0000   3E 72              MVI A,72H
10  0002   D3 33              OUT 33H        ;OUTPUT CONTROL WORD
11                        ;
12                        ; CTR=1, RL=3, M=1, BINARY COUNT
13                        ;
14  0004   3E 4B              MVI A,75       ;DECIMAL 75
15  0006   D3 31              OUT 31H        ;CTR 1, LSB
16  0008   3E 00              MVI A,00
17  000A   D3 31              OUT 31H        ;CTR 1, MSB
18                        ;
19                        ; DEVICE IS NOW INITIALIZED AND
20                        ; WAITING FOR AN EXTERNAL TRIGGER
21                        ;
22  000C                      END
```

Figure 14-15 Program to initialize the counter as a programmable one-shot.

14-8 MODE 2: RATE GENERATOR

In mode 2, the 8253 becomes a *divide-by-n* counter. The output pin of the counter goes low for one input clock period. The time between the output low-going pulses is dependent on the present count in the counter register. See

381

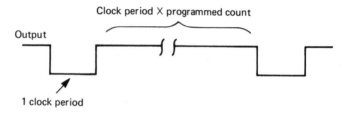

Figure 14-16   A timing diagram showing how the output of the 8253 would appear when used in mode 2 as a rate generator. The output will go to a logical 0 for one input clock period. The time between pulses is determined by the count programmed in the device.

Figure 14-16. If the count is loaded between output pulses, the present period will not be affected. A new period will occur during the next count sequence.

As an example, suppose that we wish to provide an output frequency on counter 2 that is equal to 638 hertz. This would be a period of $1/638 = 1.567$ milliseconds or 1567 microseconds. If an input clock of 1 megahertz were applied to the clock input of counter 2, then the counter would need to be programmed to 1567. This could be done in BCD or decimal. Figure 14-17 shows an example of an 8080 program that would program the 8253 for an operation such as this.

In this figure the counter is in the binary format. This requires the data to be loaded in decimal instead of hexadecimal.

```
 9   0000   3E B5          MVI  A,0B5H
10   0002   D3 33          OUT  33H        ;OUTPUT CONTROL WORD
11   0004   3E 67          MVI  A,67H
12   0006   D3 32          OUT  32H        ;CTR 2, LSB IN BCD
13   0008   3E 15          MVI  A,15H
14   000A   D3 32          OUT  32H        ;CTR 2, MSB IN BCD
15                      ;
16                      ; DEVICE IS NOW INITIALIZED AND
17                      ; THE OUTPUT FREQUENCY WILL BE
18                      ; 638 HERTZ, WITH AN INPUT FREQUENCY
19                      ; OF 1 MEGAHERTZ.
20                      ;
21   000C                  END
```

Figure 14-17   Program to initialize the 8253 as a rate generator.

14-9 MODE 3: SQUARE WAVE GENERATOR

Mode 3 is similar to mode 2 except that the output will be high for half the period and low for half. If the count is odd, the output will be high for $(n + 1)/2$ and low for $(n - 1)/2$ counts. Figure 14-18 shows an 8085 program to setup counter 0 for a square wave frequency of 10K hertz, assuming that the input clock frequency is equal to 1 megahertz.

```
10   0000   3E 36                MVI A,36H
11   0002   D3 33                OUT 33H      ;OUTPUT CONTROL WORD
12   0004   3E 64                MVI A,100
13   0006   D3 30                OUT 30H      ;CTR 0, LSB IN BINARY
14   0008   3E 00                MVI A,00
15   000A   D3 32                OUT 32H      ;CTR 0, MSB IN BINARY
16                           ;
17                           ; DEVICE IS NOW INITIALIZED AND
18                           ; THE OUTPUT FREQUENCY WILL BE
19                           ; 10 KHERTZ, WITH AN INPUT FREQUENCY
20                           ; OF 1 MEGAHERTZ.
21                           ;
22   000C                        END
```

Figure 14-18   Program to initialize the 8253 as a square wave generator with an input frequency of 1 megahertz and an output frequency of 10 kilohertz.

14-10 MODE 3: SOFTWARE-TRIGGERED STROBE

In this mode the programmer can set up the counter to give an output time out that will start when the count register is loaded. On the terminal count, when the counter equals zero the output will go low for one clock period and then it will go high again. This is shown in Figure 14-19. When the mode is set, the output will go high.

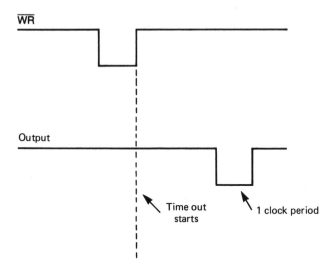

Figure 14-19   A timing diagram showing how the output of the 8253 would appear when used as a programmable software triggered strobe. When the CPU performs an I/O write operation, the time starts. At the end of the count the output will go low for one clock period.

14-11 AN EXAMPLE

As an example, let's program the device to provide a 100-millisecond, software-triggered delay at counter output 2. We will assume that the input frequency equals 1 megahertz, which requires an input clock count of 10^5. In addition, we wish to perform the count in BCD. The maximum BCD count in any register is equal to 10^4. Therefore, we will use two counters. The first will divide the frequency down to 1 kilohertz. The second will provide the software-triggered strobe. This is shown in Figure 14-20. Figure 14-21 presents an 8080 program that uses the 8253 in the manner described in this example.

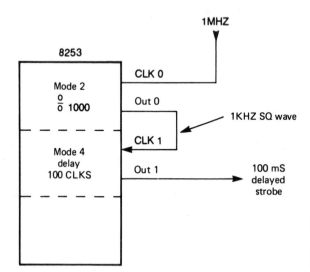

Figure 14-20   A block diagram showing how the 8253 could be used to generate a 100-millisecond-delayed software strobe.

```
 9   0000   3E 35                    MVI A,35H
10   0002   D3 33                    OUT 33H       ;OUTPUT CONTROL WORD
11   0004   3E 00                    MVI A,00
12   0006   D3 30                    OUT 30H       ;CTR 0, LSB
13   0008   3E 0A                    MVI A,10
14   000A   D3 30                    OUT 30H       ;CTR 1, MSB
15                                   ;
16                                   ; COUNTER 0 IS NOW SET UP WITH AN OUTPUT
17                                   ; FREQUENCY OF 1KHZ SQUARE WAVE WITH AN
18                                   ; INPUT FREQUENCY OF 1 MHZ.  THIS OUTPUT
19                                   ; WILL BE CONNECTED TO CTR1 CLOCK INPUT.
20                                   ;
21   000C   3E 79                    MVI A,79H
22   000E   D3 33                    OUT 33H       ;OUTPUT CONTROL WORD
23   0010   3E 00                    MVI A,00
24   0012   D3 31                    OUT 31H       ;CTR1 LSB
25   0014   3E 01                    MVI A,01
26   0016   D3 31                    OUT 31H       ;CTR1 MSB
27                                   ;
28                                   ; THE LAST OUT TO CTR1 NEED NOT BE DONE
29                                   ; UNTIL THE TIMEOUT IS WANTED.
30                                   ;
31   0018                            END
```

Figure 14-21   Program to initialize the 8253 to realize the function shown in Figure 14-20.

14-12 MODE 5: HARDWARE-TRIGGERED STROBE

With the counter in this mode, the rising edge of the trigger input will start the counter counting. The output will go low for one clock at the terminal count. The device is retriggerable, thus meaning that if the trigger input is taken low and then high during a count sequence, the sequence will start over (see Figure 14-22). Figure 14-23 shows an 8080 program to initialize the 8253 counter 1 to be used as a hardware-triggered strobe.

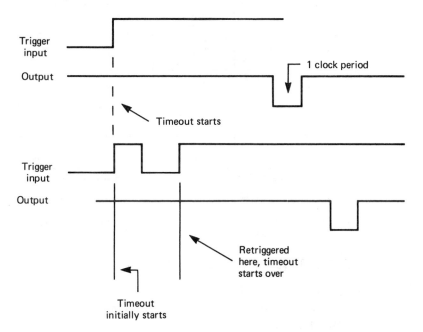

Figure 14-22 A timing diagram showing how the 8253 operates as a hardware-triggered strobe. When the external trigger input goes to a logical 1, the timer will start to time out. If the external trigger occurs again prior to the time of completing a full time-out, the timer will retrigger.

```
 9  0000   3E 7A              MVI A,07AH
10  0002   D3 33              OUT 33H      ;OUTPUT CONTROL WORD
11  0004   3E 55              MVI A,85
12  0006   D3 31              OUT 31H      ;CTR 1, LSB IN DECIMAL
13  0008   3E 00              MVI A,00
14  000A   D3 31              OUT 31H      ;CTR 1, MSB IN DECIMAL
15                          ;
16                          ; DEVICE IS NOW INITIALIZED AND
17                          ; WAITING FOR AN EXTERNAL TRIGGER
18                          ; INPUT ON TRIG1.
19                          ;
20  000C                      END
```

Figure 14-23 Program to initialize the 8253 in a hardware-triggered strobe mode of operation.

Modes \ Signal status	Low or going low	Rising	High
0	Disables counting	— —	Enables counting
1	— —	1) Indicates counting 2) Resets output after next clock	— —
2	1) Disables counting 2) Sets output immediately high	1) Reloads counter 2) Initiates counting	Enables counting
3	1) Disables counting 2) Sets output immediately high	Initiates counting	Enables counting
4	Disables counting	— —	Enables counting
5	— —	Initiates counting	— —

Figure 14-24   Table showing the different uses of the 8253 gate input pin.

14-13 USES OF THE GATE INPUT PIN

Each mode of operation for the counter chip has a different use for the gate input pin. The table shown in Figure 14-24 summarizes the operation of this particular input pin on the device.

14-14 CHAPTER SUMMARY

In this chapter we examined the important points of using the 8253 programmable timer with the 8080 CPU. We presented a block diagram of the device and a discussion of each operating mode. The 8253 can be used to serve a majority of timing functions in many system applications that use the 8080, 8085, or Z80 as a system controller. The examples in this chapter have shown that using this device is not a complicated hardware or software task.

REVIEW PROBLEMS

1. Draw a block diagram of the 8253 timer chip.

2. What is the difference between a BCD counter and a binary counter?

3. What are the maximum and minimum count rates of the 8253?

4. What is the function of the gate input?

5. Draw a schematic for connecting the 8253 to an 8085 microprocessor.

6. Repeat problem 5 for a Z80 CPU.

7. What is the function of the control word register?

8. What is the proper control word to place counter 0 in mode 0, interrupt on terminal count?

9. Write a program to set up the 8253 in the following manner:
   (a) Mode 0
   (b) Use counter 0
   (c) Input frequency of 1 megahertz
   (d) Start count = FF
   (e) Binary count

10. Show a schematic diagram that cascades two counters to give a very long delay count. The second counter will decrement each time the first counter reaches terminal count.

11. Write a program that will allow the 8253 to operate in the manner described in Problem 10.

15

USING DYNAMIC RAMS IN A MICROPROCESSOR SYSTEM

Understanding how to interface and use dynamic RAMs (DRAMs) in a microprocessor system is not a difficult task, but it does require that you have some knowledge of the internal structure of the RAM. You will see for yourself that it is possible to use dynamic RAMs successfully even if you have had no previous experience working with them. As we progress through this chapter, each signal associated with the operation of these memory devices will be thoroughly discussed. You will be introduced to and shown exactly how the terms RAS (row-address strobe), CAS (column-address strobe), and REFRESH apply to dynamic RAM systems.

This chapter contains the general information that may be applied to most dynamic RAMs and dynamic RAM systems. We will build upon the information in this chapter when we show how to interface eight 16K × 1 dynamic RAM chips to the Z80 microprocessor. This interface will be designed and discussed in the latter sections of this chapter.

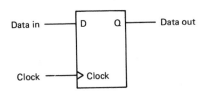

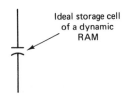

Figure 15-1   The individual storage cells in a static RAM can be thought of a D flip-flops. Once data is clocked into the cell, it remains until new data is clocked in or until power is removed from the device.

Figure 15-2   The storage cell of a dynamic RAM resembles a capacitor in its electrical operating characteristics.

15-1 DYNAMIC RAM STORAGE CELLS

Recall that RAM stands for random access memory. Two types of RAM are widely used today, *static* and *dynamic*. Data is physically stored inside the both types of RAM in a *storage cell*. However, the storage cell in a static RAM differs from the storage cell in a dynamic RAM. Static RAM stores data in a manner similar to a D flip-flop. The D flip-flop holds 1 bit of data until another bit of data is written into it or power is removed from the device. This is shown in Figure 15-1.

A dynamic RAM comprises storage cells that may be thought of electrically as capacitors. See Figure 15-2. There are many thousands of these capacitors, or storage cells, on a dynmic RAM chip. Each cell is capable of storing 1 bit of data.

Let's now focus our attention on the storage cell inside the dynamic RAM device. The capacitor that makes up this storage cell is not ideal. That is, charge placed on this capacitor will leak off given enough time. In a dynamic RAM, the charge on the capacitor represents the stored data. Therefore, the data stored in the cell can be lost. A more accurate model of the storage cell is a capacitor in parallel with a resistor, as shown in Figure 15-3.

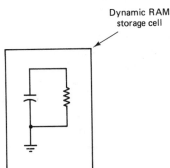

Figure 15-3   An actual storage capacitor loses its stored electrical charge. This may be represented by placing a resistor in parallel with the capacitor. If enough time passes, the charge will leak off through the resistor.

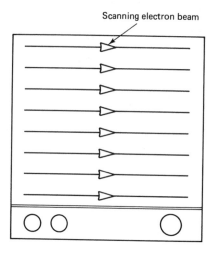

Figure 15-4 The image on a television screen must be constantly updated by the scanning electron beam, or it will fade away.

We cannot keep the capacitor from discharging. However, we do need to retain the information. These two facts present a problem: How do we keep the data in the RAM? This problem is solved by applying a technique call refreshing the RAM, or simply REFRESH. This technique is discussed in detail later in this chapter. For now, you must understand that the data in a dynamic RAM needs to be operated on in some manner to keep it from being lost.

The concept of operating on data to prevent its loss is not a new one. Consider the picture on your television screen. The picture is not a static image like a photograph. The television picture is updated constantly (refreshed) by an electron beam sweeping back and forth across the face of the picture tube. See Figure 15-4. If for some reason, the electron beam fails to scan the screen, the image will fade away. In a similar manner, the charge in a dynamic RAM cell must be constantly restored to the storage cells or data will be lost.

15-2 STORAGE CELL ORGANIZATION

In the previous section we discussed a single dynamic RAM storage cell. In this section we will examine how the thousands of storage cells are organized in a single dynamic RAM chip. Figure 15-5 illustrates how the storage cells in a dynamic RAM are grouped. We show only 16 storage cells to keep the illustration from getting needlessly complex. Once you understand how 16 storage cells are organized in a dynamic RAM, it is an easy matter to extend this small model to much larger array of cells, say 16,384.

Let's examine Figure 15-5 in detail. The 16 cells are organized in a matrix of 4 rows and 4 columns. Each cell in the matrix has a unique position specified by the intersection of a row and a column, as illustrated in Figure 15-6. Using

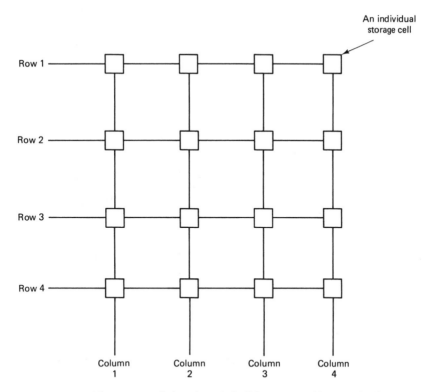

Figure 15-5   The storage cells in a dynamic RAM are grouped in a matrix of rows and columns. We have shown only 16 cells in this figure. An actual dynamic RAM device contains thousands of cells in its storage matrix.

this row and column positioning, any cell can be uniquely identified by its' row and column intersection.

External signals are used to indicate which storage cell in the internal matrix we wish to use (or access). These external signals are given the name *row-address lines* and *column-address LINES*. The row-address lines electrically inform the matrix of the proper rows to access, whereas the column-address line specify the correct columns. See Figure 15-7. You will be shown later exactly how these addresses are input to the RAM chip.

15-3 SENSE AMPLIFIERS

In the previous two sections we explained how the individual storage capacitors are grouped together to form a storage matrix. We also discussed the method used to specify the location of each cell in the matrix. Let's now examine how data is stored in and retrieved from these storage cells.

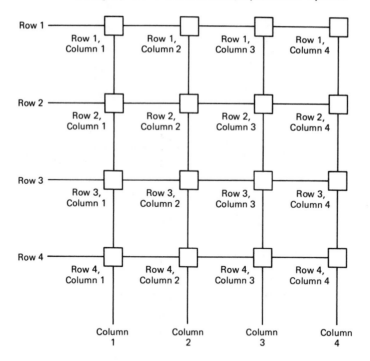

Figure 15-6   Each storage cell in the matrix has a unique position specified by the intersection of a row and a column. The identification for each cell is shown by the row and column number pictured beside it.

Due to the way in which dynamic RAMs are fabricated, the storage capacitor is not capable of providing a large current to an external load. Therefore, a circuit called a *sense amplifier* is placed at the output of the storage capacitor to increase the output current drive capability of the cell. See Figure 15-8.

The function of the sense amplifier is illustrated in Figure 15-9. Solid-state switch (SSW) S_1 is fabricated in the dynamic RAM chip to isolate the storage cell from all circuits external to that cell. S_1 is closed when the row address and column address corresponding to the storage cell are selected. Upon closing S_1, the charge stored on the capacitor flows through S_1 and resistor R_L. Current flowing through a resistor causes a voltage to be developed across it. The voltage developed across R_L is increased by the sense amplifier to a level suitable for driving an external load. This voltage is usually TTL (transistor transistor logic) compatible, which is as follows:

LOGIC 1 VOLTAGE LEVEL
2 volts to 5 volts

LOGIC 0 VOLTAGE LEVEL
0 volts to .8 volts

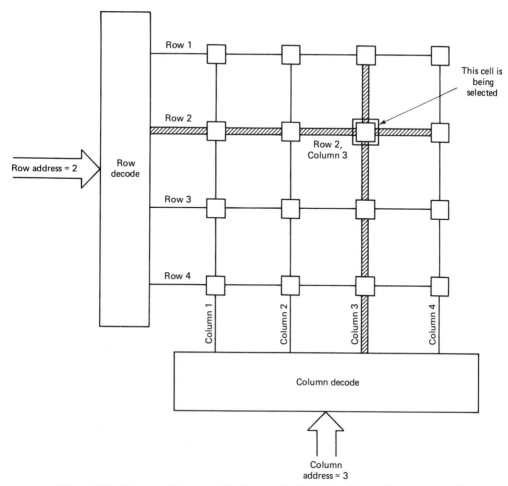

Figure 15-7 The row address specifies the row in which the cell we wish to access resides. The column address indicates in which column the cell is.

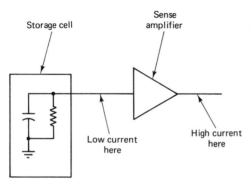

Figure 15-8 A sense amplifier is a device, internal to the dynamic RAM, that is connected to the output of a storage cell. The storage cell has a very low current output while the sense amplifier has an output capable of driving an external load.

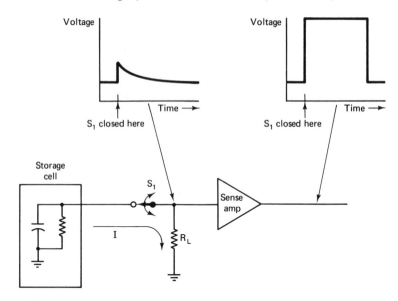

Figure 15-9 The current, I, flowing through R_1 develops voltage across R_1. This voltage is amplified by the sense amplifier to a level capable of driving a load.

When S_1 is closed, current flows from the storage cell through R_L, and the charge on the capacitor (if there is one) is lost. (*Note:* If a logic 0 were stored on the capacitor, there would be no charge present on the capacitor. If a logic 1 were stored on the capacitor, it would be electrically charged.) This type of read is known as a *destructive read* because data is destroyed simply by reading it!

The destructive read presents a problem. Data should not disappear after it is read from a storage cell. The sense amplifier, along with other circuits, is used to solve this problem. The following discussion refers to Figure 15-10. Charge flows out of the storage capacitor through R_L via the switch S_1. The voltage developed across R_L is increased by the sense amplifier and input to a latch, as shown in Figure 15-10a.

After the sense amplifier inputs the data to the latch and the charge on the storage capacitor has been depleted, switch S_1 is set to position A. This action allows the output of the latch to restore the charge on the capacitor. (See Figure 15-10b.) That is, if a logical 1 was stored in the cell, a logical 1 is rewritten to the cell. If a logical 0 was stored in the cell, a logical 0 is rewritten to the cell.

15-4 BLOCK DIAGRAM OF THE 4116 DYNAMIC RAM CHIP

Now that we have discussed the internal structure of a general dynamic RAM chip, let's turn our attention to an actual dynamic RAM, the 4116, 16K × 1 RAM. The chip has 16,384 storage cells in its' internal matrix and is commonly

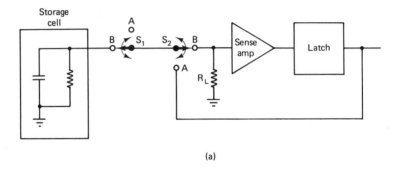

(a)

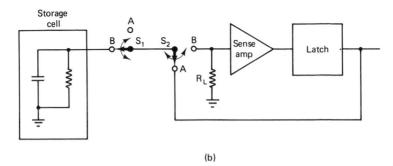

(b)

Figure 15-10   (a) During the first part of a read operation, switch S_1 is in position B, and switch S_2 is in position B. This allows the storage cell to discharge through R_1. The current flowing through R_1 develops a voltage at the input of the sense amplifier. This voltage is increased by the sense amplifier to a level capable of driving the latch; (b) during the second part of the read operation, switch S_2 is set to position A. This allows the output of the latch to rewrite the data that was in the storage cell back to the cell. In this way, data is refreshed during a read operation.

referred to as a 16K × 1 bit DRAM. This means there are 16K (16,384) unique *storage locations,* each capable of storing 1 bit of data. A location is made up of one or more storage cells. In the 4116, a location is equal to a single storage cell. Other memory devices may be described in a different manner, such as 16K × 4 RAMs. In a 16K × 4 DRAM, there are 16,384 unique storage locations, each one consisting of 4 storage cells.

A block diagram of the 4116 is shown in Figure 15-11. Notice that many of the blocks we have previously discussed are shown in this figure. For example, the storage cell matrix is similar to the storage cell matrix we introduced in Figure 15-5. The row-address decoder was shown when we examined how individual storage cells were addressed, and the sense amplifier block was discussed in Section 15-3.

Let's now examine each block in Figure 15-11 and discuss its function in the dynamic RAM. Do not be alarmed if some questions still remain at this

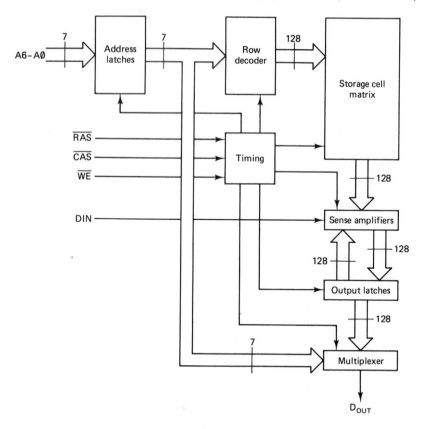

Figure 15-11   A block diagram of the 4116.

point in the chapter. This discussion is meant as an introduction only. More detail will be given as your understanding of the basics increases.

Storage Cell Matrix Block

There are 16,384 storage cells in the 4116 organized in a matrix of 128 rows and 128 columns. Each storage cell contains a storage capacitor and an SSW, which isolates the capacitor from the rest of the circuits in the dynamic RAM chip.

Address Latch Block

Seven address signals, A6–A0, are input to this block. At the proper time (determined by the timing block) an address is latched into the address latch block. The output of this block is input to both the row decode block and the column multiplexer block.

Row Decoder Block

This block has seven input lines and 128 output lines. Each address line represents a single bit in a 7-bit binary number. Therefore, 128 unique numbers can be specified ($2^7 = 128$). Each of these numbers represents a single row in the storage matrix. One output line of the row decoder is asserted when an address is applied to its input. This output line electrically informs the selected row that data is either going to be read from it or written to it.

Sense Amplifier Block

The outputs of the columns of cells in the storage matrix are tied together on a single line called a *bit line*. See Figure 15-12. There are 128 bit lines in the storage matrix which are input to 128 sense amplifiers. (*Note:* The bit line contains the output of only one storage cell at a given time.)

Data Latch Block

After data is read from a row of cells, it is written into the latches in this block. As we discussed earlier, this data is then rewritten to the selected row of storage cells.

Column Multiplexer Block

Only one of the 128 lines input to this block is switched through to the DOUT output line. The seven address signals input to the dynamic RAM chip, A6–A0, determine which one is selected.

Timing Block

The sequence in which electrical events take place within the dynamic RAM is determined by this block. It also provides control signals to most of the other blocks in the dynamic RAM. These control signals electrically inform the various circuits inside the RAM chip of the operations that are being performed. For example, the address latch is electrically instructed to latch the row address when the input signal RAS is asserted.

15-5 PINOUT OF THE 4116

In this section we discuss the external signals that are used to connect the 4116 dynamic RAM chip to a microprocessor. Figure 15-13 shows the pinout of the 4116 device. Each pin's function shown in Figure 15-13 falls into one of four categories:

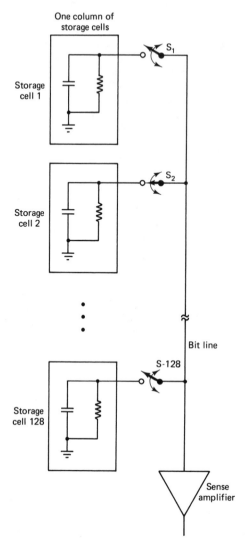

One column of
storage cells

Storage
cell 1

Storage
cell 2

Bit line

S-128

Storage
cell 128

Sense
amplifier

Figure 15-12 The outputs of 128 storage cells have their outputs tied together on a column line. Only one storage cell is connected to the sense amplifier at any given time. For example, if storage cell 2 was being accessed, switch S_2 would be closed. All other switches (S_1 and S_3–S_{128}) would be open.

1. Signals to provide power to the chip.
2. Signals to specify the row and column address.
3. Signals to get data into and out of the chip.
4. Signals to control the read, write, and refresh operations.

   Therefore, we may group the 4116's signals into the categories that follow.

POWER
$$V_{dd} = +12V, V_{cc} = +5V, V_{ss} = 0 \text{ V}, V_{bb} = -5V$$

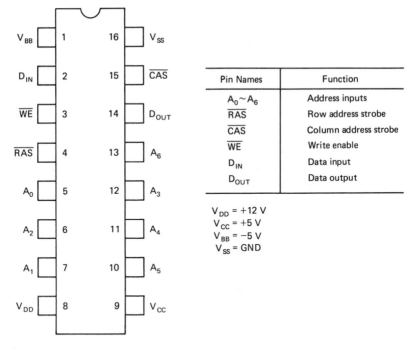

Figure 15-13 Pin configuration of the Intel 4116.

Pin Names	Function
$A_0 \sim A_6$	Address inputs
$\overline{RAS}$	Row address strobe
$\overline{CAS}$	Column address strobe
$\overline{WE}$	Write enable
D_{IN}	Data input
D_{OUT}	Data output

$V_{DD} = +12 \text{ V}$
$V_{CC} = +5 \text{ V}$
$V_{BB} = -5 \text{ V}$
$V_{SS} = \text{GND}$

ADDRESS
A6, A5, A4, A3, A2, A1, A0

DATA
DIN (Data input), DOUT (Data output)

CONTROL
$\overline{RAS}$ (row-address strobe), $\overline{CAS}$ (column-address strobe)
$\overline{WE}$ (write enable)

Do not be concerned if you are unfamiliar with the signals listed, because we will be discussing each signal and describing its function. A more detailed examination of these signals will be given later in this chapter, where a system is designed and we interface these devices to a microprocessor.

Let's now take each signal of the 4116 and discuss it's function in a dynamic RAM system.

A6–A0

A6 is the MSB address line, and A0 is the LSB. The microprocessor accesses one memory cell by outputting the row and column address on these seven lines. We show exactly how this is accomplished later in the chapter.

$\overline{\text{CAS}}$

When the microprocessor has output the column address on the A6–A0 address lines of the dynamic RAM chip, the $\overline{\text{CAS}}$ line is asserted. This action electrically informs the dynamic RAM to latch the column address.

$\overline{\text{RAS}}$

The $\overline{\text{RAS}}$ signal electrically informs the RAM chip that there is a valid row address on the A6–A0 address lines of the device. When $\overline{\text{RAS}}$ is asserted, the row address is latched internally. Note that the address lines A6–A0 are shared by the row address and the column address. We show how these lines are shared when the term multiplexing is discussed.

DIN

Data is input to a storage cell by electrically driving this line to a logical 0 or a logical 1. After the row address and column address are latched by the 4116, the microprocessor writes to the cell by placing data on this line.

DOUT

Data is read from the RAM chip via this line. After the row address and the column address are latched (strobed) into the chip, the contents of the selected storage cell are output to this line. If a logical 1 is stored in the storage cell, the voltage on the line will be between 3.4 V and 5 V. If a logical 0 is stored in the cell, the voltage on the DOUT line will be between 0 and .4 V.

$\overline{\text{WE}}$

This signal electrically informs the dynamic RAM whether the microprocessor is reading data from the RAM or writing data to the RAM. A logical 0 on this line indicates a write operation, whereas a logical 1 denotes a read operation.

15-6 WRITING DATA TO THE 4116

In the previous section we introduced you to the 4116 and the signals necessary to interface it to a microprocessor. We now focus on the write operation and explain exactly how these signals are used to write data to the RAM chip.

Writing data to the RAM occurs in a sequence of three events:

1. Generating an address. The microprocessor produces a row address and a column address. These two addresses are output by the microprocessor onto the lines A6–A0 of the 4116.

2. Decoding the address. After the row address and column address are latched into the RAM, it is decoded. This action selects a unique storage cell in the device.

3. Writing the data. The microprocessor places data on the DIN line of the 4116 and asserts the $\overline{WE}$ line of the RAM. This action writes the data present on the DIN line to the selected storage cell.

We now examine each of these events in detail. In Figure 15-14 we show 14 of the microprocessor's address lines. They are labeled MA13–MA0. The row address for the dynamic RAM is output on the address lines MA6–MA0 of the microprocessor. The column address is output on the microprocessor's address lines MA13–MA7. These 14 lines must somehow be interfaced to the 7 address lines of the dynamic RAM. See Figure 15-15.

The circuit that performs this interfacing task is called a *multiplexer*. You were introduced to the term multiplexer in Section 15-4 when we discussed the block diagram of the 4116. The multiplexer shown in Figure 15-15 inputs the row address (MA6–MA0) and the column address (MA13–MA7) and outputs each group at the proper time to the dynamic RAM address lines RA6–RA0.

Figure 15-16 illustrates that the multiplexer is equivalent to 7 SPDT switches. Transistors perform this switching function on the multiplexer chip, but for our purposes it will make the multiplexer easier to understand if we think of these transistors as being SPDT switches.

When the switches S_1–S_7 in Figure 15-16 are in position A, the address lines from the microprocessor, MA6–MA0, are switched through to the output pins of the multiplexer. This corresponds to the row address being applied to the RAM address lines.

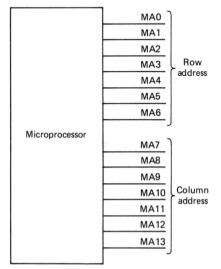

Figure 15-14 The microprocessor places the row addresses on address lines MA6–MA0. The column address is placed on the address lines MA13–MA7.

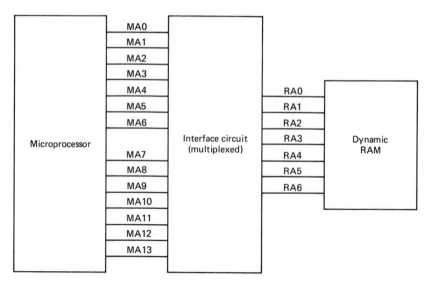

Figure 15-15 A multiplexer circuit interfaces the row address lines (MA6–MA0) and the column address lines (MA13–MA7) of the microprocessor onto the dynamic RAM's address lines, RA6–RA0.

By setting the switches to position B, the input lines from the microprocessor, MA13–MA7, are connected to the outputs of the multiplexer. This corresponds to the column address being input to the address lines of the 4116.

The switches can be electrically operated by controlling an input signal to the multiplexer called MUX. See Figure 15-17. MUX is a signal generated by the microprocessor. By controlling this input to the multiplexer, the microprocessor controls when the row address is applied to the RAM and when the column address is input to the address lines of the RAM. In Figure 15-17, a logical 0 on the MUX input line causes the microprocessor's address line MA0 to be connected to the RAM's address line RA0. When MUX is changed to a logical 1, the microprocessor's address line MA7 is switched through to the dynamic RAM's address line RA0. In this way, the microprocessor's address lines MA6–MA0 and MA13–MA7 electrically share the dynamic RAM's address lines RA6–RA0 as controlled by MUX. (Refer back to Figure 15-16.)

Figure 15-18 illustrates the timing between MUX, $\overline{\text{RAS}}$, and $\overline{\text{CAS}}$. When the row address is applied to the dynamic RAM, the $\overline{\text{RAS}}$ input to the RAM must become asserted. This action strobes the row address into an internal latch. In a similar fashion, the column address is strobed into the RAM by asserting $\overline{\text{CAS}}$ after the column address is input to the 4116's address lines.

Let's now return to our original goal of writing data to a RAM call. When the write operation begins, MUX is a logical 0, which places the microprocessor's address lines MA6–MA0 on the dynamic RAM's address lines RA6–RA0. The $\overline{\text{RAS}}$ signal now goes to a logical 0, which latches the row address into the RAM. After this occurs the MUX signal is set to a logical 1. At this

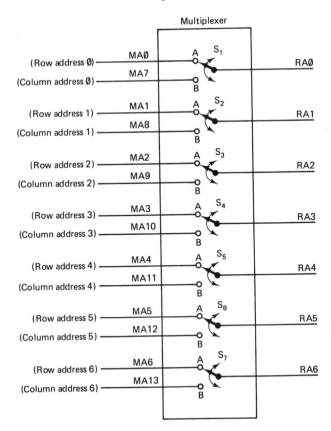

Figure 15-16 A multiplexer connects the microprocessor's address lines, MA6–MA0 (row address) to the dynamic RAM's address lines RA6–RA0 during the time that the switches S_1–S_7 are in position A. When these switches are set to position B, the multiplexer connects the microprocessor's address lines, MA13–MA7 (column address), to the dynamic RAM's address lines, RA7–RA0.

time the column address (MA13–MA7) is placed onto the RAM's address lines RA6–RA0. CAS then becomes asserted, strobing the column address into the RAM. At this point in the write operation, the correct cell in the storage matrix is selected. We now must write data to that cell.

Data is placed on the DIN line of the 4116 by the microprocessor. When the WE line of the RAM is asserted by the microprocessor, the data is strobed into the selected storage cell. See Figure 15-19. This entire operation of writing data to a storage cell in the RAM chip is known as a *write cycle*.

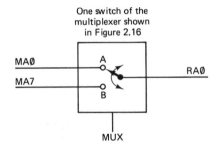

Figure 15-17 The input signal to the multiplexer, MUX, controls whether switches S_1–S_7 are set to position A or position B. The position of these switches determines whether the row address or the column address is connected to the dynamic RAM's address lines.

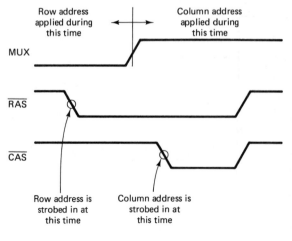

Figure 15-18 The control signal, MUX, selects between the row and column address. When MUX is a logical 0, the row address is selected, and the $\overline{RAS}$ is asserted to latch the row address into the RAM. MUX then goes to a logical 1, which selects the column address. $\overline{CAS}$ is asserted, and the column address is latched into the RAM.

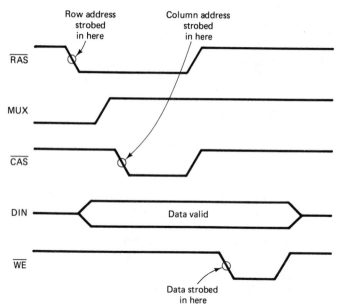

Figure 15-19 After the row address and column address has been latched into the RAM, data is placed by the microprocessor, onto the DIN line of the RAM. This data is strobed into the RAM when WE goes to a logical 0 level.

15-7 READING DATA FROM THE 4116

Data is read from the 4116 in a manner similar to the write operation. To start the read cycle, a cell in the storage matrix is addressed in exactly the same way as in a write operation. That is, the row address and the column address

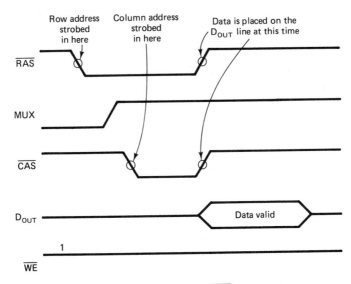

Figure 15-20   The timing of MUX, $\overline{RAS}$, and $\overline{CAS}$ is exactly the same for a read operation as a write operation, which was illustrated in Figure 15-19. The $\overline{WE}$ strobe remains at a logical 1 throughout the entire read cycle. Data is placed on the DOUT line of the RAM chip when $\overline{RAS}$ and $\overline{CAS}$ go to a logical 1.

are applied to the RAM via the multiplexer. RAS, CAS, and MUX have the same timing in the read operation as they did in the write operation. See Figure 15-20.

We stated earlier in the chapter that when a cell is selected, the entire row in which it resides is read. Selection of the proper cell is then determined by the column address. Figure 15-21 illustrates how the column address selects the data output of one cell in the accessed row. Notice that the data of an entire row of cells are rewritten to the cells through the sense amplifiers as we described in Section 15-3. This means that 128 of the 16,384 storage cells are refreshed each time a single cell is read. We make use of this fact in the next section, where refreshing the RAM is presented.

15-8 REFRESHING THE 4116

In Section 15-1, you were introduced to the concept of refreshing the dynamic RAM. At that time it was stated that refreshing is needed in a dynamic RAM system to keep data from being lost. You were then shown two types of RAM cycles, the read cycle and the write cycle, in which data is read from and written to the dynamic RAM chip. We now discuss another cycle, the *RAS-only refresh cycle,* in which data in the 4116 is restored to the storage cells on a row-by-row basis without reading from or writing to the RAM.

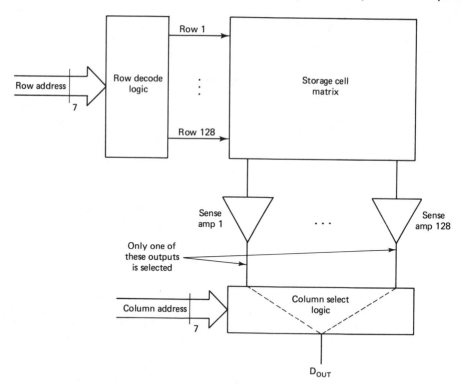

Figure 15-21   The column address selects which sense amplifier is connected to the
DOUT line of the RAM.

The RAS-only refresh cycle starts in the same way as a normal read cycle. That is, a row of cells is selected by the row address and $\overline{RAS}$ is asserted to latch the row address. At this time the row data is rewritten internally to the selected row. No column address is required because we are not accessing a particular cell. The only objective is to rewrite the row data (refresh). Since the $\overline{RAS}$ signal is the only strobe required for this operation, the cycle is designated RAS-only refresh.

Applying RAS-only refresh to a dynamic RAM memory system requires that another circuit, external to the RAM, must be used. This circuit is known as a *dynamic RAM controller* and may be designed using discrete logic gates or as an LSI chip.

Let's examine the general characteristics of dynamic RAM controllers. A dynamic RAM controller consists of circuits that perform the following functions:

1. Multiplex the row and column address output by the microprocessor onto the dynamic RAM's address lines. This was discussed in Section 15-6.

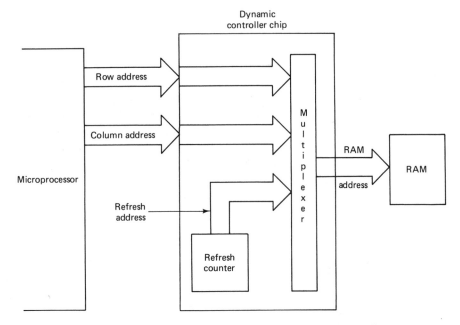

Figure 15-22 A binary counter, internal to the dynamic RAM controller, generates the refresh address. This refresh address is then multiplexed with the row address and column address. The output of the multiplexer is electrically connected to the RAM system.

2. Use the RAS-only refresh. This operation involves multiplexing another address (the refresh address) onto the dynamic RAM's address lines. Figure 15-22 illustrates how the refresh address is multiplexed with the row address and column address onto the dynamic RAM's address lines.

The refresh address is generated by the block labeled *refresh counter* in Figure 15-22. The output of this counter is a 7-bit address, CA6–CA0, which specifies the row of storage cells to be refreshed.

Up to this point in our discussion of refreshing, we have stated that the storage cells need to be refreshed periodically in order to prevent the loss of data. However we have not specified how often the refresh operation must occur. This time between refresh operations is specified by the manufacturer's data sheet for the device. This time is usually 2 milliseconds, at 25°C.

The 4116's storage matrix contains 128 rows and 128 columns, so a 128-cycle refresh operation is needed to refresh the entire storage matrix. For a 128-cycle refresh operation, the time between refresh is 2 milliseconds. Therefore, every row of the dynamic RAM's storage matrix must be refreshed every 2 milliseconds.

Two methods of refreshing RAM may be used to accomplish this task. They are known as

1. Burst refresh
2. Cycle steal refresh

Burst Refresh

Burst refresh is a technique by which the entire memory array is refreshed without interruption from the microprocessor. This refreshing is accomplished in the following manner. The refresh counter outputs the first address onto the RAM's address lines A6–A0. This address is then latched into the chip by asserting $\overline{RAS}$. See Figure 15-23. The first row of the total 128 rows contained in the RAM's storage matrix is refreshed by this action. $\overline{CAS}$ remains at a logical 1 level during this time; hence the phrase RAS-only refresh is applied. Next, the refresh counter is incremented by 1. $\overline{RAS}$ is again strobed, and the next row is refreshed. This pattern is repeated (that is, incrementing the refresh counter and strobing it into the RAM until all 128 rows of the memory device are refreshed). This takes approximately 43 microseconds to complete.

During this time, the microprocessor is prevented (held off) from accessing the dynamic RAM. That is, neither read operations nor write operations are permitted to take place with the RAM system while it is being burst refreshed.

Cycle-Stealing Refresh

The burst refresh method requires the microprocessor to stop its activity every 2 milliseconds for a time long enough to perform a refresh of the entire dynamic RAM system. This means that out of every 2 milliseconds, of time, up to 43 microseconds is used for the refresh operation; the microprocessor may not read from RAM or write to RAM during this 43 microseconds. This is perfectly acceptable in some microprocessor applications. However, other applications may require that the microprocessor not be denied access for this length of

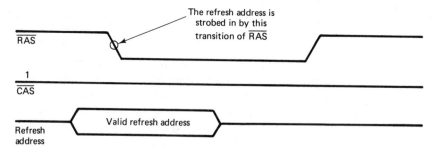

Figure 15-23 The refresh address is applied to the address lines of the RAM and strobed in by $\overline{RAS}$ going to a logic 0 level. Notice that $\overline{CAS}$ remains at a logic 1 level throughout the refresh cycle. This is known as RAS-only refresh.

time. For these applications, burst refresh is unacceptable, and cycle-stealing refresh is used.

Cycle-stealing refresh gets its name from the fact that RAM is refreshed by fitting the refresh operation into the microprocessor's normal operating cycle. That is, by making use of times when the microprocessor is not accessing memory, the refresh operation can be performed without having to stop microprocessor activity.

Designing a RAM system to accomplish this requires knowledge of the microprocessor being used and the way in which it accesses memory. You will be shown how the Z80 uses cycle-stealing refresh to interface to dynamic RAMs in a later section of this chapter.

Microprocessor instruction cycles contain times in which the microprocessor is fetching an op-code from memory and times when it is decoding the instruction. During the decoding time, the refresh counter outputs a row address onto the dynamic RAM's address bus A6–A0 and strobes it into the RAM by asserting $\overline{\text{RAS}}$. This action refreshes the addressed row in the storage matrix of the dynamic RAM. The refresh counter is then incremented to the next row address. The refresh of a single row does not affect the microprocessor's ability to access memory and therefore is *transparent* to the microprocessor. This means the cycle-steal refresh technique must output a new row address at least every (2 milliseconds)/128, which equals 15.6 microseconds. This process of refreshing rows during times when the microprocessor is not accessing memory is repeated continuously. In this way the microprocessor activity does not have to be halted for the refresh operation.

15-9 INTERFACING TO THE Z80 MICROPROCESSOR

Let's now build on the general information given in the first sections of this chapter and actually interface a dynamic RAM system to a microprocessor. We choose the Z80 microprocessor for this purpose because it has been designed for ease of interfacing with dynamic RAMs.

Multiplexing the Address Lines

A circuit line the one in Figure 15-24 can be used to perform the function of multiplexing the address inputs to the 4116. Here is how the circuit works. Address lines A0–A13 are input to the two 74LS157 digital multiplexers. A0–A6 are routed to the 4116 address inputs for the $\overline{\text{RAS}}$ active and A7–A13 are input to the 4116 during the $\overline{\text{CAS}}$. Notice the placement of the address inputs from the microprocessor to the pins of the multiplexers. Pin 1 of the 74LS157s controls the multiplexing function. We show in a later discussion how pin 1 is controlled.

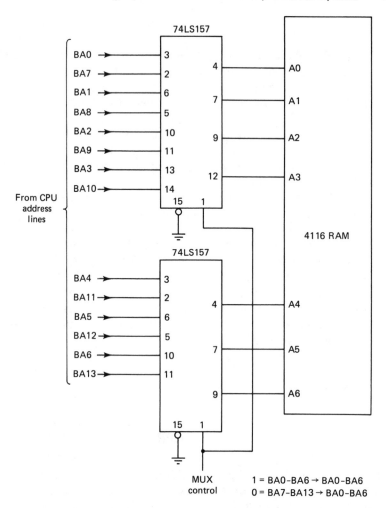

Figure 15-24 Schematic diagram showing how the 14 address inputs are applied to the 4116 device. The address lines from the microprocessor are input to the 74LS157 multiplexors. The outputs of the multiplexors are input to the RAM pins A0–A6.

Block Diagram of the Memory Array

Figure 15-25 shows the block diagram of how the dynamic RAMs will be arranged in our memory system. In viewing this diagram it is important to remember that there are many ways to interface a microprocessor with dynamic RAMs. We have chosen the technique presented here because it is straightforward and can be understood by those with only a modest amount of digital hardware experience.

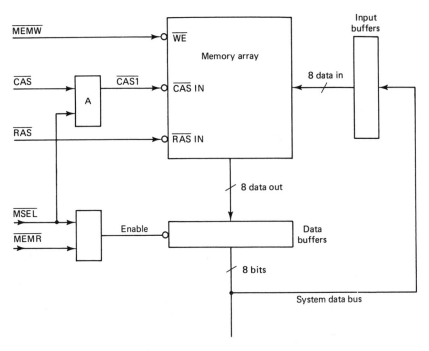

Figure 15-25   Block diagram showing the peripheral hardware associated with a dynamic RAM system.

The $\overline{RAS}$ line will be input to all memory devices in the system memory in a parallel fashion. The $\overline{CAS}$ input to the memory system may or may not be asserted during the memory operation because only the $\overline{RAS}$ input is required for RAM refresh. However, when accessing data both the $\overline{RAS}$ and $\overline{CAS}$ must be active.

The logic for generating the $\overline{CAS}$ input is shown by block A of Figure 15-25. This logic will decode the conditions of the memory-select lines and assert the $\overline{CAS}$ input to the memory at the correct time. The memory-select line is generated from the decoding of the system address lines. Our memory array will occupy address space from C000–FFFF.

Memory $\overline{WE}$ input is active logical 0 whenever the $\overline{MREQ}$ and $\overline{WR}$ outputs from the Z80 are active. These lines are active when the Z80 is performing any memory-write operation. We will show how unwanted write operations to the dynamic memory space are disabled.

Generating $\overline{RAS}$, $\overline{CAS}$, and MUX

Let's examine how the $\overline{RAS}$, $\overline{CAS}$, and MUX control lines can be generated with the Z80 microprocessor. Figure 15-26 shows how this can be accomplished with hardware. The $\overline{RAS}$ is a buffered $\overline{MREQ}$ signal and strobes the lower

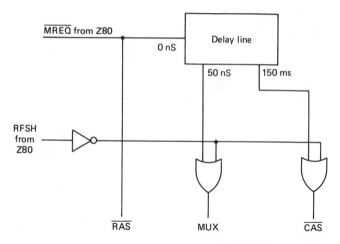

Figure 15-26   Schematic diagram showing how the $\overline{\text{RAS}}$, $\overline{\text{CAS}}$, and MUX control signals are generated.

address lines A0–A6 into the 4116. Note that the $\overline{\text{MREQ}}$ signal is active at the correct time in the memory cycle to accomplish this objective.

The MUX signal switches the address input lines from BA0–BA6 to BA7–BA13. MUX is generated from the $\overline{\text{RAS}}$ line after it has gone through a delay line. Any standard delay line may be used.

After the MUX line has switched to a logical 0, the $\overline{\text{CAS}}$ line can be asserted. Another delay line is used here. This technique for DRAM signal generation is very easy to realize with hardware.

Data Input to the RAM

Figure 15-27 shows a buffer circuit, which connects the system data bus to the 4116 input pins. Notice in this figure that the buffers are always enabled. Data from the system data bus will electrically appear at the RAM input pins at all times. This is electrically valid because the data will not be stored into the RAM until a $\overline{\text{WE}}$ signal is asserted to the device.

It is a good idea to use data buffers to isolate the RAM inputs from the entire system data bus. This will help in the debugging of a defective system. Also, the data buffers electrically isolate the RAM inputs from the glitches that normally occur on a microprocessor data bus.

Writing Data to the RAM

Let's look now at how data is written to the dynamic RAM during a memory-write cycle. The following signals must be active during a memory write: (1) $\overline{\text{MREQ}}$ and (2) $\overline{\text{WR}} = 0$. These two signals will assert the $\overline{\text{WE}}$ line to the RAM.

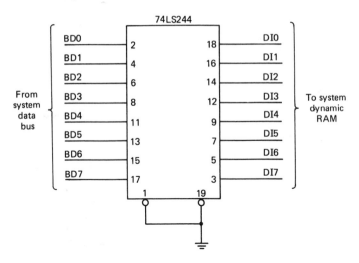

Figure 15-27 Hardware required to input the data to the RAM input lines. In some systems this input buffer may not be used. In those cases the dynamic RAM inputs will be connected directly to the system data bus lines.

The RAM space is selected when BA15 and BA14 are both logical 1s. When these conditions are met, the $\overline{CAS}$ to the RAM system will be asserted. The circuit for accomplishing the $\overline{WE}$ to the memory at the correct time is shown in Figure 15-28. The timing diagram for a memory write operation is shown in Figure 15-29a and b.

Reading Data from RAM

The circuit for enabling the data from memory onto the system data bus is shown in Figure 15-30. When the microprocessor is electrically reading data from the correct memory space, the output data buffers of Figure 15-30 are enabled.

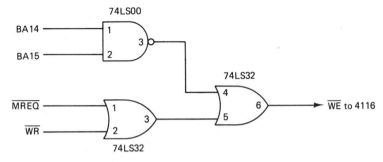

Figure 15-28 Schematic diagram showing how the $\overline{WE}$ signal to the DRAM is generated in this system application example.

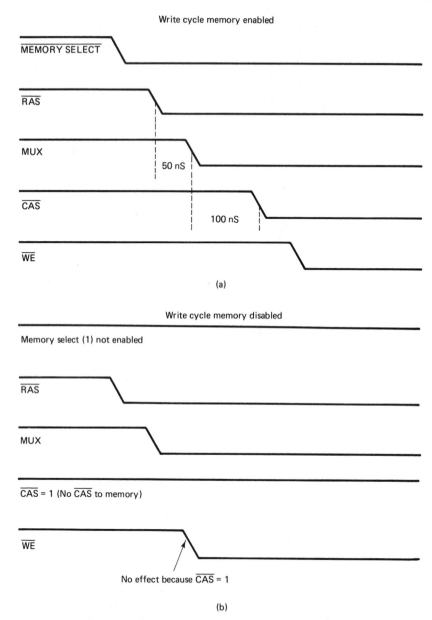

Figure 15-29 (a) Timing diagram showing the sequence of signals that will become asserted during a memory-write cycle when the dynamic RAM space is enabled; (b) timing diagram showing the sequence of signals that are asserted during a memory-write cycle when the dynamic RAM is not enabled. Notice the memory-select line is a logical 1 during this operation. Also notice no $\overline{CAS}$ signal is generated.

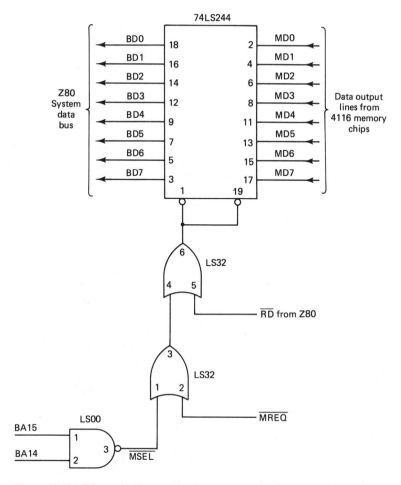

Figure 15-30 Schematic diagram showing one way the data output lines of the dynamic RAM can be placed onto the system data bus during a memory-read operation to the RAM space.

Refreshing the RAM

Up to this point in the discussion we have shown how the RAM communicates electrically with the microprocessor. Let's now discuss how the refreshing of the RAM is accomplished. It is this area of refreshing the RAM where we take advantage of the hardware designed into the Z80 CPU.

Internal to the Z80 is a counter that is called the refresh counter, or R register. The refresh counter can be set to output a refresh address on the system address bus at a specified point in time. After the refresh address has

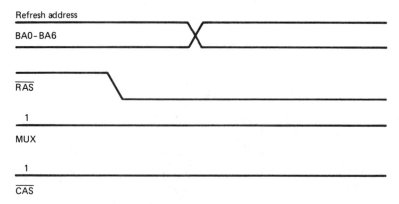

Figure 15-31 Timing diagram showing the relationship of the main dynamic RAM control signals during a memory-refresh operation.

been output, the Z80 will automatically increment the counter to output the next refresh address at the specified time.

During the op-code fetch machine cycle of an instruction cycle, the refresh address is output on the Z80 address lines A0–A6. The address is output on the Z80 address lines A0–A6. The address is output during the op-code decoding time (t4) of the Z80 cycle. Recall that this type of refresh addressing is called cycle stealing.

When the refresh address is output on the Z80, an output pin called RFSH is asserted. Figure 15-31 shows the timing required for a refresh cycle. During this time the $\overline{\text{MREQ}}$ control line is active, which will generate the $\overline{\text{RAS}}$, but the $\overline{\text{RD}}$ control line is not asserted. Logic for the $\overline{\text{CAS}}$ generation already shown will keep it from occurring during a refresh cycle.

It is implied that the Z80 will output a refresh address only when executing a program. Further, the time interval between op-code fetches must be consistent with the refresh time for the dynamic RAM.

A Complete Schematic for a Dynamic Ram System

Figure 15-32 shows a complete schematic for a dynamic RAM system for use with the Z80 microprocessor. Note the use of series resistors when interfacing the low-output impedance signals of the TTL to the high-input impedance of the 4116 DRAM. You are reminded that the hardware realization of the system can be done in many ways. The design chosen for this text was calculated to make you aware of the essential details to be considered when using dynamic RAMs.

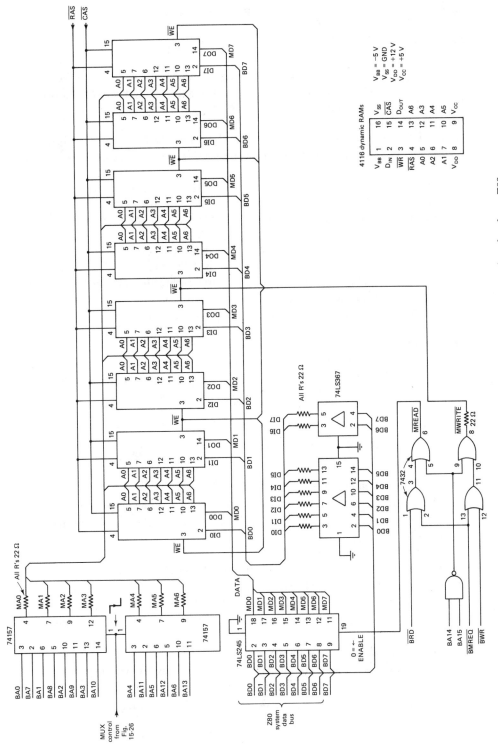

Figure 15-32 Complete schematic of a 16K X 8 dynamic RAM memory interfaced to a Z80 microprocessor.

417

15-10 CHAPTER SUMMARY

The chapter opened with an introduction to the basic memory-building block, the storage cell. We described the differences between storage cells used in static RAM and dynamic RAM. The storage cell in a dynamic RAM chip was shown to be a capacitor.

Next, the storage capacitor in a dynamic RAM chip was shown to be nonideal. We saw that data must be refreshed in the RAM to keep it from being lost.

Following our discussion of a single storage cell, we examined how these cells are organized into a marix of rows and columns. Each cell was shown to have a unique position in the matrix determined by the row and column that intersect at that cell.

Sense amplifiers were introduced next and we saw how they increase the voltage in the storage cells to a level that is TTL compatible. We also demonstrated how the sense amplifiers are used to rewrite the data in the cell back to the cell when it is read.

We then introduced an actual dynamic RAM chip, the Intel 4116, and we examined its block diagram. The pinout of the chip was shown and all signals used to interface it to a microprocessor were discussed.

The chapter next examined burst and cycle-stealing refresh techniques and the general functions of a dynamic RAM controller. Finally, a complete dynamic RAM system interface to a Z80 microprocessor was presented.

REVIEW PROBLEMS

1. Draw a simple model of an ideal dynamic RAM storage cell.
2. Modify the ideal cell of Problem 1 to reflect a more accurate model of the storage cell.
3. Give an example of a system that requires information to be refreshed or the information gets lost.
4. What is the function of the row and column addresses?
5. What is the purpose of the sense amplifier in a dynamic RAM device?
6. Explain the term destructive read.
7. The 4116 is labeled as a 16K × 1 DRAM. What does this label mean?
8. What is the difference between separate and common I/O? Which type does the 4116 DRAM use?
9. How many address lines does a 16K × 1 DRAM require?
10. How does the 4116 realize the total address lines with only seven address inputs?
11. What is the function of the $\overline{RAS}$ and $\overline{CAS}$ signals?
12. Draw a block diagram showing how the address lines from the CPU are input to the address inputs of the 4116 DRAM.

13. What is the function of the MUX signal in a dynamic RAM system?
14. What is the sequence of events required to write data to a DRAM?
15. What is the sequence of events required to read data from a DRAM?
16. When a DRAM is refreshed, only the row address needs to be cycled. Why?
17. If the refresh interval of a storage cell at room temperature for the 4116 is equal to 2 milliseconds, what is the maximum time interval between refresh addresses on the 4116 DRAM?
18. What is the difference between cycle-steal and burst refresh?
19. Which types of refreshing, cycle steal or burst, does the Z80 employ?
20. What is the function of the R register on the Z80?
21. What signals on the Z80 are active during a refresh cycle?

16

USING AN ANALOG-TO-DIGITAL CONVERTER (ADC) WITH THE 8080, 8085, AND Z80

In this chapter we show how to use the microprocessor to monitor any external instrument that outputs an analog voltage. The discussion starts off by showing a block diagram of the overall concept involved in this type of microprocessor system monitoring. From there the discussion focuses on how an analog-to-digital converter operates. The discussion is basic enough to enable the beginner to understand the principles of analog-to-digital conversion.

After the analog-to-digital converter is presented, this chapter shows the electrical connection of an analog-to-digital converter to the 8080, 8085, and Z80 microprocessors. At the conclusion of this chapter, you will have a good understanding of how devices that output an analog voltage may be monitored and controlled using a microprocessor.

16-1 BLOCK DIAGRAM OF THE PROBLEM

Figure 16-1 shows a block diagram of the overall concept that we must use. In this diagram there are three major blocks shown:

1. Block 1 is the digital monitoring device. In this case it will be an 8080, 8085, or Z80 microprocessor-based system.
2. Block 2 will transform the analog voltage output from the external device to an equivalent digital word.
3. Block 3 is the external device itself. This device is really a transducer of some sort. The transducer will output a voltage that has an amplitude dependent on the physical event being monitored.

In this chapter we will be digitally monitoring an analog event. An analog event is one that can be measured at an infinite number of possible output levels (within a given range.) An example is pressure. Pressure can be measured at any value from a few pounds per square inch to many hundreds of pounds per square inch. Further, the value can be any number of pounds per square inch between these two limits.

If the event to be monitored by the digital computer is analog, a transducer capable of accurately reporting the analog changes is necessary. For example, as the pressure being measured changes, the transducer must reflect this change in its output. When the pressure changes a little, either increasing or decreasing, the voltage output from the pressure transducer must increase or decrease. The amount of increase or decrease will depend on the transducer. See Figure 16-2.

These analog changes are in contrast to a digital transducer. One very common digital transducer is a simple switch. Such a transducer could be used

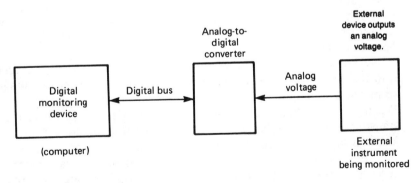

Figure 16-1   Block diagram showing an overall view of the concepts involved in monitoring an analog voltage with a digital system.

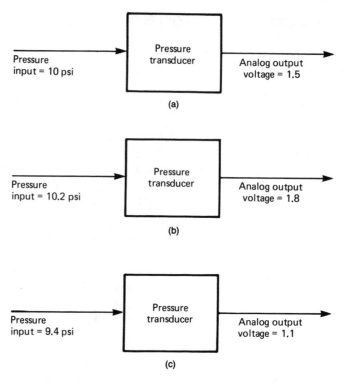

Figure 16-2 Each of these block diagrams shows the analog voltage output from a given pressure input. Note that as the pressure changes at the input of the transducer, the analog voltage must change in proportion.

for a home security system. We may use these type of transducers to indicate whether the windows and doors of our home are open or closed. Although it is true that a window or door can be wide open or just open a crack, in most cases we need not distinguish between the degrees of openness; the event we are monitoring is digital. That is, for our purposes, the door or window is either open or closed.

The transducers with which we are now concerned are analog output transducers. This means the transducer's output will vary in voltage anywhere from a low voltage, say .0 volts, to a high voltage of approximately 5.0 volts. We use this voltage range simply as an example. Typically, the outputs of transducers are variable from any negative voltage to any positive voltage. Voltage variations depend on the type of transducer used.

For the present time let us assume that an analog transducer will output a certain range of voltage. This analog voltage will be input to the block of Figure 16-1 labeled *analog-to-digital converter*. The digital output of the analog-to-digital converter is then input to the microprocessor system block.

The microprocessor will electrically input digital information being out-

put from the analog-to-digital converter. As we will see, this digital information will be equivalent to the analog voltage that is input to the analog-to-
digital converter.

16-2 THE ANALOG-TO-DIGITAL CONVERTER

Let us now define exactly what is meant by the term analog-to-digital converter
(ADC) and discuss how we can make use of this principle. In general terms, the
function of analog-to-digital conversion is to transform (or convert) an analog
input voltage into the digital output word that is its mathematical representation.

This function must be performed before a digital computer can input the
analog voltage. A device that performs this type of function is called an ADC.
The remainder of this section will be devoted to explaining how a typical
ADC operates. After this discussion, we present some examples of how the
ADC device operates in a digital system environment.

We begin the discussion by looking at Figure 16-3, a block diagram of a
typical ADC. This diagram shows general ADC inputs and outputs. Let us
concentrate on the function of each input and output shown in Figure 16-3,
beginning with the power outputs. The ADC consists of electronic circuits,
which require a source of direct current (DC) power. The power input lines are
shown at the top of Figure 16-3. Typical power connections are ±12 or 15 V, +5
V, and ground.

The ±12-V to 15-V DC supplies are used to power some of the internal

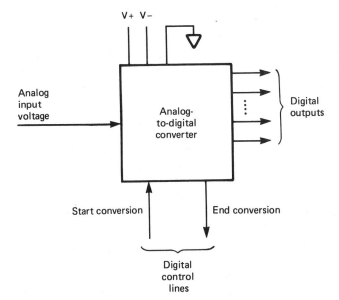

Figure 16-3 Block diagram of a typical ADC.

circuits, such as operational amplifiers and voltage comparators, located inside the ADC package. A +5-V supply will usually power the digital electronics inside the ADC package. Data sheets on any particular ADC will specify any DC power required for proper electrical operation.

In Figure 16-3 we also see the input line labeled *analog voltage*. This is the analog voltage that will be transformed into its equivalent digital word. ADCs have specified ranges of analog voltages they can accept as input. Typical ADC analog input voltage ranges are 0 to +5 V, −5 to +5 V, −10 to +10 V, and 0 to +10 V. These are just a few of the many different voltage ranges that are commonly available. Refer to the manufacturer's data sheet for exact specifications on the analog input voltage range for a particular ADC.

Some of the input voltage ranges listed are negative and positive, whereas some are only positive. If the input voltage to the ADC can be positive and negative, the ADC is said to be *bipolar*. If the input voltage range is from zero to a positive value, the ADC is said to be *unipolar*. The ADC we use later in the chapter is unipolar.

The next section of Figure 16-3 to examine shows the 8 digital output lines. The number of digital output lines on an ADC will vary. Typical numbers of output lines are 6, 8, 10, 12, and 16. In general, the greater the number of output lines, the more expensive the ADC, which is due to the greater accuracy with which the ADC can perform a conversion.

The digital output lines of the ADC will be set to a logical 1 or a logical 0, as determined by the analog input voltage. Later in the discussion, it will be shown how to calculate which digital outputs will be logical 1s and which will be logical 0s. For now, think of the digital output lines as the physical connections to the digital monitoring system. In our case, the connections are to the microprocessor system.

Finally, in Figure 16-3, we see two digital control lines, labeled *start conversion* and *end conversion*. Each line has a unique function. The start conversion input-control line is a digital signal that will electrically inform the ADC to start converting the analog input to a digital output word. This signal is generated by the digital monitoring device (in this example it is the microprocessor).

When the microprocessor system is electrically prepared to start the conversion process, the start conversion input-control line to the ADC is asserted.

The end conversion output line shown in Figure 16-3 is used to inform the digital monitoring device electrically that an analog-to-digital conversion is complete. Notice that the ADC is itself a small electronic system. It will take a short time for the ADC to adjust the digital outputs to new values based on the analog input voltage. The exact time required will vary from ADC to ADC. Typical times are between 10 and 200 microseconds.

When the microprocessor system requests an analog-to-digital conversion, it must electrically monitor the end conversion output line before it retrieves data. This monitoring will determine when the conversion is complete. Figure 16-4 shows a flowchart of how the microprocessor must communicate with the ADC.

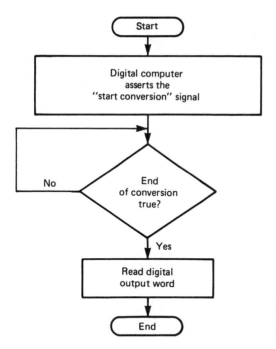

Figure 16-4 Flowchart showing the sequence of events required for electrical communication with an ADC.

In Figure 16-4 the analog-to-digital conversion will be started when the microprocessor system asserts the start conversion input line to the ADC. Once the hardware is connected, this assertion will be accomplished using software. After the conversion is started, the microprocessor system will electrically monitor the end conversion line from the ADC. When this line is true, the microprocessor will read (input) the digital output lines of the ADC. This flowchart will be realized later in this chapter, when the ADC is electrically connected to the microprocessor.

You have probably noticed that the discussion thus far has not gone into the details of how the ADC internally performs the analog-to-digital conversion. We present the ADC at the user's level, treating it as a "black box" that will perform a certain function in a certain prescribed manner. This is precisely the way ADCs are used. The electronics that actually perform the conversion are usually out of the user's control or access. This chapter, therefore, simply discusses how to make the ADC perform its prescribed function. Understanding this introduction to the ADC inputs and outputs has been our first step toward using this type of device in a microprocessor system application.

16-3 CALCULATING THE DIGITAL OUTPUTS OF THE ADC

In this section we will discuss how to determine the relationship between the logical conditions of the digital outputs and the voltage levels of the analog input. For this discussion we will use unipolar ADC. The acceptable voltage

Digital Output Line	Weight
D0	1
D1	2
D2	4
D3	8
D4	16
D5	32
D6	64
D7	128

Figure 16-5 The ADC's output lines have the same numerical weights as the microprocessor data bus lines.

input range will be from 0.0 V to +10.0 V. There will be 8 digital output lines. A block diagram of this type of device is shown in Figure 16-3.

The digital outputs will be labeled D0–D7 and will be assigned the same numerical weights that are given to bit positions in data bytes elsewhere in this text. A table of the digital output lines and the corresponding weights is shown in Figure 16-5.

As we have seen before, the minimum weight at the digital outputs will occur when all are a logical 0, and the maximum weight at the digital outputs will occur when all are a logical 1. The minimum and maximum weights are 0 and 255, respectively.

The maximum analog voltage input to the ADC is 10.0 V. If we relate the maximum and minimum input voltage to the maximum and minimum digital output weights, we have:

$$0.0 \text{ V input} = \text{weight 0 output}$$
$$10.0 \text{ V input} = \text{weight 255 output}$$

All input voltage between 0 and 10 V will have a corresponding digital weight between 0 and 255.

There are 256 different weights available for the input voltages, because there are eight digital outputs. The number of different weights available with a particular ADC is calculated by the following equation:

$$\text{Number of digital weights} = 2^{(\text{number of digital output lines})}$$

In our case we had 8 digital output lines, so the number of digital weights is 2^8, or 256. If we had an ADC with 10 digital output lines, the number of digital weights would equal 2^{10}, or 1024.

Since we know how many unique digital weights are available for the analog input voltage range, we can divide that range into equal increments. In our example we have 255 available intervals between the 256 digital weights. (One weight must be subtracted because of the 0 at the start.) Therefore, the range between 0 and 10 volts will be divided into 255 equal pieces by the following equation:

$$\frac{\text{Voltage range}}{255} = 0.03921569$$

Using this information, we can generate a list of input voltages and their corresponding digital output weights. This list is shown in Figure 16-6.

We see in Figure 16-6 that the equivalent input voltage increases in discrete steps of approximately .04 V. For example, suppose an input voltage to the ADC is equal to 1.5 V. There is not a digital output weight exactly equivalent to 1.5 V. From the list of Figure 16-6, we must choose an output weight that will yield the voltage closest to 1.5 V. We see that 1.5 V is between 1.49 V (a weight of 38) and 1.53 V (a weight of 39). The weight of the ADC digital outputs for an input voltage of 1.5 V is closest to 38, so we can use that number. However, it is not an exact representation of the analog input voltage.

The point is that the ADC can only digitally represent analog voltages between 0 and 10 V to a resolution of approximately $\pm.04$ V. This error is dependent on the number of digital output lines and the analog voltage input range. If the value of $\pm.04$ V is too great a range, then a different ADC, with more digital output lines, must be chosen.

When using the ADC, we do not want to have to look up the equivalent voltage each time a conversion takes place. It would be much easier if the voltage could be calculated based on the digital output word. This would lend itself well to computer control. These calculations can be accomplished using the following equation, where the analog voltage is equal to the digital output weight multiplied by the resolution of the ADC:

$$\text{Analog voltage} = \text{digital weight} \times (10/225)$$

For example, suppose the weight read from the ADC was equal to 125. This would equal an input voltage of

$$125 \times (10/255) = 4.9 \text{ V}$$

Remember that this voltage may actually equal 4.9 V or slightly less than 4.9 V, because of the resolution of the ADC.

In its present form, our equation is dependent upon the input voltage range and the number of digital output lines on the ADC. These values must be modified for the equation to fit a particular application.

16-4 CONNECTING THE ADC TO THE MICROPROCESSOR

Figure 16-7 shows a complete schematic for connecting an actual ADC, Analog Devices' model AD570, to an 8255, which will be connected to a microprocessor. In this section we discuss exactly how each part of Figure 16-7 operates. The AD570 is very similar in its operating characteristics to the general example we gave in Figure 16-3.

Digital Weights and Voltages

D	V	D	V	D	V	D	V	D	V
0	0.00	1	0.04	2	0.08	3	0.12	4	0.16
5	0.20	6	0.24	7	0.27	8	0.31	9	0.35
10	0.39	11	0.43	12	0.47	13	0.51	14	0.55
15	0.59	16	0.63	17	0.67	18	0.71	19	0.75
20	0.78	21	0.82	22	0.86	23	0.90	24	0.94
25	0.98	26	1.02	27	1.06	28	1.10	29	1.14
30	1.18	31	1.22	32	1.25	33	1.29	34	1.33
35	1.37	36	1.41	37	1.45	38	1.49	39	1.53
40	1.57	41	1.61	42	1.65	43	1.69	44	1.73
45	1.76	46	1.80	47	1.84	48	1.88	49	1.92
50	1.96	51	2.00	52	2.04	53	2.08	54	2.12
55	2.16	56	2.20	57	2.24	58	2.27	59	2.31
60	2.35	61	2.39	62	2.43	63	2.47	64	2.51
65	2.55	66	2.59	67	2.63	68	2.67	69	2.71
70	2.75	71	2.78	72	2.82	73	2.86	74	2.90
75	2.94	76	2.98	77	3.02	78	3.06	79	3.10
80	3.14	81	3.18	82	3.22	83	3.25	84	3.29
85	3.33	86	3.37	87	3.41	88	3.45	89	3.49
90	3.53	91	3.57	92	3.61	93	3.65	94	3.69
95	3.73	96	3.76	97	3.80	98	3.84	99	3.88
100	3.91	101	3.96	102	4.00	103	4.04	104	4.08
105	4.12	106	4.16	107	4.20	108	4.24	109	4.27
110	4.31	111	4.35	112	4.39	113	4.43	114	4.47
115	4.51	116	4.55	117	4.59	118	4.63	119	4.67
120	4.71	121	4.75	122	4.78	123	4.82	124	4.86
125	4.90	126	4.94	127	4.98	128	5.02	129	5.06
130	5.10	131	5.14	132	5.18	133	5.22	134	5.25
135	5.29	136	5.33	137	5.37	138	5.41	139	5.45
140	5.49	141	5.53	142	5.57	143	5.61	144	5.65
145	5.69	146	5.73	147	5.76	148	5.80	149	5.84
150	5.88	151	5.92	152	5.96	153	6.00	154	6.04
155	6.08	156	6.12	157	6.16	158	6.20	159	6.24
160	6.27	161	6.31	162	6.35	163	6.39	164	6.43
165	6.47	166	6.51	167	6.55	168	6.59	169	6.63
170	6.67	171	6.71	172	6.75	173	6.78	174	6.82
175	6.86	176	6.90	177	6.94	178	6.98	179	7.02
180	7.06	181	7.10	182	7.14	183	7.18	184	7.22
185	7.25	186	7.29	187	7.33	188	7.37	189	7.41
190	7.45	191	7.49	192	7.53	193	7.57	194	7.61
195	7.65	196	7.69	197	7.73	198	7.76	199	7.80
200	7.84	201	7.88	202	7.92	203	7.96	204	8.00
205	8.04	206	8.08	207	8.12	208	8.16	209	8.20
210	8.24	211	8.27	212	8.31	213	8.35	214	8.39
215	8.43	216	8.47	217	8.51	218	8.55	219	8.59
220	8.63	221	8.67	222	8.71	223	8.75	224	8.78
225	8.82	226	8.86	227	8.90	228	8.94	229	8.98
230	9.02	231	9.06	232	9.10	233	9.14	234	9.18
235	9.22	236	9.25	237	9.29	238	9.33	239	9.37
240	9.41	241	9.45	242	9.49	243	9.53	244	9.57
245	9.61	246	9.65	247	9.69	248	9.73	249	9.76
250	9.80	251	9.84	252	9.88	253	9.92	254	9.96
255	10.00								

D - DIGITAL WEIGHT
V - VOLTAGE

Figure 16-6 List of digital output weights and corresponding analog input voltages they represent.

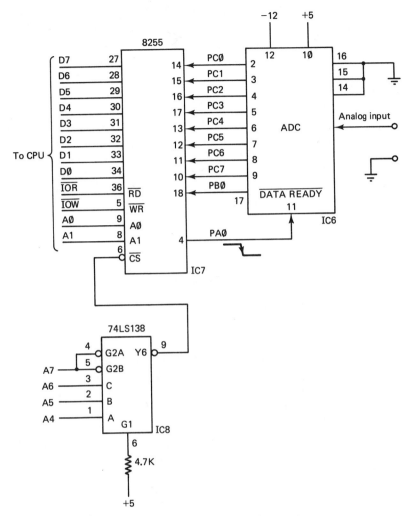

Figure 16-7 Schematic diagram showing one way the AD570 ADC can be connected to a microprecessor system using an 8255 as the port interface.

The ADC is labeled IC6 in Figure 16-7. The device is powered by +12 V, −5 V, and ground lines.

On the right side of the ADC (IC6) is the analog input voltage to be converted. This is the voltage input from an external source, between 0 and 10 V. The digital outputs of the ADC are input to an 8255 PIO, which is set up as an input port and labeled IC7. We discussed how this type of device operated in Chapter 12.

The data outputs of the 8255 are connected directly to the microprocessor data bus lines, D0–D7. When the CPU electrically reads the data at the ADC outputs, the 8255 will become enabled and input data to the microprocessor.

In order to start the analog-to-digital conversion, the input pin 11 of IC6 in Figure 16-7 must be set to a logical 1 and then reset to a logical 0. This is the *start conversion* input signal for this particular ADC. When input pin 11 is reset to a logical 0, the analog-to-digital conversion will start. This is graphically shown by the waveform next to the signal line at IC6.

Input pin 11 of the ADC is connected to the PA0 output line of the 8255 (IC7). This output line is acting as a strobe input to the ADC. Whenever the user wishes to set the PA0 output to a logical 1, the D0 bit is set to a logical 1 and output to port PA0, which will be port code 60H.

We can perform all the steps necessary to start the conversion by using the OUT instruction in assembly language. For example, the instruction

```
MVI A,01
OUT 60H
```

will set the PA0 output of the 8255 a logical 1, and the instruction

```
MVI A,00
OUT 60H
```

will set the PA0 output of the 8255 to a logical 0.

The important point here is that we can control an individual hardware line via a software instruction. Specifically, we can assume that the input pin 11 of IC6 can be set to a logical 1 or a logical 0 under software control.

To know when the analog-to-digital conversion is complete, the microprocessor must electrically monitor the output pin 17 of IC6. This pin is the end conversion signal for the ADC. Output pin 17 of IC6 in Figure 16-7 is connected to input pin PB0 of the 8255. Pin 17 of the ADC will be a logical 0 when the analog-to-digital conversion is complete. At all other times, it will be a logical 1.

When the microprocessor executes an IN instruction to the correct address, port B of the 8255 will become enabled. This will allow the logical state of the output pin 11 of IC6 to be enabled onto the microprocessor data bus. At that time the microprocessor will electrically input the logical level into an assembly language program. This type of input monitoring of a device is called *polling*. The microprocessor constantly monitors the input pin, waiting for the conditions to become true.

An alternative to polling would be to configure the ADC such that it interrupts the microprocessor when the end conversion line is true.

We have now seen how the end conversion signal is monitored by the microprocessor. We have also seen how the microprocessor will set the input pin 11 of IC6 to a logical 1 or a logical 0 under control of software.

Each of these events will be accomplished using either an IN or OUT instruction. A major part of both of these instructions is the address. We now discuss how the address of the OUT or IN instructions is related to the schematic diagram given in Figure 16-7.

We will be running our program on a system with the 8255 PIO connected with the following port definitions:

PORT NUMBER	DEFINITION
63H	control port for the 8255
62H	port C ADC output lines
61H	port B end conversion PB0
60H	port A start conversion PA0

With these hardware definitions we see that the PIO must be set up with PORT A = OUTPUT, PORT B = INPUT, and PORT C = INPUT. To start the conversion, the PA0 output line of the 8255 will be made to toggle from a logical 0 to a logical 1 and back to a logical 0 again.

Once the conversion is started, PB0 will be monitored for the end of conversion. The end of conversion is true when PB0 goes to a logical 0 state. At that time the microprocessor can read the digitized value of the analog voltage by inputting the data at port C.

We have now discussed the hardware action taking place in the schematic shown in Figure 16-7 in some detail to relate hardware events to software instructions. The discussion was meant to show how the hardware of the interface operates.

16-5 SOFTWARE FOR ANALOG-TO-DIGITAL CONVERSION

Now that we have the ADC connected to the microprocessor, let's control it with the software. We will show an assembly language program required to input a digital value from the ADC. A flowchart for the steps of this program is shown in Figure 16-8. The intent is to show you how it can be done, not the only way to do it. Once you see how it is accomplished, you can easily adapt the details to your own application and system.

The following discussion will realize the flowchart of Figure 16-8 with 8085 programming statements. It is assumed that the ADC I/O device is located at addresses specified in the previous section.

Step 1.   The first step in the program is to initialize the 8255 with the correct configuration, which is PA = OUTPUT, PB = INPUT, and PC = INPUT. Referring back to Section 12.5, we see that the correct hexadecimal byte to write to the control port will be 8BH (or 1000 1011). We can write this to port 63H using the following two instructions:

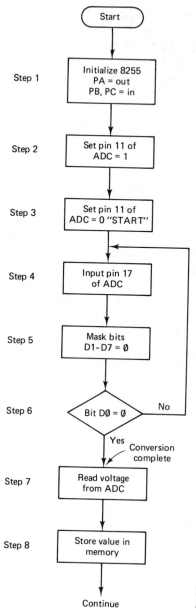

Figure 16-8 Flowchart for inputting and controlling the ADC of Figure 16-7.

```
MVI A,8BH      data to A
OUT 63H        output to 8255 control port
```

Step 2.   The next step is to write a logical 1 to the ADC pin 11 getting ready to start the conversion process. We can accomplish this using an OUT instruction and setting the output data equal to 01H.

```
MVI A,01H     data to A reg
OUT 60H       data to port PA
```

Step 3.   In this step the start conversion input line to the ADC will be set to a logical 0, starting the conversion process. This can be accomplished in the same way as in Step 2, except that the data is now a 0 instead of a 1:

```
DCR A         set A = 0
OUT 60H
```

The analog-to-digital conversion has now started.

Step 4.   We must now read the logical level of pin 17 of IC6 to determine if the conversion is complete. This can be accomplished using the following instruction:

```
IN 61H        read PB
```

Step 5.   Notice that when we read PB, the A register will be the summation of all the data lines, PB0–PB7. However, as we see in Figure 16-7, we are physically controlling the logical level of only one input line during execution of the IN instruction. The other input lines can be a logical 1 or a logical 0. Further, the other input lines may not show the same value during each IN instruction. The actual logical level will depend on many factors, most of which are out of the user's control.

Therefore, we would like logically to ignore data lines PB1–PB7 during the IN instruction. Although we cannot electrically do this, we can do it using software. In the system software we can ignore data lines PB1–PB7 by masking them off and setting them to a known value. In this case the known value will be a logical 0. This can be accomplished as follows:

```
ANI A,01H     set all bits PB1-PB7=0
```

Step 6.   We now have all unused bits set to a known state. All that remains is to test whether bit PB0 is a logical 1 or a logical 0. If bit PB0 is a logical 1, the conversion is not over and we must go back to Step 4 and input the test byte from port B again. If bit PB0 is a logical 0, we are finished with the conversion:

```
JNZ STEP 4    go back if not done
```

Step 7.   If the conversion is complete, the value of A will be equal to 0. The digital outputs of the ADC can now be read by the microprocessor. This is done by the following statement:

```
IN 62H        read ADC value into A
```

The A register will be equal to the combined weight of the input data lines.

```
;STEP 1
      MVI   A,8BH
      OUT   63H
;
;STEP 2
      MVI   A,01H
      OUT   60H        ;pin 11 = 1

;STEP 3
      DCR   A
      OUT   60H        ;pin 11 = 0

;STEP 4
      IN    61H        ;read pin 17

;STEP 5
      ANI   A,01H      ;mask D1-D7
;
;STEP 6
      JNZ   STEP_4     ;check for EOC

;STEP 7
      IN    62H        ;read ADC value

;STEP 8
      STA   VALUE_1    ;save in memory

      END
```

Figure 16-9 Assembly language program to realize the flow charts of Figure 16-8.

Step 8.   The final operation is to operate on the input voltage value to perform whatever function you desire. In our case we will simply store the voltage value into memory and go back to get another reading.

```
            STA VALUE_1       save in memory
```

This last statement simply stored the data into memory. However, your program could have operated on the data in any manner you desire. For example the input data could have been the speed of a motor, which then has to be displayed. Perhaps the motor is rotating too fast and its speed must be reduced. Maybe the value input to the A register is the temperature of a sample that you are running and experimenting on. Regardless of the overall application, the preceeding steps are the basic ones required to input data from typical ADCs.

If we now put the entire set of instructions together, they appear as shown in Figure 16-9. If the program were used in a system, the statements of Figure 16-9 would most likely appear in an assembly language subroutine. This would allow access to the ADC at any time during a program by executing a CALL statement.

16-6 SUMMARY

In this chapter we discussed the details of using an analog-to-digital converter. The discussion started by explaining the need for analog-to-digital conversion. Next we discussed conceptually how it is accomplished. We deliberately did not attempt to explain the details of how an ADC operates internally.

Next we connected a real ADC, the AD570, to CPU. Actual circuits were designed and discussed. It was shown how to calculate the input voltage based on the digital word or weight read from the ADC.

The main focus of this chapter was to show how to perform analog-to-digital conversion in a microprocessor-based system. Whatever quantity you wish to measure in your particular application, if it requires an analog interface to your microprocessor, you can use the information shown in this chapter.

REVIEW PROBLEMS

1. What is the function of an ADC?
2. List four analog events that occur in the real world.
3. List four digital events that occur in the real world.
4. Draw a block diagram of a typical ADC.
5. What is the difference between a unipolar and a bipolar ADC?
6. If an ADC is 8 bits, and has an input voltage range of 0–5 V, what is its voltage resolution?
7. Assume you have an ADC that is specified as unipolar, 0–10 V, 10 bits. What is the digital output word for the following analog inputs?
   (a) 2.4 V
   (b) 5.6 V
   (c) 9.7 V
   (d) 1.35 V
8. Using the ADC in Problem 7, calculate the input voltage based on the following digital output words. Assume the output words are written in hexadecimal.
   (a) 035
   (b) 3AC
   (c) 14F
   (d) 0F1
9. Using the block diagram of the ADC from Problem 4, connect the device to the buses of an 8085 microprocessor.
10. Write an 8085 assembly language program that will control the ADC as connected in Problem 9. Your program should start the conversion, monitor the end conversion line, read the value from the ADC, and place the ADC value in memory. Assume the output from the ADC is equal to 8 bits.
11. If the ADC in Problem 10 had an output word of 10 bits, show how the hardware for interfacing it to the 8085 system buses would differ from the hardware connection given in Problem 9.
12. Repeat Problem 10 for the new ADC connections given in Problem 11.
13. List five microprocessor system applications where an ADC may be used.

17

USING A DIGITAL-TO-ANALOG CONVERTER

Certain peripheral devices that you can use in your microprocessor application require an analog input voltage in order to operate. A microprocessor system is completely digital, which presents an interfacing problem. What we need is a means of producing an analog voltage from a digital source or controller. Such a process is called digital-to-analog conversion and uses a digital-to-analog converter (DAC). Figure 17-1 shows a block diagram of this concept.

One type of external device that might require an analog input voltage is a DC motor. Most DC motors have the operating characteristic of rotating faster when a higher voltage is applied to the input. This type of external device, along with its analog input is shown in the block diagram of Figure 17-2.

We have shown only a single example of an external device that would require an analog voltage, but there are many other applications using devices that require analog voltages for control. For instance, communications and recording systems utilize digital-to-analog conversion and analog-to-digital conversion at their inputs and outputs. Electronic music and waveform-generation

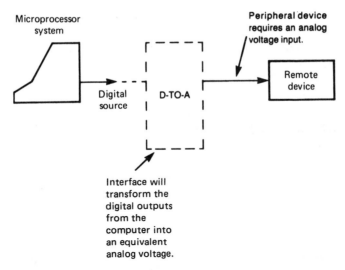

Figure 17-1   Block diagram showing the interfacing problem that occurs when a completely digital source needs to control a peripheral device that requires an analog voltage input. A DAC will solve this interfacing problem.

systems also use digital-to-analog conversion as a part of the overall system. These systems are growing in popularity. In short, any device whose output (speed, musical pitch, brightness, and so on) is variable may need digital-to-analog conversion to be controlled by digital electronics.

The intention of this chapter is to show how you can control and set an analog output voltage from a digital source, such as a microprocessor. Our discussions cover the basics of digital-to-analog conversion. We design and discuss the hardware and software necessary to allow the microprocessor automatically to control an analog output voltage. The circuits shown are inexpensive and available from many suppliers. These circuits show one way to achieve digital-to-analog conversion. Nothing in the chapter is left up to the reader; no

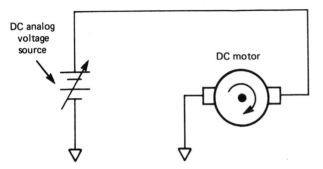

Figure 17-2   Diagram showing a DC motor being controlled by an analog voltage. When the voltage increases, the motor will increase in speed. This is only one type of peripheral device that can be controlled by an analog input voltage.

important detail is omitted. All pin numbers, connecting lines, and types of devices are labeled.

If your microprocessor application requires an analog output voltage to control one or more pieces of peripheral hardware, then this chapter will help you get the job done.

17-1 DEFINITION OF DIGITAL-TO-ANALOG CONVERSION

Before we design and discuss a digital-to-analog conversion interface in detail, let us examine the overall concept. In the introduction to this chapter a very informal definition was given. This section brings that definition into sharper focus.

Figure 17-3 shows a general block diagram of a DAC. In this diagram the block labeled B1 is the electrical interface to the digital source that will input data to and control the converter. In our case the source is the digital output lines from the microprocessor. Block B1 controls the block labeled B2 in Figure 17-3.

Block B2 is the precision scaling network. The job of the scaling network

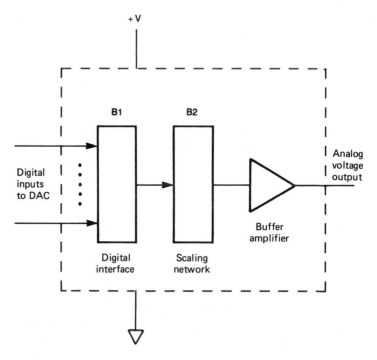

Figure 17-3 Block diagram of a general DAC. Most DACs will have a block diagram similar to this one.

is to determine electrically how much of the reference voltage or current will appear at the converter output. For example, suppose the reference voltage is equal to 8.00 volts. The scaling network is capable of setting the output voltage to an exact portion of the reference, say 2.04 V. The reference could also be a smaller voltage than the output device requires, such as 1.5 V. In this case, the scaling network would select a portion of the 1.5 V, and this would be amplified before being output from the DAC.

Digital inputs to the DAC originating at the microprocessor will electrically control the scaling network, determining how much of the reference will be applied to the output. By setting the proper digital information on the DAC inputs, we could force the DAC output voltage to be equal to exactly half of the 8.00-volt reference, or 4.00 V. We cannot give a formal, general definition of what voltage will appear at the output as a function of the digital inputs because every DAC may have different specifications.

Using what we know about digital-to-analog conversion, let us take an example and show how to set the analog output voltage as a function of the digital inputs. Suppose we have a DAC that will output 10.00 V when all the digital inputs are set to a logical 1; that is, suppose we have a DAC whose reference voltage is 10.0 V. (We use this as an illustration only. Every DAC has a unique set of specifications.) Let us further assume that the DAC will output .0 V when all the digital inputs are logical 0s. Figure 17-4 shows these two conditions. A DAC with this type of voltage output specification is unipolar. The output voltage will swing in one direction (toward one electrical pole) only. In this case the swing is from .0 V to +10.0 V. The term unipolar also describes an ADC with the same voltage characteristics.

Next we need to know how many digital inputs the DAC has. We will assume that our hypothetical converter has 10 digital inputs. There are therefore 2^{10}, or 1024, different input combinations that can be used. See Figure 17-5. From this number we can calculate the minimum voltage swing, the increment by which the voltage will change for each unique digital combinations on the input.

To calculate how much the DAC output voltage will change when the digital input combination changes by only 1 LSB, we can use the following formula:

1. Minimum output voltage change = maximum output voltage change ÷ maximum number of states.
2. Maximum output voltage change = maximum DAC output voltage ($V_{out\,max}$) − Minimum DAC output voltage ($V_{out\,min}$) $V_{out\,max} - V_{out\,min.}$
3. $V_{out\,min}$ = .00 V, $V_{out\,max}$ = 10.00 V; maximum output voltage change = 10.0 − 0.0 = 10.0.
4. Maximum number of unique digital combinations with 10 inputs = 1024. With 1024 endpoints, there are 1023 intervals from 0 to 10 V. Therefore, we have actually 1023 states that produce a voltage.

(a)

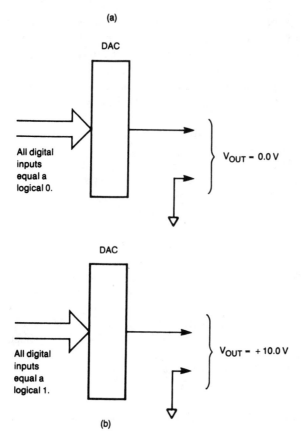

DAC

All digital inputs equal a logical 0.

$V_{OUT} = 0.0\,V$

DAC

All digital inputs equal a logical 1.

$V_{OUT} = +10.0\,V$

(b)

Figure 17-4 (a) When all digital inputs to the DAC are a logical 0, the output voltage will equal .00 V; (b) when all digital inputs to the DAC are a logical 1, the output voltage will equal 10.00 V. These are the extreme ranges of this DAC's output voltage swing.

10 digital inputs

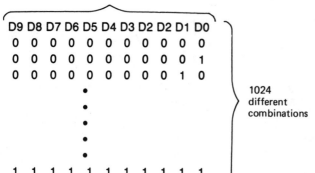

D9	D8	D7	D6	D5	D4	D3	D2	D2	D1	D0
0	0	0	0	0	0	0	0	0	0	0
0	0	0	0	0	0	0	0	0	0	1
0	0	0	0	0	0	0	0	0	1	0
1	1	1	1	1	1	1	1	1	1	1

1024 different combinations

Figure 17-5 With 10 digital inputs to the DAC, there are 1024 different combinations that may be applied. The number 1024 is derived from raising 2 to the tenth power.

5. Minimum voltage swing = 10.0 V/1023 = .009775 V, or 10 mV.

The preceding calculations mean the output voltage will change 10 milli-volts for each digital input bit that changes. See Figure 17-6. This DAC would be described as a unipolar, 10-bit voltage DAC. (The term *voltage* indicates that the output is a voltage and not a current.) Suppose our hypothetical DAC

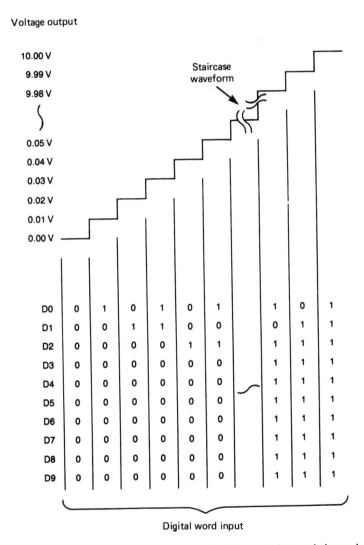

Figure 17-6   The relationship between the digital input values and the analog output voltage for a DAC with 10 inputs and an output voltage swing between .00 and +10.00 V.

had 12 inputs instead of 10. Using the formulas given, the minimum voltage change at the output would be equal to the following:

6. Minimum voltage change = $(10.0 \text{ V})/(4096 - 1) = 10/4095 = .0024$ V. Let us now suppose that this same DAC had only 6 inputs. The minimum voltage change at the output pin would be as follows:

7. Minimum voltage change = $10.0 \text{ V}/(64 - 1) = 10.0/63 = .158$ V.

Notice that for a given output voltage range, the greater the number of digital inputs, the smaller the minimum voltage change at the DAC output. See Figure 17-7. In general, the greater the number of digital input lines, the better the output voltage resolution; that is, the DAC can resolve a smaller increment of voltage. This means we can come closer to obtaining the "smooth" staircase waveform at the top of Figure 17-7.

Most commercial DACs are specified to be accurate only to $\pm\frac{1}{2}$, or 1 LSB. This means when you connect your DAC to the system and you obtain an output voltage, you will get the calculated voltage plus or minus the resolution (1 LSB) or one-half the resolution ($\frac{1}{2}$ LSB).

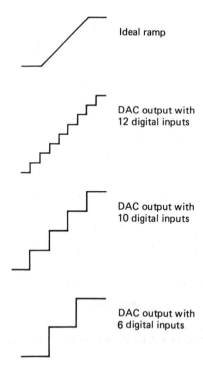

Ideal ramp

DAC output with 12 digital inputs

DAC output with 10 digital inputs

DAC output with 6 digital inputs

Figure 17-7 Different staircase wave forms for the analog voltage output of the DAC. As the digital input word changes by a single count, the output will increase by the minimum voltage change. Notice that as the number of digital inputs to the DAC increases, the staircase more closely resembles the ideal ramp.

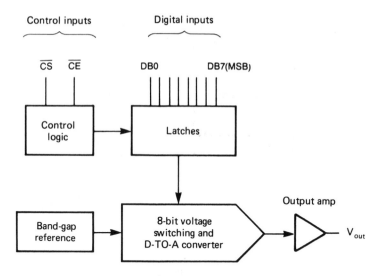

Figure 17-8 Block diagram for the AD558 DAC. Compare this to the general block diagram given in Figure 17-2.

17-2 AN ACTUAL DAC

To illustrate the points we have covered about digital-to-analog conversion, let us examine a DAC that is available off the shelf. The DAC chosen for this example is the AD558, manufactured by Analog Devices. A block diagram of the AD558 is shown in Figure 17-8; let us discuss it in detail.

In this figure we see the digital input block, labeled DB0–DB7. These digital inputs connect to the data lines of the microprocessor system. This DAC has built-in latches that allow the data output from the CPU to be strobed and electrically stored without the use of an external latch. See Figure 17-9. This is a very useful feature of the AD558 because it reduces the total number of parts necessary to complete the interface. We will show exactly how to connect the data lines of the CPU to the DAC digital inputs in a later section of this chapter.

The next block shown in Figure 17-8 is the control logic. This logic block provides the strobe signal to the input storage latches at the correct time. The CPU outputs a strobe pulse to the control logic block.

Another important block of Figure 17-8 is the band-gap reference. This is the internal reference voltage for the DAC. With the reference voltage inside the package, we do not have to provide an external precision reference for the DAC. This is useful because it means that standard power supplies may be used to power the device. Most digital electronic systems have power supplies to suit this application.

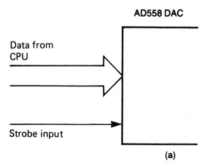

(a)

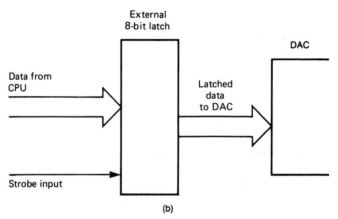

(b)

Figure 17-9   (a) The AD558 will connect directly to the CPU data lines. There is an internal 8-bit storage latch on the device; (b) other DACs may require an external storage latch to capture the logical state of the data lines at exactly the correct instant in time.

A large block of Figure 17-8 is the 8-bit voltage switching and DAC. This block is the scaling network that applies a portion of the reference voltage to the output. In this device the voltage is output to the outside world from an amplifier, labeled *output amp* in Figure 17-8. At this point in the discussion you should compare the block diagram of Figure 17-8 to the general block diagram that was given in Figure 17-2.

There are two modes of operation for the AD558. One mode allows the output voltage to swing between .00 V and +2.50 volts. The second mode allows the output voltage to swing between .00 V and +10.24 V. Figure 17-10 shows how to connect the pins of the AD558 for an output voltage swing in each of the two modes. The 10.24-V mode requires the V_{cc} input voltage pin to have

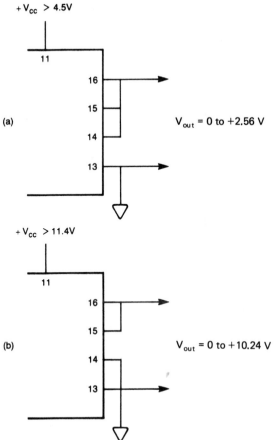

+V_cc > 4.5V

11

16

15

(a) V_{out} = 0 to +2.56 V

14

13

+V_cc > 11.4V

11

16

15

(b) V_{out} = 0 to +10.24 V

14

13

Figure 17-10 (a) Wiring diagram for using the AD558 with a maximum output voltage of 2.56 V, (b) wiring diagram for using the AD558 with a maximum of 10.24 V.

a potential between 11.4 and 16.5 V. The 2.56-V mode requires a minimum potential of 4.6 V.

17-3 CONNECTING THE DAC TO THE MICROPROCESSOR SYSTEM BUSES

Let's now discuss how to connect the AD558 to the buses of a microprocessor system for automatic control. The information presented can be applied to connecting a microprocessor to most other DACs as well. Figure 17-11 shows the complete schematic diagram for connecting the AD558 to a CPU.

We see in Figure 17-11 that the data bus lines from the CPU D0–D7 are connected directly to the data input pins of the DAC device. We can do this because of the AD558's built-in 8-bit latch, which will store the logical conditions of the data lines at the correct time. Other DACs may not have this

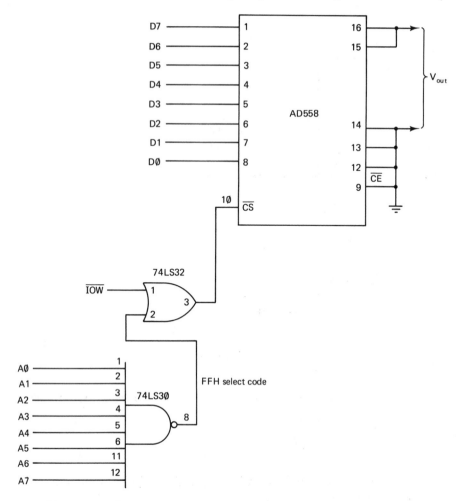

Figure 17-11 Complete schematic showing the connection of the AD558 to the CPU. The AD558 employs an internal data latch, which allows it to be connected directly to the system data bus.

feature. DACs without built-in storage latches require an external 8-bit latch, used as shown in Figure 17-12.

To strobe the data into the DAC circuit shown in Figure 17-11 or 17-12, the CPU will execute an OUT instruction. Timing considerations for latching the data during an OUT instruction were discussed in Chapter 8. When the OUT instruction is executed, the data specified by the instruction is placed at the AD558 data inputs via the eight data lines.

At this time the port-write strobe line goes to a logical 0. When this occurs, the $\overline{CS}$ input pin 10 of the AD558 is set to a logical 0. After the OUT instruction is

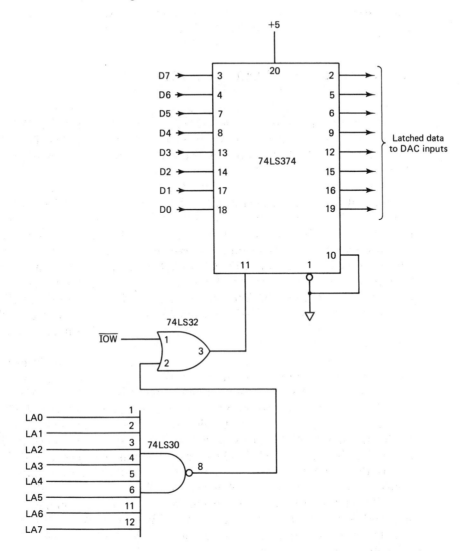

Figure 17-12 Schematic diagram showing how a 74LS374 8-bit latch may be used on DACs that do not have an internal latch like the AD558.

complete, the $\overline{CS}$ input line to the AD558 is set to a logical 1 because the $\overline{IOW}$ control line as well as the port-select line go to a logical 1. The operation of the port-select line and $\overline{IOW}$ control line was discussed in Chapter 8.

When the $\overline{CS}$ input line to the AD558 goes to a logical 1, the data at the input pins is strobed into the internal 8-bit latch. The output voltage at pin 16 depends on the data written to the DAC during the execution of the OUT instruction. By using the OUT instruction and the circuit shown in Figure 17-11, we can set any digital input combination between 0 and 255 at the DAC input pins. Now we need to know how to calculate exactly what data should be written to set any voltage at the DAC output.

17-4 SETTING ANY OUTPUT VOLTAGE ON THE DAC

We assume that the DAC is connected to an 8085 system that can control the digital inputs to the DAC as discussed in Section 17-3. The DAC is physically connected to the system, as shown in Figure 17-11. If the 8085 can control the digital inputs to the DAC, how can we calculate the digital byte needed to set the correct voltage output?

In Section 17-1 we discussed briefly how to calculate the minimum voltage change that will occur at the output of the DAC. That discussion assumed there were 10 digital inputs to the device, but the AD558 has only 8 inputs. Therefore, the equations given in Section 17-1 need to be modified slightly to handle the differences in the two examples. We could interface a 10-bit DAC to the 8085 using two output ports. DACs or ADCs handling any number of bits can be used with any 8-bit microprocessor. The microprocessor will simply use two I/O ports to write the complete data word to the DAC. A microprocessor does not have to support 16-bit I/O operations to use a DAC with an input word greater than 8 bits. We have used 8-bit DACs and ADCs here to illustrate the concepts of ADC and DAC.

Our objective in this section is to calculate the total weight of the digital inputs required to output a certain voltage from the DAC. For example, suppose we wish the DAC's output voltage to be equal to 4.8 V. The microprocessor will be programmed to output a certain combination of logical 1s and 0s to obtain the correct DAC output voltage. The digital output byte in this case would be equal to 120. Let us go through the steps necessary to perform this calculation. We use the AD558 DAC as the example. However, you can easily modify the equation given to fit any DAC with a different number of digital inputs and a different output voltage swing.

The first step is to calculate the minimum voltage change at the DAC output. This may be accomplished using the procedure outlined in Section 17-1 and the appropriate specifications for the AD558.

$$\text{Minimum voltage swing} \quad = \quad \frac{\text{maximum voltage swing}}{\text{total number of digital input combinations} - 1}$$

The maximum voltage swing on the AD558 will be equal to 10.24 V. There are 8 digital inputs to the AD558. (That is why it is called an 8-bit DAC.) With 8 digital inputs there are a total of 256 different input combinations. One of these combinations is used to set the DAC to its lowest value, so we use 255 ($256 - 1$) combinations to generate the output voltage swing.

$$\text{Minimum voltage swing} = \frac{10.24}{255} = .040 \text{ V} \quad (40 \text{ mV})$$

This means that if the digital input byte to the DAC were 00000001, the output voltage would be 40 millivolts.

Given the digital input word, we can calculate the DAC output voltage by the equations

$$V_{out} = (\text{digital input word}) \times .040 \text{ V}$$

For example, let us suppose we have a digital input word equal to 65. The output voltage is $65 \times .040 = 2.60$ V.

However, we are interested in the converse of the preceding case: Given a needed voltage, what digital input word is required by the DAC? To calculate this we use the relationship between the voltage desired and the known voltage output change for each digital input change. The analog output voltage for the DAC will change .040 V for each digital count. Therefore, we can divide the voltage needed or desired at the DAC output by .040. The result will be the digital value needed to produce the output voltage.

For example, suppose we wished the DAC output voltage to be equal to 8.00 V. To find out what digital word is needed by the DAC we set up the relationship:

$$\text{Digital word} = \frac{V_{out} \text{ wanted}}{.040}$$

In our example the equation is

$$\text{Digital word} = \frac{8.00}{.040} = 200$$

If we input the digital weight of 200 to the DAC data input lines, the DAC output voltage will be equal to 8.00 V.

Let us suppose that we wished the DAC to output a voltage equal to 5.25 V. To calculate the digital word, we would use the equation

$$\text{Digital word} = \frac{5.25}{.040} = 131.25$$

Notice that the digital word is not an integer but is a real number. To control the DAC with the CPU, we must output digital words as integers: 1, 2, 234, 179, and so on. The exact voltage we require at the output of the DAC cannot be obtained. We must choose whether we wish the output voltage to be a little larger than 5.25 or a little less than 5.25. See Figure 17-13.

We must choose between the integer numbers 131 and 132. Neither of

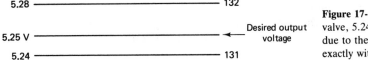

Figure 17-13 You must choose which valve, 5.24 or 5.28, to output. This is due to the fact that 5.25 cannot be set exactly with the DAC.

these two numbers will give the exact output voltage of 5.25 V. The number 131 will be an output voltage equal to $131 \times .040 = 5.24$ V. The number 132 will give an output voltage equal to $132 \times .040 = 5.28$ V. It is clear that the integer 131 gives an output voltage value closer to 5.25 V than does 132.

As a general rule, we may round to the closest whole number and use the rounded value as the digital word to be output. If the decimal part of the required digital word is less than but not equal to .5, use the smaller integer. If the decimal part of the digital word is greater than or equal to .5, use the greater integer as the output word.

To illustrate, let us suppose the required digital word was equal to 156.34. The decimal part of this word is equal to .34 so we will use the smaller integer, 156, as the output word. If the required digital word was equal to 156.67, we would use the greater integer, 157, as the output word.

The same type of calculations for the digital input word and the rounding of the word to integer values may be applied to a DAC with any number of input lines and any output voltage swing.

17-5 CONTROLLING THE DAC WITH SOFTWARE

In this section we present an assembly language program that will allow the AD558 to be programmed to any output voltage between .00 and 10.24 V. The program will input a byte from switches that are at an input port. The program is shown in Figure 17-14.

You see that the program shown in Figure 17-14 is very short. The first line is used to get a byte 00–FF from input port 60H. This byte is used to write to the DAC. Once the byte is input, a single OUT instruction writes the data to the device.

This program demonstrates how easy it is to control a DAC with the system software. Once you have the hardware operating, the software simply outputs the correct byte to the DAC inputs to set any voltage you choose.

```
IN   60H     ;Get 8 bits to output
OUT  0FFH    ;Output data to DAC
```

Figure 17-14 A simple two-line program to control the DAC output voltage. The binary word to write to the DAC comes from the setting of the switches at input port 60H.

17-6 CHAPTER SUMMARY

In this chapter we began by explaining the nature of digital-to-analog conversion. We showed that the number of unique output voltages of a converter depends on the number of digital input lines and the converter's range of output voltage levels. Also shown in this chapter was a general block diagram of a

DAC. When you encounter a DAC for the first time, it will be helpful to keep in mind this general block diagram.

From this general discussion of the DAC, a formula was derived for calculating the minimum voltage change for a DAC with any number of digital inputs and any maximum output voltage swing. We then discussed how to determine what digital word to write to the DAC to set a desired output voltage.

The chapter finished by showing a sample program for controlling a DAC with the 8085 CPU. DACs are quite popular and have many uses in microprocessor systems. If you understand the information given in this chapter, then using and controlling the peripheral devices that use DACs will be a much easier task.

REVIEW PROBLEMS

1. Write a definition for the term digital-to-analog conversion.
2. Draw a block diagram showing the typical components that comprise a general DAC.
3. List four microprocessor system applications where a DAC might be used.
4. Assume the DAC given in the block diagram of Problem 2 has an 8-bit input word. Design an interface to an 8085 microprocessor which controls the DAC.
5. Define the terms unipolar and bipolar as they refer to a DAC.
6. What is meant by the term 8-bit DAC?
7. If you are using a unipolar, 0–10 V, 12-bit DAC, what is the change in output voltage when the input word changes by one count, that is, the LSB goes from a logical 0 to a logical 1 or from a logical 1 to a logical 0?
8. Assume you have an 8-bit, unipolar DAC, 0–2.55 V. What is the output voltage given the following digital input words? Assume all input values are in hexadecimal.
    (a) 3F
    (b) 05
    (c) 2A
    (d) CD
    (e) 48
9. Given an 8-bit, unipolar, 0–10.24 V DAC, what is the required input word to set the output voltage equal to or slightly less than the following voltages?
    (a) 10.00 V
    (c) 0.34 V
    (c) 8.43 V
    (d) 2.76 V
    (e) 3.94 V
10. Depending on the desired DAC output voltage, the value can sometimes be set exactly and other times it must be set slightly less than or slightly greater than the desired output value. Explain this.

Appendix A

DATA SHEETS

8080

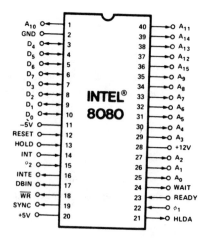

Courtesy of Intel Corporation.

Instructions for the 8080 require from one to five machine cycles for complete execution. The 8080 sends out 8 bit of status information on the data bus at the beginning of each machine cycle (during SYNC time). The following table defines the status information.

STATUS INFORMATION DEFINITION

Symbols	Data Bus Bit	Definition
INTA*	D_0	Acknowledge signal for INTERRUPT request. Signal should be used to gate a restart instruction onto the data bus when DBIN is active.
$\overline{WO}$	D_1	Indicates that the operation in the current machine cycle will be a WRITE memory or OUTPUT function ($\overline{WO}$ = 0). Otherwise, a READ memory or INPUT operation will be executed.
STACK	D_2	Indicates that the address bus holds the pushdown stack address from the Stack Pointer.
HLTA	D_3	Acknowledge signal for HALT instruction.
OUT	D_4	Indicates that the address bus contains the address of an output device and the data bus will contain the output data when $\overline{WR}$ is active.
M_1	D_5	Provides a signal to indicate that the CPU is in the fetch cycle for the first byte of an instruction.
INP*	D_6	Indicates that the address bus contains the address of an input device and the input data should be placed on the data bus when DBIN is active.
MEMR*	D_7	Designates that the data bus will be used for memory read data.

*These three status bits can be used to control the flow of data onto the 8080 data bus.

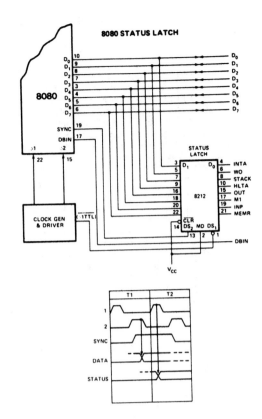

STATUS WORD CHART

Data Bus Bit	Status Information	① INSTRUCTION FETCH	② MEMORY READ	③ MEMORY WRITE	④ STACK READ	⑤ STACK WRITE	⑥ INPUT READ	⑦ OUTPUT WRITE	⑧ INTERRUPT ACKNOWLEDGE	⑨ HALT ACKNOWLEDGE	⑩ INTERRUPT ACKNOWLEDGE WHILE HALT
D_0	INTA	0	0	0	0	0	0	0	1	0	1
D_1	$\overline{WO}$	1	1	0	1	0	1	0	1	1	1
D_2	STACK	0	0	0	1	1	0	0	0	0	0
D_3	HLTA	0	0	0	0	0	0	0	0	1	1
D_4	OUT	0	0	0	0	0	0	1	0	0	0
D_5	M_1	1	0	0	0	0	0	0	1	0	1
D_6	INP	0	0	0	0	0	1	0	0	0	0
D_7	MEMR	1	1	0	1	0	0	0	0	1	0

TYPE OF MACHINE CYCLE

(N) STATUS WORD

Table 4-1. 8080 Status Bit Definitions

Courtesy of Intel Corporation.

The events that take place during the T3 state are determined by the kind of machine cycle in progress. In a FETCH machine cycle, the processor interprets the data on its data bus as an instruction. During a MEMORY READ or a STACK READ, data on this bus is interpreted as a data word. The processor outputs data on this bus during a MEMORY WRITE machine cycle. During I/O operations, the processor may either transmit or receive data, depending on whether an OUTPUT or an INPUT operation is involved.

Figure 4–7 illustrates the timing that is characteristic of a data input operation. As shown, the low-to-high transition of ϕ_2 during T2 clears status information from the processor's data lines, preparing these lines for the receipt of incoming data. The data presented to the processor must have stabilized prior to both the "ϕ_1—data set-up" interval (t_{DS1}), that precedes the falling edge of the ϕ_1 pulse defining state T3, and the "ϕ_2—data set-up" interval (t_{DS2}), that precedes the rising edge of ϕ_2 in state T3. This same

data must remain stable during the "data hold" interval (t_{DH}) that occurs following the rising edge of the ϕ_2 pulse. Data placed on these lines by memory or by other external devices will be sampled during T3.

During the input of data to the processor, the 8080 generates a DBIN signal which should be used externally to enable the transfer. Machine cycles in which DBIN is available include: FETCH, MEMORY READ, STACK READ, and INTERRUPT. DBIN is initiated by the rising edge of ϕ_2 during state T2 and terminated by the corresponding edge of ϕ_2 during T3. Any TW phases intervening between T2 and T3 will therefore extend DBIN by one or more clock periods.

Figure 4-7 shows the timing of a machine cycle in which the processor outputs data. Output data may be destined either for memory or for peripherals. The rising edge of ϕ_2 within state T2 clears status information from the CPU's data lines, and loads in the data which is to be output to external devices. This substitution takes place within the

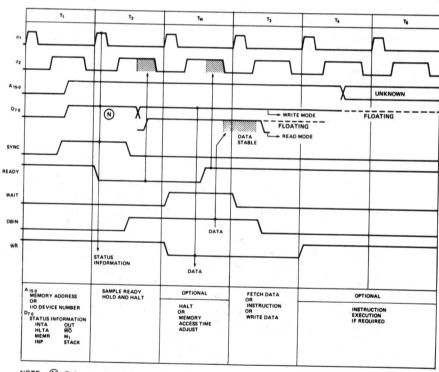

NOTE: (N) Refer to Status Word Chart on Page 4-6.

Figure 4–5. Basic 8080 Instruction Cycle

Courtesy of Intel Corporation.

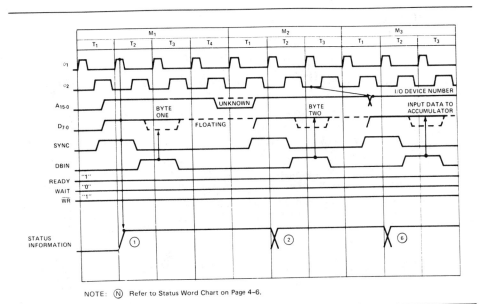

NOTE: (N) Refer to Status Word Chart on Page 4-6.

Figure 4-6. Input Instruction Cycle

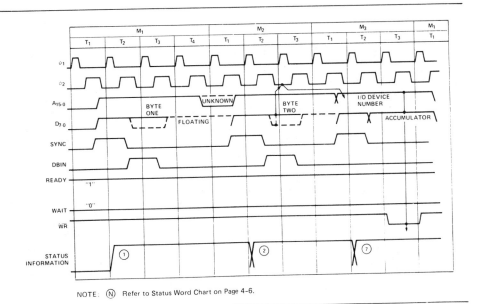

NOTE: (N) Refer to Status Word Chart on Page 4-6.

Figure 4-7. Output Instruction Cycle   Courtesy of Intel Corporation.

8085

intel®

8085A/8085A-2
SINGLE CHIP 8-BIT N-CHANNEL MICROPROCESSORS

■ **Single +5V Power Supply**

■ **100% Software Compatible with 8080A**

■ **1.3 μs Instruction Cycle (8085A); 0.8 μs (8085A-2)**

■ **On-Chip Clock Generator (with External Crystal, LC or RC Network)**

■ **On-Chip System Controller; Advanced Cycle Status Information Available for Large System Control**

■ **Four Vectored Interrupt Inputs (One is non-Maskable) Plus an 8080A-compatible interrupt**

■ **Serial In/Serial Out Port**

■ **Decimal, Binary and Double Precision Arithmetic**

■ **Direct Addressing Capability to 64k Bytes of Memory**

The Intel® 8085A is a complete 8 bit parallel Central Processing Unit (CPU). Its instruction set is 100% software compatible with the 8080A microprocessor, and it is designed to improve the present 8080A's performance by higher system speed. Its high level of system integration allows a minimum system of three IC's [8085A (CPU), 8156 (RAM/IO) and 8355/8755A (ROM/PROM/IO)] while maintaining total system expandability. The 8085A-2 is a faster version of the 8085A.

The 8085A incorporates all of the features that the 8224 (clock generator) and 8228 (system controller) provided for the 8080A, thereby offering a high level of system integration.

The 8085A uses a multiplexed data bus. The address is split between the 8 bit address bus and the 8 bit data bus. The on-chip address latches of 8155/8156/8355/8755A memory products allow a direct interface with the 8085A.

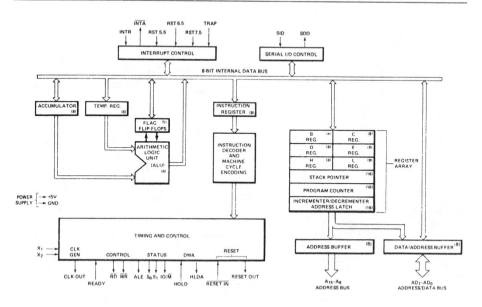

Figure 1. 8085A CPU Functional Block Diagram

8085A/8085A-2

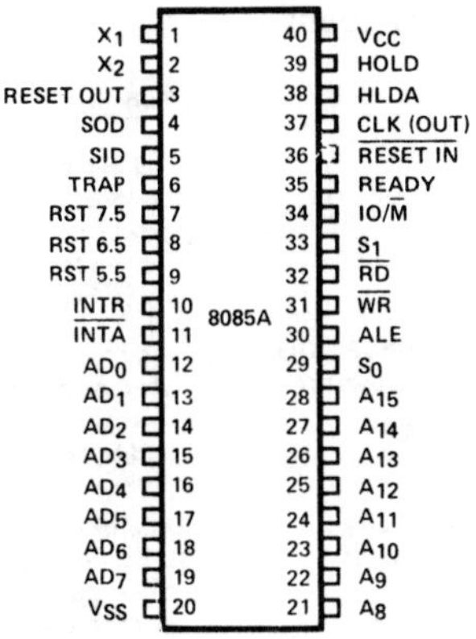

Figure 2. 8085A Pinout Diagram

8085A FUNCTIONAL PIN DEFINITION

The following describes the function of each pin:

Symbol	Function
A_8–A_{15} **(Output, 3-state)**	Address Bus: The most significant 8 bits of the memory address or the 8 bits of the I/O address, 3-stated during Hold and Halt modes and during RESET.
AD_{0-7} **(Input/Output, 3-state)**	Multiplexed Address/Data Bus: Lower 8 bits of the memory address (or I/O address) appear on the bus during the first clock cycle (T state) of a machine cycle. It then becomes the data bus during the second and third clock cycles.
ALE **(Output)**	Address Latch Enable: It occurs during the first clock state of a machine cycle and enables the address to get latched into the on-chip latch of peripherals. The falling edge of ALE is set to guarantee setup and hold times for the address information. The falling edge of ALE can also be used to strobe the status information. ALE is never 3-stated.

S_0, S_1, and $IO/\overline{M}$
(Output)

Machine cycle status:

$IO/\overline{M}$	S_1	S_0	Status
0	0	1	Memory write
0	1	0	Memory read
1	0	1	I/O write
1	1	0	I/O read
0	1	1	Opcode fetch
1	1	1	Interrupt Acknowledge
*	0	0	Halt
*	X	X	Hold
*	X	X	Reset

* = 3-state (high impedance)
X = unspecified

Symbol	Function
	S_1 can be used as an advanced $R/\overline{W}$ status. $IO/\overline{M}$,S_0 and S_1 become valid at the beginning of a machine cycle and remain stable throughout the cycle. The falling edge of ALE may be used to latch the state of these lines.
$\overline{RD}$ **(Output, 3-state)**	READ control: A low level on $\overline{RD}$ indicates the selected memory or I/O device is to be read and that the Data Bus is available for the data transfer, 3-stated during Hold and Halt modes and during RESET.
$\overline{WR}$ **(Output, 3-state)**	WRITE control: A low level on $\overline{WR}$ indicates the data on the Data Bus is to be written into the selected memory or I/O location. Data is set up at the trailing edge of $\overline{WR}$. 3-stated during Hold and Halt modes and during RESET.
READY **(Input)**	If READY is high during a read or write cycle, it indicates that the memory or peripheral is ready to send or receive data. If READY is low, the cpu will wait an integral number of clock cycles for READY to go high before completing the read or write cycle.
HOLD **(Input)**	HOLD indicates that another master is requesting the use of the address and data buses. The cpu, upon receiving the hold request, will relinquish the use of the bus as soon as the completion of the current bus transfer. Internal processing can continue. The processor can regain the bus only after the HOLD is removed. When the HOLD is acknowledged, the Address, Data, $\overline{RD}$, $\overline{WR}$, and $IO/\overline{M}$ lines are 3-stated.
HLDA **(Output)**	HOLD ACKNOWLEDGE: Indicates that the cpu has received the HOLD request and that it will relinquish the bus in the next clock cycle. HLDA goes low after the Hold request is removed. The cpu takes the bus one half clock cycle after HLDA goes low.
INTR **(Input)**	INTERRUPT REQUEST: is used as a general purpose interrupt. It is sampled only during the next to the last clock cycle of an instruction and during Hold and Halt states. If it is active, the Program Counter (PC) will be inhibited from incrementing and an $\overline{INTA}$ will be issued. During this cycle a RESTART or CALL instruction can be inserted to jump to the interrupt service routine. The INTR is enabled and disabled by software. It is disabled by Reset and immediately after an interrupt is accepted.

8085A/8085A-2

8085A FUNCTIONAL PIN DESCRIPTION (Continued)

Symbol	Function
$\overline{\text{INTA}}$ (Output)	INTERRUPT ACKNOWLEDGE: Is used instead of (and has the same timing as) $\overline{\text{RD}}$ during the Instruction cycle after an INTR is accepted. It can be used to activate the 8259 Interrupt chip or some other interrupt port.
RST 5.5 RST 6.5 RST 7.5 (Inputs)	RESTART INTERRUPTS: These three inputs have the same timing as INTR except they cause an internal RE-START to be automatically inserted. The priority of these interrupts is ordered as shown in Table 1. These interrupts have a higher priority than INTR. In addition, they may be individually masked out using the SIM instruction.
TRAP (Input)	Trap interrupt is a nonmaskable RE-START interrupt. It is recognized at the same time as INTR or RST 5.5-7.5. It is unaffected by any mask or Interrupt Enable. It has the highest priority of any interrupt. (See Table 1.)
RESET IN (Input)	Sets the Program Counter to zero and resets the Interrupt Enable and HLDA flip-flops. The data and address buses and the control lines are 3-stated during RESET and because of the asynchronous nature of RESET, the processor's internal registers and flags may be altered by RESET with unpredictable results. $\overline{\text{RESET IN}}$ is a

Symbol	Function
	Schmitt-triggered input, allowing connection to an R-C network for power-on RESET delay. The cpu is held in the reset condition as long as $\overline{\text{RESET IN}}$ is applied.
RESET OUT (Output)	Indicates cpu is being reset. Can be used as a system reset. The signal is synchronized to the processor clock and lasts an integral number of clock periods.
X_1, X_2 (Input)	X_1 and X_2 are connected to a crystal, LC, or RC network to drive the internal clock generator. X_1 can also be an external clock input from a logic gate. The input frequency is divided by 2 to give the processor's internal operating frequency.
CLK (Output)	Clock Output for use as a system clock. The period of CLK is twice the X_1, X_2 input period.
SID (Input)	Serial input data line. The data on this line is loaded into accumulator bit 7 whenever a RIM instruction is executed.
SOD (Output)	Serial output data line. The output SOD is set or reset as specified by the SIM instruction.
V_{CC}	+5 volt supply.
V_{SS}	Ground Reference.

TABLE 1. INTERRUPT PRIORITY, RESTART ADDRESS, AND SENSITIVITY

Name	Priority	Address Branched To (1) When Interrupt Occurs	Type Trigger
TRAP	1	24H	Rising edge AND high level until sampled.
RST 7.5	2	3CH	Rising edge (latched).
RST 6.5	3	34H	High level until sampled.
RST 5.5	4	2CH	High level until sampled.
INTR	5	See Note (2).	High level until sampled.

NOTES:

(1) The processor pushes the PC on the stack before branching to the indicated address.
(2) The address branched to depends on the instruction provided to the cpu when the interrupt is acknowledged.

Courtesy of Intel Corporation.

Z8400
Z80® CPU Central
Processing Unit

Zilog

Product
Specification

April 1985

FEATURES

- The instruction set contains 158 instructions. The 78 instructions of the 8080A are included as a subset; 8080A software compatibility is maintained.

- Eight MHz, 6 MHz, 4 MHz, and 2.5 MHz clocks for the Z80H, Z80B, Z80A, and Z80 CPU result in rapid instruction execution with consequent high data throughput.

- The extensive instruction set includes string, bit, byte, and word operations. Block searches and block transfers, together with indexed and relative addressing, result in the most powerful data handling capabilities in the microcomputer industry.

- The Z80 microprocessors and associated family of peripheral controllers are linked by a vectored interrupt system. This system may be daisy-chained to allow implementation of a priority interrupt scheme. Little, if any, additional logic is required for daisy-chaining.

- Duplicate sets of both general-purpose and flag registers are provided, easing the design and operation of system software through single-context switching, background-foreground programming, and single-level interrupt processing. In addition, two 16-bit index registers facilitate program processing of tables and arrays.

- There are three modes of high speed interrupt processing: 8080 similar, non-Z80 peripheral device, and Z80 Family peripheral with or without daisy chain.

- On-chip dynamic memory refresh counter.

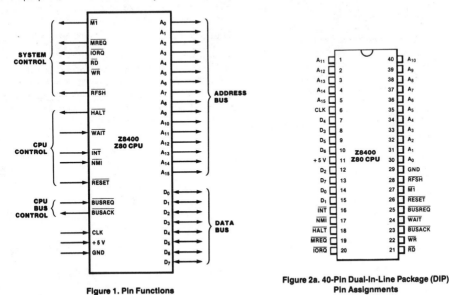

Figure 1. Pin Functions

**Figure 2a. 40-Pin Dual-In-Line Package (DIP)
Pin Assignments**

2001-001,002

PIN DESCRIPTIONS

A$_0$-A$_{15}$. *Address Bus* (output, active High, 3-state). A$_0$-A$_{15}$ form a 16-bit address bus. The Address Bus provides the address for memory data bus exchanges (up to 64K bytes) and for I/O device exchanges.

BUSACK. *Bus Acknowledge* (output, active Low). Bus Acknowledge indicates to the requesting device that the CPU address bus, data bus, and control signals MREQ, IORQ, RD, and WR have entered their high-impedance states. The external circuitry can now control these lines.

BUSREQ. *Bus Request* (input, active Low). Bus Request has a higher priority than NMI and is always recognized at the end of the current machine cycle. BUSREQ forces the CPU address bus, data bus, and control signals MREQ, IORQ, RD, and WR to go to a high-impedance state so that other devices can control these lines. BUSREQ is normally wired-OR and requires an external pullup for these applications. Extended BUSREQ periods due to extensive DMA operations can prevent the CPU from properly refreshing dynamic RAMs.

D$_0$-D$_7$. *Data Bus* (input/output, active High, 3-state). D$_0$-D$_7$ constitute an 8-bit bidirectional data bus, used for data exchanges with memory and I/O.

Halt. *Halt State* (output, active Low). HALT indicates that the CPU has executed a Halt instruction and is awaiting either a nonmaskable or a maskable interrupt (with the mask enabled) before operation can resume. While halted, the CPU executes NOPs to maintain memory refresh.

INT. *Interrupt Request* (input, active Low). Interrupt Request is generated by I/O devices. The CPU honors a request at the end of the current instruction if the internal software-controlled interrupt enable flip-flop (IFF) is enabled. INT is normally wired-OR and requires an external pullup for these applications.

IORQ. *Input/Output Request* (output, active Low, 3-state). IORQ indicates that the lower half of the address bus holds a valid I/O address for an I/O read or write operation. IORQ is also generated concurrently with M1 during an interrupt acknowledge cycle to indicate that an interrupt response vector can be placed on the data bus.

M1. *Machine Cycle One* (output, active Low). M1, together with MREQ, indicates that the current machine cycle is the opcode fetch cycle of an instruction execution. M1, together with IORQ, indicates an interrupt acknowledge cycle.

MREQ. *Memory Request* (output, active Low, 3-state). MREQ indicates that the address bus holds a valid address for a memory read or memory write operation.

NMI. *Non-Maskable Interrupt* (input, negative edge-triggered). NMI has a higher priority than INT. NMI is always recognized at the end of the current instruction, independent of the status of the interrupt enable flip-flop, and automatically forces the CPU to restart at location 0066H.

RD. *Read* (output, active Low, 3-state). RD indicates that the CPU wants to read data from memory or an I/O device. The addressed I/O device or memory should use this signal to gate data onto the CPU data bus.

RESET. *Reset* (input, active Low). RESET initializes the CPU as follows: it resets the interrupt enable flip-flop, clears the PC and Registers I and R, and sets the interrupt status to Mode 0. During reset time, the address and data bus go to a high-impedance state, and all control output signals go to the inactive state. Note that RESET must be active for a minimum of three full clock cycles before the reset operation is complete.

RFSH. *Refresh* (output, active Low). RFSH, together with MREQ, indicates that the lower seven bits of the system's address bus can be used as a refresh address to the system's dynamic memories.

WAIT. *Wait* (input, active Low). WAIT indicates to the CPU that the addressed memory or I/O devices are not ready for a data transfer. The CPU continues to enter a Wait state as long as this signal is active. Extended WAIT periods can prevent the CPU from refreshing dynamic memory properly.

WR. *Write* (output, active Low, 3-state). WR indicates that the CPU data bus holds valid data to be stored at the addressed memory or I/O location.

CPU TIMING

The Z80 CPU executes instructions by proceeding through a specific sequence of operations:

■ Memory read or write

■ I/O device read or write

■ Interrupt acknowledge

The basic clock period is referred to as a T time or cycle, and three or more T cycles make up a machine cycle (M1, M2 or M3 for instance). Machine cycles can be extended either by the CPU automatically inserting one or more Wait states or by the insertion of one or more Wait states by the user.

Instruction Opcode Fetch. The CPU places the contents of the Program Counter (PC) on the address bus at the start of the cycle (Figure 5). Approximately one-half clock cycle later, $\overline{MREQ}$ goes active. When active, $\overline{RD}$ indicates that the memory data can be enabled onto the CPU data bus.

The CPU samples the $\overline{WAIT}$ input with the falling edge of clock state T_2. During clock states T_3 and T_4 of an $\overline{M1}$ cycle, dynamic RAM refresh can occur while the CPU starts decoding and executing the instruction. When the Refresh Control signal becomes active, refreshing of dynamic memory can take place.

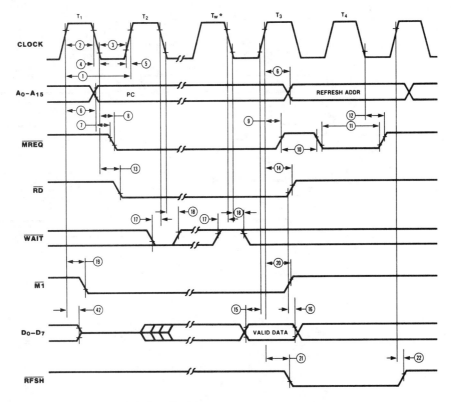

*T_W = Wait cycle added when necessary for slow anciliary devices.

Figure 5. Instruction Opcode Fetch

2001-006

Memory Read or Write Cycles. Figure 6 shows the timing of memory read or write cycles other than an opcode fetch (M1) cycle. The MREQ and RD signals function exactly as in the fetch cycle. In a memory write cycle, MREQ also becomes active when the address bus is stable. The WR line is active when the data bus is stable, so that it can be used directly as an R/W pulse to most semiconductor memories.

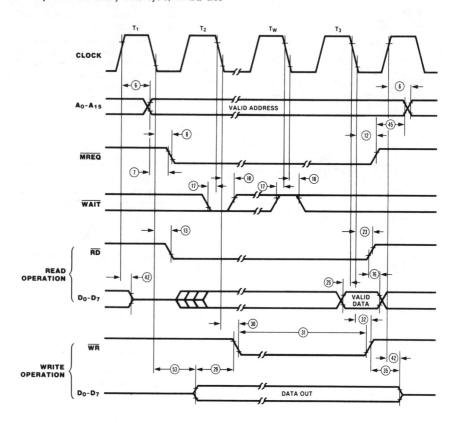

Figure 6. Memory Read or Write Cycles

Input or Output Cycles. Figure 7 shows the timing for an I/O read or I/O write operation. During I/O operations, the CPU automatically inserts a single Wait state (T_{WA}). This extra Wait state allows sufficient time for an I/O port to decode the address from the port address lines.

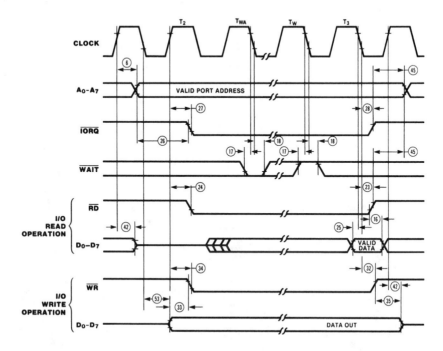

T_{WA} = One wait cycle automatically inserted by CPU.

Figure 7. Input or Output Cycles

Interrupt Request/Acknowledge Cycle. The CPU samples the interrupt signal with the rising edge of the last clock cycle at the end of any instruction (Figure 8). When an interrupt is accepted, a special $\overline{M1}$ cycle is generated.

During this $\overline{M1}$ cycle, $\overline{IORQ}$ becomes active (instead of $\overline{MREQ}$) to indicate that the interrupting device can place an 8-bit vector on the data bus. The CPU automatically adds two Wait states to this cycle.

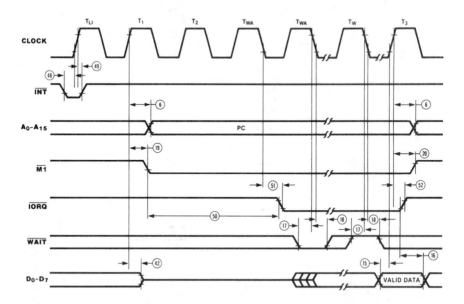

Z80 CPU

NOTES: 1) T_{LI} = Last state of any instruction cycle.
2) T_{WA} = Wait cycle automatically inserted by CPU.

Figure 8. Interrupt Request/Acknowledge Cycle

2001-009

Non-Maskable Interrupt Request Cycle. $\overline{\text{NMI}}$ is sampled at the same time as the maskable interrupt input $\overline{\text{INT}}$ but has higher priority and cannot be disabled under software control. The subsequent timing is similar to that of a normal memory read operation except that data put on the bus by the memory is ignored. The CPU instead executes a restart (RST) operation and jumps to the $\overline{\text{NMI}}$ service routine located at address 0066H (Figure 9).

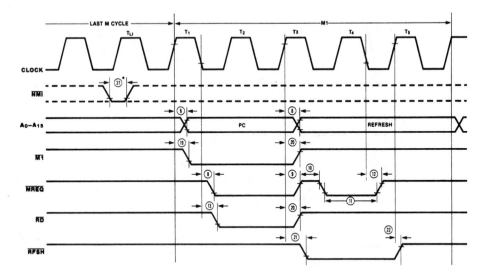

*Although $\overline{\text{NMI}}$ is an asynchronous input, to guarantee its being recognized on the following machine cycle, $\overline{\text{NMI}}$'s falling edge must occur no later than the rising edge of the clock cycle preceding the last state of any instruction cycle (T_{LI}).

Figure 9. Non-Maskable Interrupt Request Operation

2001-010

Bus Request/Acknowledge Cycle. The CPU samples BUSREQ with the rising edge of the last clock period of any machine cycle (Figure 10). If BUSREQ is active, the CPU sets its address, data, and MREQ, IORQ, RD, and WR lines to a high-impedance state with the rising edge of the next clock pulse. At that time, any external device can take control of these lines, usually to transfer data between memory and I/O devices.

Z80 CPU

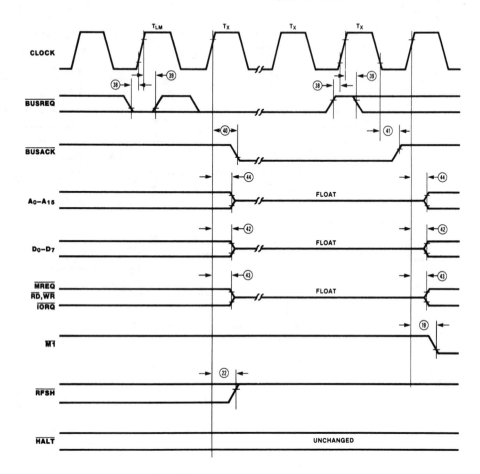

NOTES: 1) T_{LM} = Last state of any M cycle.
2) T_X = An arbitrary clock cycle used by requesting device.

Figure 10. Z-BUS Request/Acknowledge Cycle

Halt Acknowledge Cycle. When the CPU receives a HALT instruction, it executes NOP states until either an INT or NMI input is received. When in the Halt state, the HALT output is active and remains so until an interrupt is received (Figure 11). INT will also force a Halt exit.

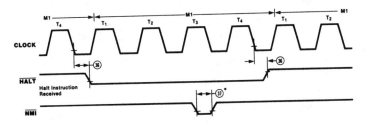

* Although NMI is an asynchronous input, to guarantee its being recognized on the following machine cycle, MNI's falling edge must occur no later than the rising edge of the clock cycle preceding the last state of any instruction cycle (T_LI).

Figure 11. Halt Acknowledge Cycle

Reset Cycle. RESET must be active for at least three clock cycles for the CPU to properly accept it. As long as RESET remains active, the address and data buses float, and the control outputs are inactive. Once RESET goes inactive, two internal T cycles are consumed before the CPU resumes normal processing operation. RESET clears the PC register, so the first opcode fetch will be to location 0000H (Figure 12).

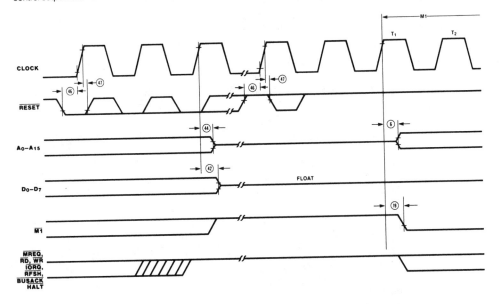

Figure 12. Reset Cycle

2001-012,013

AC CHARACTERISTICS†

Number	Symbol	Parameter	Z80 CPU Min	Z80 CPU Max	Z80A CPU Min	Z80A CPU Max	Z80B CPU Min	Z80B CPU Max	Z80H CPU Min	Z80H CPU Max
1	TcC	Clock Cycle Time	400*		250*		165*		125*	
2	TwCh	Clock Pulse Width (High)	180	2000	110	2000	65	2000	55	2000
3	TwCl	Clock Pulse Width (Low)	180	2000	110	2000	65	2000	55	2000
4	TfC	Clock Fall Time		30		30		20		10
5	TrC	Clock Rise Time		30		30		20		10
6	TdCr(A)	Clock ↑ to Address Valid Delay		145		110		90		80
7	TdA(MREQf)	Address Valid to $\overline{\text{MREQ}}$ ↓ Delay	125*		65*		35*		20*	
8	TdCf(MREQf)	Clock ↓ to $\overline{\text{MREQ}}$ ↓ Delay		100		85		70		60
9	TdCr(MREQr)	Clock ↑ to $\overline{\text{MREQ}}$ ↑ Delay		100		85		70		60
10	TwMREQh	$\overline{\text{MREQ}}$ Pulse Width (High)	170*		110*		65*		45*	
11	TwMREQl	$\overline{\text{MREQ}}$ Pulse Width (Low)	360*		220*		135*		100*	
12	TdCf(MREQr)	Clock ↓ to $\overline{\text{MREQ}}$ ↑ Delay		100		85		70		60
13	TdCf(RDf)	Clock ↓ to $\overline{\text{RD}}$ ↓ Delay		130		95		80		70
14	TdCr(RDr)	Clock ↑ to $\overline{\text{RD}}$ ↑ Delay		100		85		70		60
15	TsD(Cr)	Data Setup Time to Clock ↑	50		35		30		30	
16	ThD(RDr)	Data Hold Time to $\overline{\text{RD}}$ ↑		0		0		0		0
17	TsWAIT(Cf)	$\overline{\text{WAIT}}$ Setup Time to Clock ↓	70		70		60		50	
18	ThWAIT(Cf)	$\overline{\text{WAIT}}$ Hold Time after Clock ↓		0		0		0		0
19	TdCr(M1f)	Clock ↑ to $\overline{\text{M1}}$ ↓ Delay		130		100		80		70
20	TdCr(M1r)	Clock ↑ to $\overline{\text{M1}}$ ↑ Delay		130		100		80		70
21	TdCr(RFSHf)	Clock ↑ to $\overline{\text{RFSH}}$ ↓ Delay		180		130		110		95
22	TdCr(RFSHr)	Clock ↑ to $\overline{\text{RFSH}}$ ↑ Delay		150		120		100		85
23	TdCf(RDr)	Clock ↓ to $\overline{\text{RD}}$ ↑ Delay		110		85		70		60
24	TdCr(RDf)	Clock ↑ to $\overline{\text{RD}}$ ↓ Delay		100		85		70		60
25	TsD(Cf)	Data Setup to Clock ↓ during M_2, M_3, M_4, or M_5 Cycles	60		50		40		30	
26	TdA(IORQf)	Address Stable prior to $\overline{\text{IORQ}}$ ↓	320*		180*		110*		75*	
27	TdCr(IORQf)	Clock ↑ to $\overline{\text{IORQ}}$ ↓ Delay		90		75		65		55
28	TdCf(IORQr)	Clock ↓ to $\overline{\text{IORQ}}$ ↑ Delay		110		85		70		60
29	TdD(WRf)	Data Stable prior to $\overline{\text{WR}}$ ↓	190*		80*		25*		5*	
30	TdCf(WRf)	Clock ↓ to $\overline{\text{WR}}$ ↓ Delay		90		80		70		60
31	TwWR	$\overline{\text{WR}}$ Pulse Width	360*		220*		135*		100*	
32	TdCf(WRr)	Clock ↓ to $\overline{\text{WR}}$ ↑ Delay		100		80		70		60
33	TdD(WRf)	Data Stable prior to $\overline{\text{WR}}$ ↓	20*		-10*		-55*		55*	
34	TdCr(WRf)	Clock ↑ to $\overline{\text{WR}}$ ↓ Delay		80		65		60		55
35	TdWRr(D)	Data Stable from $\overline{\text{WR}}$ ↑	120*		60*		30*		15*	
36	TdCf(HALT)	Clock ↓ to $\overline{\text{HALT}}$ ↑ or ↓		300		300		260		225
37	TwNMI	$\overline{\text{NMI}}$ Pulse Width	80		80		70		60*	
38	TsBUSREQ(Cr)	$\overline{\text{BUSREQ}}$ Setup Time to Clock ↑	80		50		50		40	

*For clock periods other than the minimums shown, calculate parameters using the table on the following page. Calculated values above assumed TrC = TfC = 20 ns.

†Units in nanoseconds (ns).

AC CHARACTERISTICS† (Continued)

Number	Symbol	Parameter	Z80 CPU Min	Z80 CPU Max	Z80A CPU Min	Z80A CPU Max	Z80B CPU Min	Z80B CPU Max	Z80H CPU Min	Z80H CPU Max
39	ThBUSREQ(Cr)	BUSREQ Hold Time after Clock ↑	0		0		0		0	
40	TdCr(BUSACKf)	Clock ↑ to BUSACK ↓ Delay		120		100		90		80
41	TdCf(BUSACKr)	Clock ↓ to BUSACK ↑ Delay		110		100		90		80
42	TdCr(Dz)	Clock ↑ to Data Float Delay		90		90		80		70
43	TdCr(CTz)	Clock ↑ to Control Outputs Float Delay (MREQ, IORQ, RD, and WR)		110		80		70		60
44	TdCr(Az)	Clock ↑ to Address Float Delay		110		90		80		70
45	TdCTr(A)	MREQ ↑, IORQ ↑, RD ↑, and WR ↑ to Address Hold Time	160*		80*		35*		20*	
46	TsRESET(Cr)	RESET to Clock ↑ Setup Time	90		60		60		45	
47	ThRESET(Cr)	RESET to Clock ↑ Hold Time		0		0		0		0
48	TsINTf(Cr)	INT to Clock ↑ Setup Time	80		80		70		55	
49	ThINTr(Cr)	INT to Clock ↑ Hold Time		0		0		0		0
50	TdM1f(IORQf)	M1 ↓ to IORQ ↓ Delay	920*		565*		365*		270*	
51	TdCf(IORQf)	Clock ↓ to IORQ ↓ Delay		110		85		70		60
52	TdCf(IORQr)	Clock ↑ IORQ ↑ Delay		100		85		70		60
53	TdCf(D)	Clock ↓ to Data Valid Delay		230		150		130		115

*For clock periods other than the minimums shown, calculate parameters using the following table. Calculated values above assumed TrC = TfC = 20 ns.
†Units in nanoseconds (ns).

FOOTNOTES TO AC CHARACTERISTICS

Number	Symbol	General Parameter	Z80	Z80A	Z80B	Z80H
1	TcC	TwCh + TwCl + TrC + TfC				
7	TdA(MREQf)	TwCh + TfC	− 75	− 65	− 50	− 45
10	TwMREQh	TwCh + TfC	− 30	− 20	− 20	− 20
11	TwMREQl	TcC	− 40	− 30	− 30	− 25
26	TdA(IORQf)	TcC	− 80	− 70	− 55	− 50
29	TdD(WRf)	TcC	− 210	− 170	− 140	− 120
31	TwWR	TcC	− 40	− 30	− 30	− 25
33	TdD(WRf)	TwCl + TrC	− 180	− 140	− 140	− 120
35	TdWRr(D)	TwCl + TrC	− 80	− 70	− 55	− 50
45	TdCTr(A)	TwCl + TrC	− 40	− 50	− 50	− 45
50	TdM1f(IORQf)	2TcC + TwCh + TfC	− 80	− 65	− 50	− 45

AC Test Conditions:
$V_{IH} = 2.0$ V $V_{OH} = 1.5$ V
$V_{IL} = 0.8$ V $V_{OL} = 1.5$ V
$V_{IHC} = V_{CC} - 0.6$ V   FLOAT = ±0.5 V
$V_{ILC} = 0.45$ V

ABSOLUTE MAXIMUM RATINGS

Voltages on all pins with respect to ground . . .0.3V to +7V
Operating Ambient
Temperature . . . . .See Ordering Information
Storage Temperature . . . . −65°C to +150°C

Stresses greater than those listed under Absolute Maximum Ratings may cause permanent damage to the device. This is a stress rating only; operation of the device at any condition above these indicated in the operational sections of these specifications is not implied. Exposure to absolute maximum rating conditions for extended periods may affect device reliability.

STANDARD TEST CONDITIONS

The DC Characteristics and Capacitance sections below apply for the following standard test conditions, unless otherwise noted. All voltages are referenced to GND (0V). Positive current flows into the referenced pin.

Available operating temperature ranges are:

- S = 0°C to +70°C, +4.75V $\leqslant$ V_{CC} $\leqslant$ +5.25V

- E = −40°C to +85°C, +4.75V $\leqslant$ V_{CC} $\leqslant$ +5.25V

- M = −55°C to +125°C, +4.5V $\leqslant$ V_{CC} $\leqslant$ +5.25V

The Ordering Information section lists temperature ranges and product numbers. Package drawings are in the Package Information section in this book. Refer to the Literature List for additional documentation.

All ac parameters assume a load capacitance of 100 pf. Add 15 ns delay for each 50 pf increase in load up to a maximum of 200 pf for the data bus. AC timing measurements are referenced to 1.5 volts (except for clock, which is referenced to the 10% and 90% points).

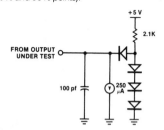

Z80 CPU

DC CHARACTERISTICS
All parameters are tested unless otherwise noted.

Symbol	Parameter	Min	Max	Unit	Test Condition
V_{ILC}	Clock Input Low Voltage	−0.3	0.45	V	
V_{IHC}	Clock Input High Voltage	V_{CC} −.6	V_{CC} +.3	V	
V_{IL}	Input Low Voltage	−0.3	0.8	V	
V_{IH}	Input High Voltage	2.0[1]	V_{CC}	V	
V_{OL}	Output Low Voltage		0.4	V	I_{OL} = 2.0 mA
V_{OH}	Output High Voltage	2.4[1]		V	I_{OH} = −250 μA
I_{CC}	Power Supply Current		200	mA	Note 3
I_{LI}	Input Leakage Current		10	μA	V_{IN} = 0 to V_{CC}
I_{LO}	3-State Output Leakage Current in Float	−10	10[2]	μA	V_{OUT} = 0.4 to V_{CC}

1. For military grade parts, refer to the Z80 Military Electrical Specification.
2. A_{15}-A_0, D_7-D_0, MREQ, IORQ, RD, and WR.
3. Measurements made with outputs floating.

CAPACITANCE
Guaranteed by design and characterization.

Symbol	Parameter	Min	Max	Unit
C_{CLOCK}	Clock Capacitance		35	pf
C_{IN}	Input Capacitance		5	pf
C_{OUT}	Output Capacitance		15	pf

NOTES:
T_A = 25°C, f = 1 MHz.
Unmeasured pins returned to ground.

Appendix B

INSTRUCTION SETS

8080/8085

CHAPTER 5
THE INSTRUCTION SET

5.1 WHAT THE INSTRUCTION SET IS

A computer, no matter how sophisticated, can do only what it is instructed to do. A program is a sequence of instructions, each of which is recognized by the computer and causes it to perform an operation. Once a program is placed in memory space that is accessible to your CPU, you may run that same sequence of instructions as often as you wish to solve the same problem or to do the same function. The set of instructions to which the 8085A CPU will respond is permanently fixed in the design of the chip.

Each computer instruction allows you to initiate the performance of a specific operation. The 8085A implements a group of instructions that move data between registers, between a register and memory, and between a register and an I/O port. It also has arithmetic and logic instructions, conditional and unconditional branch instructions, and machine control instructions. The CPU recognizes these instructions only when they are coded in binary form.

5.2 SYMBOLS AND ABBREVIATIONS:

The following symbols and abbreviations are used in the subsequent description of the 8085A instructions:

SYMBOLS	MEANING
accumulator	Register A
addr	16-bit address quantity
data	8-bit quantity
data 16	16-bit data quantity
byte 2	The second byte of the instruction
byte 3	The third byte of the instruction
port	8-bit address of an I/O device
r,r1,r2	One of the registers A,B,C, D,E,H,L

DDD,SSS   The bit pattern designating one of the registers A,B,C,D, E,H,L (DDD = destination, SSS = source):

DDD or SSS	REGISTER NAME
111	A
000	B
001	C
010	D
011	E
100	H
101	L

rp   One of the register pairs:

B represents the B,C pair with B as the high-order register and C as the low-order register;

D represents the D,E pair with D as the high-order register and E as the low-order register;

H represents the H,L pair with H as the high-order register and L as the low-order register;

SP represents the 16-bit stack pointer register.

RP   The bit pattern designating one of the register pairs B,D,H,SP:

RP	REGISTER PAIR
00	B-C
01	D-E
10	H-L
11	SP

rh   The first (high-order) register of a designated register pair.

rl   The second (low-order) register of a designated register pair.

Courtesy of Intel Corporation.

THE INSTRUCTION SET

PC	16-bit program counter register (PCH and PCL are used to refer to the high-order and low-order 8 bits respectively).
SP	16-bit stack pointer register (SPH and SPL are used to refer to the high-order and low-order 8 bits respectively).
r_m	Bit m of the register r (bits are number 7 through 0 from left to right).
LABEL	16-bit address of subroutine.
	The condition flags:
Z	Zero
S	Sign
P	Parity
CY	Carry
AC	Auxiliary Carry
( )	The contents of the memory location or registers enclosed in the parentheses.
←	"Is transferred to"
∧	Logical AND
⊻	Exclusive OR
∨	Inclusive OR
+	Addition
−	Two's complement subtraction
•	Multiplication
↔	"Is exchanged with"
‾‾‾	The one's complement (e.g., $\overline{(A)}$)
n	The restart number 0 through 7
NNN	The binary representation 000 through 111 for restart number 0 through 7 respectively.

The instruction set encyclopedia is a detailed description of the 8085A instruction set. Each instruction is described in the following manner:

1. The MCS-85 macro assembler format, consisting of the instruction mnemonic and operand fields, is printed in **BOLDFACE** on the first line.

2. The name of the instruction is enclosed in parentheses following the mnemonic.

3. The next lines contain a symbolic description of what the instruction does.

4. This is followed by a narrative description of the operation of the instruction.

5. The boxes describe the binary codes that comprise the machine instruction.

6. The last four lines contain information about the execution of the instruction. The number of machine cycles and states required to execute the instruction are listed first. If the instruction has two possible execution times, as in a conditional jump, both times are listed, separated by a slash. Next, data addressing modes are listed if applicable. The last line lists any of the five flags that are affected by the execution of the instruction.

5.3 INSTRUCTION AND DATA FORMATS

Memory used in the MCS-85 system is organized in 8-bit bytes. Each byte has a unique location in physical memory. That location is described by one of a sequence of 16-bit binary addresses. The 8085A can address up to 64K (K = 1024, or 2^{10}; hence, 64K represents the decimal number 65,536) bytes of memory, which may consist of both random-access, read-write memory (RAM) and read-only memory (ROM), which is also random-access.

Data in the 8085A is stored in the form of 8-bit binary integers:

DATA WORD

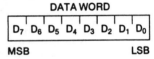

| D_7 | D_6 | D_5 | D_4 | D_3 | D_2 | D_1 | D_0 |

MSB   LSB

When a register or data word contains a binary number, it is necessary to establish the order in which the bits of the number are written. In the Intel 8085A, BIT 0 is referred to as the **Least Significant Bit (LSB)**, and BIT 7 (of an 8-bit number) is referred to as the **Most Significant Bit (MSB)**.

An 8085A program instruction may be one, two or three bytes in length. Multiple-byte instructions must be stored in successive memory locations; the address of the first byte is always used as the address of the instruction. The exact instruction format will depend on the particular operation to be executed.

Single Byte Instructions

| D_7 | | | | | | | D_0 | Op Code |

Two-Byte Instructions

Byte One | D_7 | | | | | | | D_0 | Op Code

Byte Two | D_7 | | | | | | | D_0 | Data or Address

Courtesy of Intel Corporation.

THE INSTRUCTION SET

Three-Byte Instructions

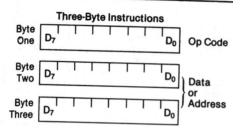

Byte One D_7 ... D_0 Op Code

Byte Two D_7 ... D_0 } Data or Address

Byte Three D_7 ... D_0 }

5.4 ADDRESSING MODES:

Often the data that is to be operated on is stored in memory. When multi-byte numeric data is used, the data, like instructions, is stored in successive memory locations, with the least significant byte first, followed by increasingly significant bytes. The 8085A has four different modes for addressing data stored in memory or in registers:

- Direct — Bytes 2 and 3 of the instruction contain the exact memory address of the data item (the low-order bits of the address are in byte 2, the high-order bits in byte 3).

- Register — The instruction specifies the register or register pair in which the data is located.

- Register Indirect — The instruction specifies a register pair which contains the memory address where the data is located (the high-order bits of the address are in the first register of the pair, the low-order bits in the second).

- Immediate — The instruction contains the data itself. This is either an 8-bit quantity or a 16-bit quantity (least significant byte first, most significant byte second).

Unless directed by an interrupt or branch institution, the execution of instructions proceeds through consecutively increasing memory locations. A branch instruction can specify the address of the next instruction to be executed in one of two ways:

- Direct — The branch instruction contains the address of the next instruction to be executed. (Except for the 'RST' instruction, byte 2 contains the low-order address and byte 3 the high-order address.)

- Register Indirect — The branch instruction indicates a register-pair which contains the address of the next instruction to be executed. (The high-order bits of the address are in the first register of the pair, the low-order bits in the second.)

The RST instruction is a special one-byte call instruction (usually used during interrupt sequences). RST includes a three-bit field; program control is transferred to the instruction whose address is eight times the contents of this three-bit field.

5.5 CONDITION FLAGS:

There are five condition flags associated with the execution of instructions on the 8085A. They are Zero, Sign, Parity, Carry, and Auxiliary Carry. Each is represented by a 1-bit register (or flip-flop) in the CPU. A flag is set by forcing the bit to 1; it is reset by forcing the bit to 0.

Unless indicated otherwise, when an instruction affects a flag, it affects it in the following manner:

Zero: If the result of an instruction has the value 0, this flag is set; otherwise it is reset.

Sign: If the most significant bit of the result of the operation has the value 1, this flag is set; otherwise it is reset.

Parity: If the modulo 2 sum of the bits of the result of the operation is 0, (i.e., if the result has even parity), this flag is set; otherwise it is reset (i.e., if the result has odd parity).

Carry: If the instruction resulted in a carry (from addition), or a borrow (from subtraction or a comparison) out of the high-order bit, this flag is set; otherwise it is reset.

Auxiliary Carry: If the instruction caused a carry out of bit 3 and into bit 4 of the resulting value, the auxiliary carry is set; otherwise it is reset. This flag is affected by single-precision additions, subtractions, increments, decrements, comparisons, and logical operations, but is principally used with additions and increments preceding a DAA (Decimal Adjust Accumulator) instruction.

THE INSTRUCTION SET

5.6 INSTRUCTION SET ENCYCLOPEDIA

In the ensuing dozen pages, the complete 8085A instruction set is described, grouped in order under five different functional headings, as follows:

1. **Data Transfer Group** — Moves data between registers or between memory locations and registers. Includes moves, loads, stores, and exchanges. (See below.)

2. **Arithmetic Group** — Adds, subtracts, increments, or decrements data in registers or memory. (See page 5-13.)

3. **Logic Group** — ANDs, ORs, XORs, compares, rotates, or complements data in registers or between memory and a register. (See page 5-16.)

4. **Branch Group** — Initiates conditional or unconditional jumps, calls, returns, and restarts. (See page 5-20.)

5. **Stack, I/O, and Machine Control Group** — Includes instructions for maintaining the stack, reading from input ports, writing to output ports, setting and reading interrupt masks, and setting and clearing flags. (See page 5-22.)

The formats described in the encyclopedia reflect the assembly language processed by Intel-supplied assembler, used with the Intellec® development systems.

5.6.1 Data Transfer Group

This group of instructions transfers data to and from registers and memory. **Condition flags are not affected by any instruction in this group.**

MOV r1, r2   (Move Register)
(r1) ← (r2)
The content of register r2 is moved to register r1.

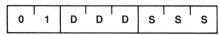

Cycles:	1
States:	4 (8085), 5 (8080)
Addressing:	register
Flags:	none

MOV r, M   (Move from memory)
(r) ← ((H) (L))
The content of the memory location, whose address is in registers H and L, is moved to register r.

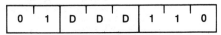

Cycles:	2
States:	7
Addressing:	reg. indirect
Flags:	none

MOV M, r   (Move to memory)
((H) (L)) ← (r)
The content of register r is moved to the memory location whose address is in registers H and L.

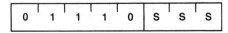

Cycles:	2
States:	7
Addressing:	reg. indirect
Flags:	none

MVI r, data   (Move Immediate)
(r) ← (byte 2)
The content of byte 2 of the instruction is moved to register r.

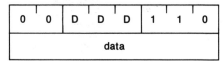

Cycles:	2
States:	7
Addressing:	immediate
Flags:	none

MVI M, data   (Move to memory immediate)
((H) (L)) ← (byte 2)
The content of byte 2 of the instruction is moved to the memory location whose address is in registers H and L.

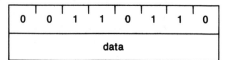

Cycles:	3
States:	10
Addressing:	immed./reg. indirect
Flags:	none

*All mnemonics copyrighted © Intel Corporation 1976.   5-4

Courtesy of Intel Corporation.

THE INSTRUCTION SET

LXI rp, data 16 (Load register pair immediate)
(rh) ← (byte 3),
(rl) ← (byte 2)
Byte 3 of the instruction is moved into the high-order register (rh) of the register pair rp. Byte 2 of the instruction is moved into the low-order register (rl) of the register pair rp.

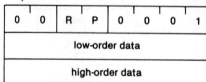

0	0	R	P	0	0	0	1
low-order data								
high-order data								

Cycles: 3
States: 10
Addressing: immediate
Flags: none

LDA addr   (Load Accumulator direct)
(A) ← ((byte 3)(byte 2))
The content of the memory location, whose address is specified in byte 2 and byte 3 of the instruction, is moved to register A.

0	0	1	1	1	0	1	0
low-order addr								
high-order addr								

Cycles: 4
States: 13
Addressing: direct
Flags: none

STA addr   (Store Accumulator direct)
((byte 3)(byte 2)) ← (A)
The content of the accumulator is moved to the memory location whose address is specified in byte 2 and byte 3 of the instruction.

0	0	1	1	0	0	1	0
low-order addr								
high-order addr								

Cycles: 4
States: 13
Addressing: direct
Flags: none

LHLD addr   (Load H and L direct)
(L)←((byte 3)(byte 2))
(H)←((byte 3)(byte 2)+1)
The content of the memory location, whose address is specified in byte 2 and byte 3 of the instruction, is moved to register L. The content of the memory location at the succeeding address is moved to register H.

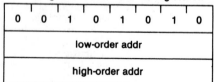

0	0	1	0	1	0	1	0
low-order addr								
high-order addr								

Cycles: 5
States: 16
Addressing: direct
Flags: none

SHLD addr   (Store H and L direct)
((byte 3)(byte 2))←(L)
((byte 3)(byte 2)+1)←(H)
The content of register L is moved to the memory location whose address is specified in byte 2 and byte 3. The content of register H is moved to the succeeding memory location.

0	0	1	0	0	0	1	0
low-order addr								
high-order addr								

Cycles: 5
States: 16
Addressing: direct
Flags: none

LDAX rp   (Load accumulator indirect)
(A) ← ((rp))
The content of the memory location, whose address is in the register pair rp, is moved to register A. Note: only register pairs rp = B (registers B and C) or rp = D (registers D and E) may be specified.

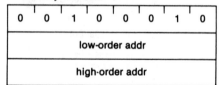

0	0	R	P	1	0	1	0

Cycles: 2
States: 7
Addressing: reg. indirect
Flags: none

Courtesy of Intel Corporation.

THE INSTRUCTION SET

STAX rp   (Store accumulator indirect)
((rp)) ← (A)
The content of register A is moved to the memory location whose address is in the register pair rp. Note: only register pairs rp = B (registers B and C) or rp = D (registers D and E) may be specified.

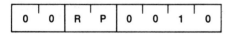

| 0 | 0 | R | P | 0 | 0 | 1 | 0 |

Cycles: 2
States: 7
Addressing: reg. indirect
Flags: none

XCHG   (Exchange H and L with D and E)
(H) ↔ (D)
(L) ↔ (E)
The contents of registers H and L are exchanged with the contents of registers D and E.

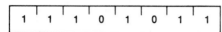

| 1 | 1 | 1 | 0 | 1 | 0 | 1 | 1 |

Cycles: 1
States: 4
Addressing: register
Flags: none

5.6.2 Arithmetic Group

This group of instructions performs arithmetic operations on data in registers and memory.

Unless indicated otherwise, all instructions in this group affect the Zero, Sign, Parity, Carry, and Auxiliary Carry flags according to the standard rules.

All subtraction operations are performed via two's complement arithmetic and set the carry flag to one to indicate a borrow and clear it to indicate no borrow.

ADD r   (Add Register)
(A) ← (A) + (r)
The content of register r is added to the content of the accumulator. The result is placed in the accumulator.

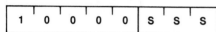

| 1 | 0 | 0 | 0 | 0 | S | S | S |

Cycles: 1
States: 4
Addressing: register
Flags: Z,S,P,CY,AC

ADD M   (Add memory)
(A) ← (A) + ((H) (L))
The content of the memory location whose address is contained in the H and L registers is added to the content of the accumulator. The result is placed in the accumulator.

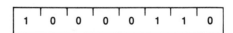

| 1 | 0 | 0 | 0 | 0 | 1 | 1 | 0 |

Cycles: 2
States: 7
Addressing: reg. indirect
Flags: Z,S,P,CY,AC

ADI data   (Add immediate)
(A) ← (A) + (byte 2)
The content of the second byte of the instruction is added to the content of the accumulator. The result is placed in the accumulator.

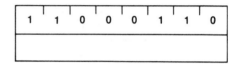

| 1 | 1 | 0 | 0 | 0 | 1 | 1 | 0 |
| | | | | | | | |

Cycles: 2
States: 7
Addressing: immediate
Flags: Z,S,P,CY,AC

ADC r   (Add Register with carry)
(A) ← (A) + (r) + (CY)
The content of register r and the content of the carry bit are added to the content of the accumulator. The result is placed in the accumulator.

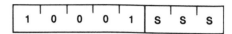

| 1 | 0 | 0 | 0 | 1 | S | S | S |

Cycles: 1
States: 4
Addressing: register
Flags: Z,S,P,CY,AC

THE INSTRUCTION SET

ADC M   (Add memory with carry)
(A) ← (A) + ((H) (L)) + (CY)
The content of the memory location whose address is contained in the H and L registers and the content of the CY flag are added to the accumulator. The result is placed in the accumulator.

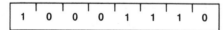

    Cycles: 2
    States: 7
  Addressing: reg. indirect
    Flags: Z,S,P,CY,AC

ACI data   (Add immediate with carry)
(A) ← (A) + (byte 2) + (CY)
The content of the second byte of the instruction and the content of the CY flag are added to the contents of the accumulator. The result is placed in the accumulator.

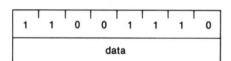

    Cycles: 2
    States: 7
  Addressing: immediate
    Flags: Z,S,P,CY,AC

SUB r   (Subtract Register)
(A) ← (A) − (r)
The content of register r is subtracted from the content of the accumulator. The result is placed in the accumulator.

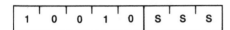

    Cycles: 1
    States: 4
  Addressing: register
    Flags: Z,S,P,CY,AC

SUB M   (Subtract memory)
(A) ← (A) − ((H) (L))
The content of the memory location whose address is contained in the H and L registers is subtracted from the content of the accumulator. The result is placed in the accumulator.

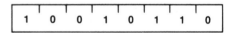

    Cycles: 2
    States: 7
  Addressing: reg. indirect
    Flags: Z,S,P,CY,AC

SUI data   (Subtract immediate)
(A) ← (A) − (byte 2)
The content of the second byte of the instruction is subtracted from the content of the accumulator. The result is placed in the accumulator.

    Cycles: 2
    States: 7
  Addressing: immediate
    Flags: Z,S,P,CY,AC

SBB r   (Subtract Register with borrow)
(A) ← (A) − (r) − (CY)
The content of register r and the content of the CY flag are both subtracted from the accumulator. The result is placed in the accumulator.

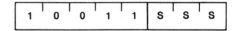

    Cycles: 1
    States: 4
  Addressing: register
    Flags: Z,S,P,CY,AC

Courtesy of Intel Corporation.

THE INSTRUCTION SET

SBB M　　　(Subtract memory with borrow)
(A) ← (A) − ((H) (L)) − (CY)
The content of the memory location whose address is contained in the H and L registers and the content of the CY flag are both subtracted from the accumulator. The result is placed in the accumulator.

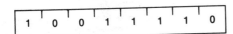

| 1 | 0 | 0 | 1 | 1 | 1 | 1 | 0 |

Cycles:　　2
States:　　7
Addressing:　reg. indirect
Flags:　　Z,S,P,CY,AC

SBI data　　(Subtract immediate with borrow)
(A) ← (A) − (byte 2) − (CY)
The contents of the second byte of the instruction and the contents of the CY flag are both subtracted from the accumulator. The result is placed in the accumulator.

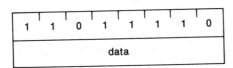

| 1 | 1 | 0 | 1 | 1 | 1 | 1 | 0 |

| data |

Cycles:　　2
States:　　7
Addressing:　immediate
Flags:　　Z,S,P,CY,AC

INR r　　　(Increment Register)
(r) ← (r) + 1
The content of register r is incremented by one. Note: All condition flags **except CY** are affected.

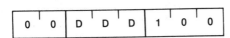

| 0 | 0 | D | D | D | 1 | 0 | 0 |

Cycles:　　1
States:　　4 (8085), 5 (8080)
Addressing:　register
Flags:　　Z,S,P,AC

INR M　　　(Increment memory)
((H) (L)) ← ((H) (L)) + 1
The content of the memory location whose address is contained in the H and L registers is incremented by one. Note: All condition flags **except CY** are affected.

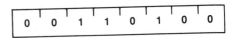

| 0 | 0 | 1 | 1 | 0 | 1 | 0 | 0 |

Cycles:　　3
States:　　10
Addressing:　reg. indirect
Flags:　　Z,S,P,AC

DCR r　　　(Decrement Register)
(r) ← (r) − 1
The content of register r is decremented by one. Note: All condition flags **except CY** are affected.

| 0 | 0 | D | D | D | 1 | 0 | 1 |

Cycles:　　1
States:　　4 (8085), 5 (8080)
Addressing:　register
Flags:　　Z,S,P,AC

DCR M　　　(Decrement memory)
((H) (L)) ← ((H) (L)) − 1
The content of the memory location whose address is contained in the H and L registers is decremented by one. Note: All condition flags **except CY** are affected.

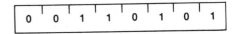

| 0 | 0 | 1 | 1 | 0 | 1 | 0 | 1 |

Cycles:　　3
States:　　10
Addressing:　reg. indirect
Flags:　　Z,S,P,AC

*All mnemonics copyrighted © Intel Corporation 1976.　　5-8

Courtesy of Intel Corporation.

THE INSTRUCTION SET

INX rp   (Increment register pair)
(rh) (rl) ← (rh) (rl) + 1
The content of the register pair rp is incremented by one. Note: **No condition flags are affected.**

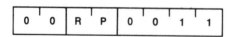

0	0	R	P	0	0	1	1

Cycles: 1
States: 6 (8085), 5 (8080)
Addressing: register
Flags: none

DCX rp   (Decrement register pair)
(rh) (rl) ← (rh) (rl) − 1
The content of the register pair rp is decremented by one. Note: **No condition flags are affected.**

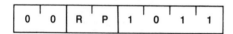

0	0	R	P	1	0	1	1

Cycles: 1
States: 6 (8085), 5 (8080)
Addressing: register
Flags: none

DAD rp   (Add register pair to H and L)
(H) (L) ← (H) (L) + (rh) (rl)
The content of the register pair rp is added to the content of the register pair H and L. The result is placed in the register pair H and L. Note: **Only the CY flag is affected.** It is set if there is a carry out of the double precision add; otherwise it is reset.

0	0	R	P	1	0	0	1

Cycles: 3
States: 10
Addressing: register
Flags: CY

DAA   (Decimal Adjust Accumulator)
The eight-bit number in the accumulator is adjusted to form two four-bit Binary-Coded-Decimal digits by the following process:

1. If the value of the lease significant 4 bits of the accumulator is greater than 9 **or** if the AC flag is set, 6 is added to the accumulator.
2. If the value of the most significant 4 bits of the accumulator is now greater than 9, **or** if the CY flag is set, 6 is added to the most significant 4 bits of the accumulator.

NOTE: All flags are affected.

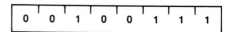

0	0	1	0	0	1	1	1

Cycles: 1
States: 4
Flags: Z,S,P,CY,AC

5.6.3 Logical Group

This group of instructions performs logical (Boolean) operations on data in registers and memory and on condition flags.

Unless indicated otherwise, all instructions in this group affect the Zero, Sign, Parity, Auxiliary Carry, and Carry flags according to the standard rules.

ANA r   (AND Register)
(A) ← (A) ∧ (r)
The content of register r is logically ANDed with the content of the accumulator. The result is placed in the accumulator. **The CY flag is cleared and AC is set (8085). The CY flag is cleared and AC is set to the OR'ing of bits 3 of the operands (8080).**

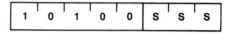

1	0	1	0	0	S	S	S

Cycles: 1
States: 4
Addressing: register
Flags: Z,S,P,CY,AC

Courtesy of Intel Corporation.

THE INSTRUCTION SET

ANA M   (AND memory)
(A) ← (A) ∧ ((H) (L))
The contents of the memory location whose address is contained in the H and L registers is logically ANDed with the content of the accumulator. The result is placed in the accumulator. **The CY flag is cleared and AC is set (8085). The CY flag is cleared and AC is set to the OR'ing of bits 3 of the operands (8080).**

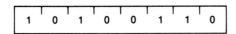

| 1 | 0 | 1 | 0 | 0 | 1 | 1 | 0 |

Cycles: 2
States: 7
Addressing: reg. indirect
Flags: Z,S,P,CY,AC

ANI data   (AND immediate)
(A) ← (A) ∧ (byte 2)
The content of the second byte of the instruction is logically ANDed with the contents of the accumulator. The result is placed in the accumulator. **The CY flag is cleared and AC is set (8085). The CY flag is cleared and AC is set to the OR'ing of bits 3 of the operands (8080).**

| 1 | 1 | 1 | 0 | 0 | 1 | 1 | 0 |
| data |

Cycles: 2
States: 7
Addressing: immediate
Flags: Z,S,P,CY,AC

XRA r   (Exclusive OR Register)
(A) ← (A) ∀ (r)
The content of register r is exclusive-OR'd with the content of the accumulator. The result is placed in the accumulator. **The CY and AC flags are cleared.**

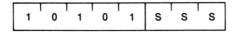

| 1 | 0 | 1 | 0 | 1 | S | S | S |

Cycles: 1
States: 4
Addressing: register
Flags: Z,S,P,CY,AC

XRA M   (Exclusive OR Memory)
(A) ← (A) ∀ ((H) (L))
The content of the memory location whose address is contained in the H and L registers is exclusive-OR'd with the content of the accumulator. The result is placed in the accumulator. **The CY and AC flags are cleared.**

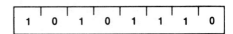

| 1 | 0 | 1 | 0 | 1 | 1 | 1 | 0 |

Cycles: 2
States: 7
Addressing: reg. indirect
Flags: Z,S,P,CY,AC

XRI data   (Exclusive OR immediate)
(A) ← (A) ∀ (byte 2)
The content of the second byte of the instruction is exclusive-OR'd with the content of the accumulator. The result is placed in the accumulator. **The CY and AC flags are cleared.**

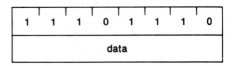

| 1 | 1 | 1 | 0 | 1 | 1 | 1 | 0 |
| data |

Cycles: 2
States: 7
Addressing: immediate
Flags: Z,S,P,CY,AC

ORA r   (OR Register)
(A) ← (A) V (r)
The content of register r is inclusive-OR'd with the content of the accumulator. The result is placed in the accumulator. **The CY and AC flags are cleared.**

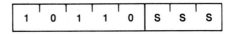

| 1 | 0 | 1 | 1 | 0 | S | S | S |

Cycles: 1
States: 4
Addressing: register
Flags: Z,S,P,CY,AC

Courtesy of Intel Corporation.

THE INSTRUCTION SET

ORA M   (OR memory)
(A) ← (A) V ((H) (L))
The content of the memory location whose
address is contained in the H and L
registers is inclusive-OR'd with the content
of the accumulator. The result is placed in
the accumulator. **The CY and AC flags are
cleared.**

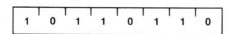

1	0	1	1	0	1	1	0

        Cycles: 2
        States: 7
    Addressing: reg. indirect
        Flags: Z,S,P,CY,AC

ORI data   (OR Immediate)
(A) ← (A) V (byte 2)
The content of the second byte of the instruction is inclusive-OR'd with the content
of the accumulator. The result is placed in
the accumulator. **The CY and AC flags are
cleared..**

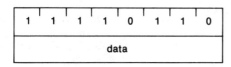

1	1	1	1	0	1	1	0
			data				

        Cycles: 2
        States: 7
    Addressing: immediate
        Flags: Z,S,P,CY,AC

CMP r   (Compare Register)
(A) − (r)
The content of register r is subtracted from
the accumulator. The accumulator remains
unchanged. The condition flags are set as
a result of the subtraction. **The Z flag is set
to 1 if (A) = (r). The CY flag is set to 1 if (A)
< (r).**

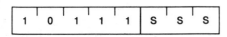

1	0	1	1	1	S	S	S

        Cycles: 1
        States: 4
    Addressing: register
        Flags: Z,S,P,CY,AC

CMP M   (Compare memory)
(A) − ((H) (L))
The content of the memory location whose
address is contained in the H and L
registers is subtracted from the accumulator. The accumulator remains unchanged. The condition flags are set as a
result of the subtraction. **The Z flag is set
to 1 if (A)=((H) (L)). The CY flag is set to 1 if
(A)< ((H) (L)).**

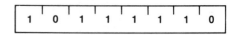

1	0	1	1	1	1	1	0

        Cycles: 2
        States: 7
    Addressing: reg. indirect
        Flags: Z,S,P,CY,AC

CPI data   (Compare immediate)
(A) − (byte 2)
The content of the second byte of the instruction is subtracted from the accumulator. The condition flags are set by
the result of the subtraction. **The Z flag is
set to 1 if (A)=(byte 2). The CY flag is set to
1 if (A)< (byte 2).**

1	1	1	1	1	1	1	0
			data				

        Cycles: 2
        States: 7
    Addressing: immediate
        Flags: Z,S,P,CY,AC

RLC   (Rotate left)
$(A_{n+1}) ← (A_n)$;$(A_0) ← (A_7)$
$(CY) ← (A_7)$
The content of the accumulator is rotated
left one position. The low order bit and the
CY flag are both set to the value shifted out
of the high order bit position. **Only the CY
flag is affected.**

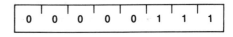

0	0	0	0	0	1	1	1

        Cycles: 1
        States: 4
        Flags: CY

Courtesy of Intel Corporation.

THE INSTRUCTION SET

RRC   (Rotate right)
$(A_n) \leftarrow (A_{n+1}); (A_7) \leftarrow (A_0)$
$(CY) \leftarrow (A_0)$
The content of the accumulator is rotated right one position. The high order bit and the CY flag are both set to the value shifted out of the low order bit position. **Only the CY flag is affected.**

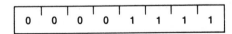

	0	0	0	0	1	1	1	1	

Cycles: 1
States: 4
Flags: CY

RAL   (Rotate left through carry)
$(A_{n+1}) \leftarrow (A_n); (CY) \leftarrow (A_7)$
$(A_0) \leftarrow (CY)$
The content of the accumulator is rotated left one position through the CY flag. The low order bit is set equal to the CY flag and the CY flag is set to the value shifted out of the high order bit. **Only the CY flag is affected.**

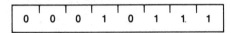

	0	0	0	1	0	1	1	1	

Cycles: 1
States: 4
Flags: CY

RAR   (Rotate right through carry)
$(A_n) \leftarrow (A_{n+1}); (CY) \leftarrow (A_0)$
$(A_7) \leftarrow (CY)$
The content of the accumulator is rotated right one position through the CY flag. The high order bit is set to the CY flag and the CY flag is set to the value shifted out of the low order bit. **Only the CY flag is affected.**

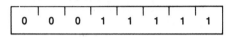

	0	0	0	1	1	1	1	1	

Cycles: 1
States: 4
Flags: CY

CMA   (Complement accumulator)
$(A) \leftarrow (\overline{A})$
The contents of the accumulator are complemented (zero bits become 1, one bits become 0). **No flags are affected.**

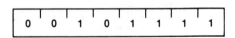

	0	0	1	0	1	1	1	1	

Cycles: 1
States: 4
Flags: none

CMC   (Complement carry)
$(CY) \leftarrow (\overline{CY})$
The CY flag is complemented. **No other flags are affected.**

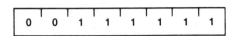

	0	0	1	1	1	1	1	1	

Cycles: 1
States: 4
Flags: CY

STC   (Set carry)
$(CY) \leftarrow 1$
The CY flag is set to 1. **No other flags are affected.**

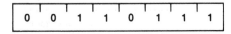

	0	0	1	1	0	1	1	1	

Cycles: 1
States: 4
Flags: CY

Courtesy of Intel Corporation.

THE INSTRUCTION SET

5.6.4 Branch Group

This group of instructions alter normal sequential program flow.

Condition flags are not affected by any instruction in this group.

The two types of branch instructions are unconditional and conditional. Unconditional transfers simply perform the specified operation on register PC (the program counter). Conditional transfers examine the status of one of the four processor flags to determine if the specified branch is to be executed. The conditions that may be specified are as follows:

CONDITION		CCC
NZ —	not zero (Z = 0)	000
Z —	zero (Z = 1)	001
NC —	no carry (CY = 0)	010
C —	carry (CY = 1)	011
PO —	parity odd (P = 0)	100
PE —	parity even (P = 1)	101
P —	plus (S = 0)	110
M —	minus (S = 1)	111

JMP addr   (Jump)
(PC) ← (byte 3) (byte 2)
Control is transferred to the instruction whose address is specified in byte 3 and byte 2 of the current instruction.

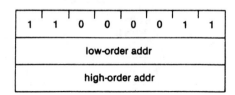

1	1	0	0	0	0	1	1
low-order addr							
high-order addr							

          Cycles: 3
          States: 10
     Addressing: immediate
          Flags: none

Jcondition addr (Conditional jump)
   If (CCC),
   (PC) ← (byte 3) (byte 2)
   If the specified condition is true, control is transferred to the instruction whose address is specified in byte 3 and byte 2 of the current instruciton; otherwise, control continues sequentially.

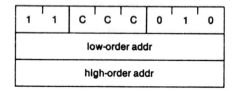

1	1	C	C	C	0	1	0
low-order addr							
high-order addr							

          Cycles: 2/3 (8085), 3 (8080)
          States: 7/10 (8085), 10 (8080)
     Addressing: immediate
          Flags: none

CALL addr   (Call)
   ((SP) − 1) ← (PCH)
   ((SP) − 2) ← (PCL)
   (SP) ← (SP) − 2
   (PC) ← (byte 3) (byte 2)
   The high-order eight bits of the next instruction address are moved to the memory location whose address is one less than the content of register SP. The low-order eight bits of the next instruction address are moved to the memory location whose address is two less than the content of register SP. The content of register SP is decremented by 2. Control is transferred to the instruction whose address is specified in byte 3 and byte 2 of the current instruction.

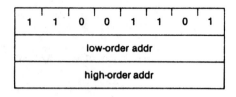

1	1	0	0	1	1	0	1
low-order addr							
high-order addr							

          Cycles: 5
          States: 18 (8085), 17 (8080)
     Addressing: immediate/
                      reg. indirect
          Flags: none

THE INSTRUCTION SET

Ccondition addr (Condition call)
If (CCC),
    ((SP) − 1) ← (PCH)
    ((SP) − 2) ← (PCL)
    (SP) ← (SP) − 2
    (PC) ← (byte 3) (byte 2)
If the specified condition is true, the actions specified in the CALL instruction (see above) are performed; otherwise, control continues sequentially.

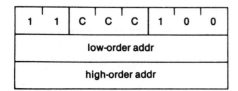

1	1	C	C	C	1	0	0
low-order addr
high-order addr

Cycles: 2/5 (8085), 3/5 (8080)
States: 9/18 (8085), 11/17 (8080)
Addressing: immediate/
              reg. indirect
Flags: none

RET   (Return)
(PCL) ← ((SP));
(PCH) ← ((SP) + 1);
(SP) ← (SP) + 2;
The content of the memory location whose address is specified in register SP is moved to the low-order eight bits of register PC. The content of the memory location whose address is one more than the content of register SP is moved to the high-order eight bits of register PC. The content of register SP is incremented by 2.

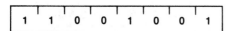

1	1	0	0	1	0	0	1

Cycles: 3
States: 10
Addressing: reg. indirect
Flags: none

Rcondition   (Conditional return)
If (CCC),
    (PCL) ← ((SP))
    (PCH) ← ((SP) + 1)
    (SP) ← (SP) + 2
If the specified condition is true, the actions specified in the RET instruction (see above) are performed; otherwise, control continues sequentially.

1	1	C	C	C	0	0	0

Cycles: 1/3
States: 6/12 (8085), 5/11 (8080)
Addressing: reg. indirect
Flags: none

RST n   (Restart)
((SP) − 1) ← (PCH)
((SP) − 2) ← (PCL)
(SP) ← (SP) − 2
(PC) ← 8 * (NNN)
The high-order eight bits of the next instruction address are moved to the memory location whose address is one less than the content of register SP. The low-order eight bits of the next instruction address are moved to the memory location whose address is two less than the content of register SP. The content of register SP is decremented by two. Control is transferred to the instruction whose address is eight times the content of NNN.

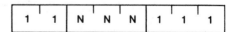

1	1	N	N	N	1	1	1

Cycles: 3
States: 12 (8085), 11 (8080)
Addressing: reg. indirect
Flags: none

15	14	13	12	11	10	9	8	7	6	5	4	3	2	1	0
0	0	0	0	0	0	0	0	0	0	N	N	N	0	0	0

Program Counter After Restart

THE INSTRUCTION SET

PCHL   (Jump H and L indirect —
                  move H and L to PC)
(PCH) ← (H)
(PCL) ← (L)
The content of register H is moved to the
high-order eight bits of register PC. The
content of register L is moved to the low-
order eight bits of register PC.

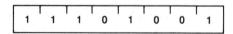

1	1	1	0	1	0	0	1

        Cycles: 1
        States: 6 (8085), 5 (8080)
    Addressing: register
         Flags: none

5.6.5   Stack, I/O, and Machine Control Group

This group of instructions performs I/O, manipu-
lates the Stack, and alters internal control
flags.

Unless otherwise specified, **condition flags are
not affected by any instructions in this group.**

PUSH rp   (Push)
((SP) − 1) ← (rh)
((SP) − 2) ← (rl)
((SP) ← (SP) − 2

The content of the high-order register of
register pair rp is moved to the memory
location whose address is one less than
the content of register SP. The content of
the low-order register of register pair rp is
moved to the memory location whose address is two less than the content of
register SP. The content of register SP is
decremented by 2. **Note: Register pair rp =
SP may not be specified.**

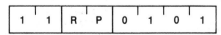

1	1	R	P	0	1	0	1

        Cycles: 3
        States: 12 (8085), 11 (8080)
    Addressing: reg. indirect
         Flags: none

PUSH PSW   (Push processor status word)
((SP) − 1) ← (A)
((SP) − 2)$_0$ ← (CY) , ((SP) − 2)$_1$ ← X
((SP) − 2)$_2$ ← (P) , ((SP) − 2)$_3$ ← X
((SP) − 2)$_4$ ← (AC), ((SP) − 2)$_5$ ← X
((SP) − 2)$_6$ ← (Z) , ((SP) − 2)$_7$ ← (S)
(SP) ← (SP) − 2   X: Undefined.

5–15

The content of register A is moved to the
memory location whose address is one
less than register SP. The contents of the
condition flags are assembled into a processor status word and the word is moved
to the memory location whose address is
two less than the content of register SP.
The content of register SP is decremented
by two.

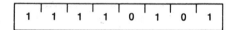

1	1	1	1	0	1	0	1

        Cycles: 3
        States: 12 (8085), 11 (8080)
    Addressing: reg. indirect
         Flags: none

FLAG WORD

D_7	D_6	D_5	D_4	D_3	D_2	D_1	D_0
S	Z	X	AC	X	P	X	CY

X: undefined

POP rp   (Pop)
(rl) ← ((SP))
(rh) ← ((SP) + 1)
(SP) ← (SP) + 2

The content of the memory location, whose
address is specified by the content of
register SP, is moved to the low-order
register of register pair rp. The content of
the memory location, whose address is one
more than the content of register SP, is
moved to the high-order register of register
rp. The content of register SP is incremented by 2. **Note: Register pair rp =
SP may not be specified.**

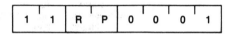

1	1	R	P	0	0	0	1

        Cycles: 3
        States: 10
    Addressing: reg.indirect
         Flags: none

THE INSTRUCTION SET

POP PSW   (Pop processor status word)

$(CY) \leftarrow ((SP))_0$
$(P) \leftarrow ((SP))_2$
$(AC) \leftarrow ((SP))_4$
$(Z) \leftarrow ((SP))_6$
$(S) \leftarrow ((SP))_7$
$(A) \leftarrow ((SP) + 1)$
$(SP) \leftarrow (SP) + 2$

The content of the memory location whose address is specified by the content of register SP is used to restore the condition flags. The content of the memory location whose address is one more than the content of register SP is moved to register A. The content of register SP is incremented by 2.

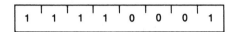

```
Cycles:      3
States:      10
Addressing:  reg. indirect
Flags:       Z,S,P,CY,AC
```

XTHL   (Exchange stack top with H and L)

$(L) \leftrightarrow ((SP))$
$(H) \leftrightarrow ((SP) + 1)$

The content of the L register is exchanged with the content of the memory location whose address is specified by the content of register SP. The content of the H register is exchanged with the content of the memory location whose address is one more than the content of register SP.

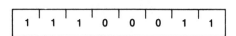

```
Cycles:      5
States:      16 (8085), 18 (8080)
Addressing:  reg. indirect
Flags:       none
```

SPHL   (Move HL to SP)

$(SP) \leftarrow (H) (L)$

The contents of registers H and L (16 bits) are moved to register SP.

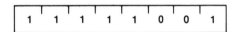

```
Cycles:      1
States:      6 (8085), 5 (8080)
Addressing:  register
Flags:       none
```

IN port   (Input)

$(A) \leftarrow (data)$

The data placed on the eight bit bidirectional data bus by the specified port is moved to register A.

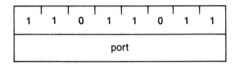

```
Cycles:      3
States:      10
Addressing:  direct
Flags:       none
```

OUT port   (Output)

$(data) \leftarrow (A)$

The content of register A is placed on the eight bit bi-directional data bus for transmission to the specified port.

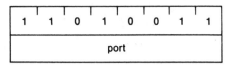

```
Cycles:      3
States:      10
Addressing:  direct
Flags:       none
```

THE INSTRUCTION SET

EI   (Enable interrupts)

The interrupt system is enabled following the execution of the next instruction. Interrupts are not recognized during the EI instruction.

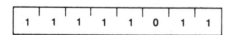

Cycles: 1
States: 4
Flags: none

NOTE: Placing an EI instruction on the bus in response to INTA during an INA cycle is prohibited. (8085)

DI   (Disable interrupts)

The interrupt system is disabled immediately following the execution of the DI instruction. Interrupts are not recognized during the DI instruction.

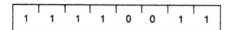

Cycles: 1
States: 4
Flags: none

NOTE: Placing a DI instruction on the bus in response to INTA during an INA cycle is prohibited. (8085)

HLT   (Halt)

The processor is stopped. The registers and flags are unaffected. (8080) A second ALE is generated during the execution of HLT to strobe out the Halt cycle status information. (8085)

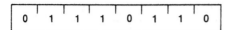

Cycles: 1 + (8085), 1 (8080)
States: 5 (8085), 7 (8080)
Flags: none

NOP   (No op)

No operation is performed. The registers and flags are unaffected.

*All mnemonics copyrighted © Intel Corporation 1976.   5-17

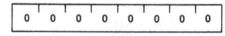

Cycles: 1
States: 4
Flags: none

RIM   (Read Interrupt Masks) (8085 only)

The RIM instruction loads data into the accumulator relating to interrupts and the serial input. This data contains the following information:

- Current interrupt mask status for the RST 5.5, 6.5, and 7.5 hardware interrupts (1 = mask disabled)
- Current interrupt enable flag status (1 = interrupts enabled) except immediately following a TRAP interrupt. (See below.)
- Hardware interrupts pending (i.e., signal received but not yet serviced), on the RST 5.5, 6.5, and 7.5 lines.
- Serial input data.

Immediately following a TRAP interrupt, the RIM instruction must be executed as a part of the service routine if you need to retrieve current interrupt status later. Bit 3 of the accumulator is (in this special case only) loaded with the interrupt enable (IE) flag status that existed prior to the TRAP interrupt. Following an RST 5.5, 6.5, 7.5, or INTR interrupt, the interrupt flag flip-flop reflects the current interrupt enable status. Bit 6 of the accumulator (I7.5) is loaded with the status of the RST 7.5 flip-flop, which is always set (edge-triggered) by an input on the RST 7.5 input line, even when that interrupt has been previously masked. (See SIM instruction.)

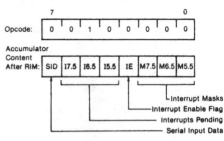

Cycles: 1
States: 4
Flags: none

Courtesy of Intel Corporation.

THE INSTRUCTION SET

SIM   (Set Interrupt Masks) (8085 only)

The execution of the SIM instruction uses the contents of the accumulator (which must be previously loaded) to perform the following functions:

- Program the interrupt mask for the RST 5.5, 6.5, and 7.5 hardware interrupts.
- Reset the edge-triggered RST 7.5 input latch.
- Load the SOD output latch.

To program the interrupt masks, first set accumulator bit 3 to 1 and set to 1 any bits 0, 1, and 2, which disable interrupts RST 5.5, 6.5, and 7.5, respectively. Then do a SIM instruction. If accumulator bit 3 is 0 when the SIM instruction is executed, the interrupt mask register will not change. If accumulator bit 4 is 1 when the SIM instruction is executed, the RST 7.5 latch is then reset. RST 7.5 is distinguished by the fact that its latch is always set by a rising edge on the RST 7.5 input pin, even if the jump to service routine is inhibited by masking. This latch remains high until cleared by a RESET IN, by a SIM Instruction with accumulator bit 4 high, or by an internal processor acknowledge to an RST 7.5 interrupt subsequent to the removal of the mask (by a SIM instruction). The RESET IN signal always sets all three RST mask bits.

If accumulator bit 6 is at the 1 level when the SIM instruction is executed, the state of accumulator bit 7 is loaded into the SOD latch and thus becomes available for interface to an external device. The SOD latch is unaffected by the SIM instruction if bit 6 is 0. SOD is always reset by the RESET IN signal.

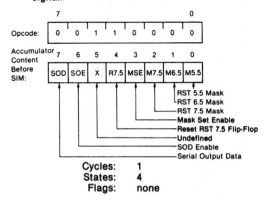

```
Opcode:        0   0   1   1   0   0   0   0
```

Cycles: 1
States: 4
Flags: none

Courtesy of Intel Corporation.

8085A

8080A/8085A INSTRUCTION SET INDEX
Table 5-1

Instruction		Code	Bytes	T States 8085A	T States 8080A	Machine Cycles
ACI	DATA	CE data	2	7	7	F R
ADC	REG	1000 1SSS	1	4	4	F
ADC	M	8E	1	7	7	F R
ADD	REG	1000 0SSS	1	4	4	F
ADD	M	86	1	7	7	F R
ADI	DATA	C6 data	2	7	7	F R
ANA	REG	1010 0SSS	1	4	4	F
ANA	M	A6	1	7	7	F R
ANI	DATA	E6 data	2	7	7	F R
CALL	LABEL	CD addr	3	18	17	S R R W W*
CC	LABEL	DC addr	3	9/18	11/17	S R•/S R R W W*
CM	LABEL	FC addr	3	9/18	11/17	S R•/S R R W W*
CMA		2F	1	4	4	F
CMC		3F	1	4	4	F
CMP	REG	1011 1SSS	1	4	4	F
CMP	M	BE	1	7	7	F R
CNC	LABEL	D4 addr	3	9/18	11/17	S R•/S R R W W*
CNZ	LABEL	C4 addr	3	9/18	11/17	S R•/S R R W W*
CP	LABEL	F4 addr	3	9/18	11/17	S R•/S R R W W*
CPE	LABEL	EC addr	3	9/18	11/17	S R•/S R R W W*
CPI	DATA	FE data	2	7	7	F R
CPO	LABEL	E4 addr	3	9/18	11/17	S R•/S R R W W*
CZ	LABEL	CC addr	3	9/18	11/17	S R•/S R R W W*
DAA		27	1	4	4	F
DAD	RP	00RP 1001	1	10	10	F B B
DCR	REG	00SS S101	1	4	5	F*
DCR	M	35	1	10	10	F R W
DCX	RP	00RP 1011	1	6	5	S*
DI		F3	1	4	4	F
EI		FB	1	4	4	F
HLT		76	1	5	5	F B
IN	PORT	DB data	2	10	10	F R I
INR	REG	00SS S100	1	4	5	F*
INR	M	34	1	10	10	F R W.
INX	RP	00RP 0011	1	6	5	S*
JC	LABEL	DA addr	3	7/10	10	F R/F R R†
JM	LABEL	FA addr	3	7/10	10	F R/F R R†
JMP	LABEL	C3 addr	3	10	10	F R R
JNC	LABEL	D2 addr	3	7/10	10	F R/F R R†
JNZ	LABEL	C2 addr	3	7/10	10	F R/F R R†
JP	LABEL	F2 addr	3	7/10	10	F R/F R R†
JPE	LABEL	EA addr	3	7/10	10	F R/F R R†
JPO	LABEL	E2 addr	3	7/10	10	F R/F R R†
JZ	LABEL	CA addr	3	7/10	10	F R/F R R†
LDA	ADDR	3A addr	3	13	13	F R R R
LDAX	RP	000X 1010	1	7	7	F R
LHLD	ADDR	2A addr	3	16	16	F R R R R

Instruction		Code	Bytes	T States 8085A	T States 8080A	Machine Cycles
LXI	RP,DATA16	00RP 0001 data16	3	10	10	F R R
MOV	REG,REG	01DD 0SSS	1	4	5	F*
MOV	M,REG	0111 0SSS	1	7	7	F W
MOV	REG,M	01DD D110	1	7	7	F R
MVI	REG,DATA	00DD D110 data	2	7	7	F R
MVI	M,DATA	36 data	2	10	10	F R W
NOP		00	1	4	4	F
ORA	REG	1011 0SSS	1	4	4	F
ORA	M	B6	1	7	7	F R
ORI	DATA	F6 data	2	7	7	F R
OUT	PORT	D3 data	2	10	10	F R O
PCHL		E9	1	6	5	S*
POP	RP	11RP 0001	1	10.	10	F R R
PUSH	RP	11RP 0101	1	12	11	S W W*
RAL		17	1	4	4	F
RAR		1F	1	4	4	F
RC		D8	1	6/12	5/11	S/S R R*
RET		C9	1	10	10	F R R
RIM (8085A only)		20	1	4	–	F
RLC		07	1	4	4	F
RM		F8	1	6/12	5/11	S/S R R*
RNC		D0	1	6/12	5/11	S/S R R*
RNZ		C0	1	6/12	5/11	S/S R R*
RP		F0	1	6/12	5/11	S/S R R*
RPE		E8	1	6/12	5/11	S/S R R*
RPO		E0	1	6/12	5/11	S/S R R*
RRC		0F	1	4	4	F
RST	N	11XX X111	1	12	11	S W W*
RZ		C8	1	6/12	5/11	S/S R R*
SBB	REG	1001 1SSS	1	4	4	F
SBB	M	9E	1	7	7	F R
SBI	DATA	DE data	2	7	7	F R
SHLD	ADDR	22 addr	3	16	16	F R R W W
SIM (8085A only)		30	1	4	–	F
SPHL		F9	1	6	5	S*
STA	ADDR	32 addr	3	13	13	F R R W
STAX	RP	000X 0010	1	7	7	F W
STC		37	1	4	4	F
SUB	REG	1001 0SSS	1	4	4	F
SUB	M	96	1	7	7	F R
SUI	DATA	D6 data	2	7	7	F R
XCHG		EB	1	4	4	F
XRA	REG	1010 1SSS	1	4	4	F
XRA	M	AE	1	7	7	F R
XRI	DATA	EE data	2	7	7	F R
XTHL		E3	1	16	18	F R R W W

Machine cycle types:

F	Four clock period instr fetch
S	Six clock period instr fetch
R	Memory read
I	I/O read
W	Memory write
O	I/O write
B	Bus idle
X	Variable or optional binary digit
DDD	Binary digits identifying a destination register
SSS	Binary digits identifying a source register
RP	Register Pair

B = 000, C = 001, D = 010 Memory = 110
E = 011, H = 100, L = 101 A = 111
BC = 00, HL = 10
DE = 01, SP = 11

*Five clock period instruction fetch with 8080A.

†The longer machine cycle sequence applies regardless of condition evaluation with 8080A.

•An extra READ cycle (R) will occur for this condition with 8080A.

Courtesy of Intel Corporation.

8085A

8085A CPU INSTRUCTIONS IN OPERATION CODE SEQUENCE
Table 5-2

OP CODE	MNEMONIC	OP CODE	MNEMONIC	OP CODE	MNEMONIC	OP CODE	MNEMONIC	OP CODE	MNEMONIC	OP CODE	MNEMONIC
00	NOP	2B	DCX H	56	MOV D,M	81	ADD C	AC	XRA H	D7	RST 2
01	LXI B,D16	2C	INR L	57	MOV D,A	82	ADD D	AD	XRA L	D8	RC
02	STAX B	2D	DCR L	58	MOV E,B	83	ADD E	AE	XRA M	D9	–
03	INX B	2E	MVI L,D8	59	MOV E,C	84	ADD H	AF	XRA A	DA	JC Adr
04	INR B	2F	CMA	5A	MOV E,D	85	ADD L	B0	ORA B	DB	IN D8
05	DCR B	30	SIM	5B	MOV E,E	86	ADD M	B1	ORA C	DC	CC Adr
06	MVI B,D8	31	LXI SP,D16	5C	MOV E,H	87	ADD A	B2	ORA D	DD	–
07	RLC	32	STA Adr	5D	MOV E,L	88	ADC B	B3	ORA E	DE	SBI D8
08	–	33	INX SP	5E	MOV E,M	89	ADC C	B4	ORA H	DF	RST 3
09	DAD B	34	INR M	5F	MOV E,A	8A	ADC D	B5	ORA L	E0	RPO
0A	LDAX B	35	DCR M	60	MOV H,B	8B	ADC E	B6	ORA M	E1	POP H
0B	DCX B	36	MVI M,D8	61	MOV H,C	8C	ADC H	B7	ORA A	E2	JPO Adr
0C	INR C	37	STC	62	MOV H,D	8D	ADC L	B8	CMP B	E3	XTHL
0D	DCR C	38	–	63	MOV H,E	8E	ADC M	B9	CMP C	E4	CPO Adr
0E	MVI C,D8	39	DAD SP	64	MOV H,H	8F	ADC A	BA	CMP D	E5	PUSH H
0F	RRC	3A	LDA Adr	65	MOV H,L	90	SUB B	BB	CMP E	E6	ANI D8
10	–	3B	DCX SP	66	MOV H,M	91	SUB C	BC	CMP H	E7	RST 4
11	LXI D,D16	3C	INR A	67	MOV H,A	92	SUB D	BD	CMP L	E8	RPE
12	STAX D	3D	DCR A	68	MOV L,B	93	SUB E	BE	CMP M	E9	PCHL
13	INX D	3E	MVI A,D8	69	MOV L,C	94	SUB H	BF	CMP A	EA	JPE Adr
14	INR D	3F	CMC	6A	MOV L,D	95	SUB L	C0	RNZ	EB	XCHG
15	DCR D	40	MOV B,B	6B	MOV L,E	96	SUB M	C1	POP B	EC	CPE Adr
16	MVI D,D8	41	MOV B,C	6C	MOV L,H	97	SUB A	C2	JNZ Adr	ED	–
17	RAL	42	MOV B,D	6D	MOV L,L	98	SBB B	C3	JMP Adr	EE	XRI D8
18	–	43	MOV B,E	6E	MOV L,M	99	SBB C	C4	CNZ Adr	EF	RST 5
19	DAD D	44	MOV B,H	6F	MOV L,A	9A	SBB D	C5	PUSH B	F0	RP
1A	LDAX D	45	MOV B,L	70	MOV M,B	9B	SBB E	C6	ADI D8	F1	POP PSW
1B	DCX D	46	MOV B,M	71	MOV M,C	9C	SBB H	C7	RST 0	F2	JP Adr
1C	INR E	47	MOV B,A	72	MOV M,D	9D	SBB L	C8	RZ	F3	DI
1D	DCR E	48	MOV C,B	73	MOV M,E	9E	SBB M	C9	RET Adr	F4	CP Adr
1E	MVI E,D8	49	MOV C,C	74	MOV M,H	9F	SBB A	CA	JZ	F5	PUSH PSW
1F	RAR	4A	MOV C,D	75	MOV M,L	A0	ANA B	CB	–	F6	ORI D8
20	RIM	4B	MOV C,E	76	HLT	A1	ANA C	CC	CZ Adr	F7	RST 6
21	LXI H,D16	4C	MOV C,H	77	MOV M,A	A2	ANA D	CD	CALL Adr	F8	RM
22	SHLD Adr	4D	MOV C,L	78	MOV A,B	A3	ANA E	CE	ACI D8	F9	SPHL
23	INX H	4E	MOV C,M	79	MOV A,C	A4	ANA H	CF	RST 1	FA	JM Adr
24	INR H	4F	MOV C,A	7A	MOV A,D	A5	ANA L	D0	RNC	FB	EI
25	DCR H	50	MOV D,B	7B	MOV A,E	A6	ANA M	D1	POP D	FC	CM Adr
26	MVI H,D8	51	MOV D,C	7C	MOV A,H	A7	ANA A	D2	JNC Adr	FD	–
27	DAA	52	MOV D,D	7D	MOV A,L	A8	XRA B	D3	OUT D8	FE	CPI D8
28	–	53	MOV D,E	7E	MOV A,M	A9	XRA C	D4	CNC Adr	FF	RST 7
29	DAD H	54	MOV D,H	7F	MOV A,A	AA	XRA D	D5	PUSH D		
2A	LHLD Adr	55	MOV D,L	80	ADD B	AB	XRA E	D6	SUI D8		

D8 = constant, or logical/arithmetic expression that evaluates to an 8-bit data quantity.

D16 = constant, or logical/arithmetic expression that evaluates to a 16-bit data quantity.

Adr = 16-bit address.

Courtesy of Intel Corporation.

8085A

8085A INSTRUCTION SET SUMMARY BY FUNCTIONAL GROUPING
Table 5-3

Mnemonic	Description	D7	D6	D5	D4	D3	D2	D1	D0	Page
MOVE, LOAD, AND STORE										
MOVr1 r2	Move register to register	0	1	D	D	D	S	S	S	5-4
MOV M.r	Move register to memory	0	1	1	1	0	S	S	S	5-4
MOV r.M	Move memory to register	0	1	D	D	D	1	1	0	5-4
MVI r	Move immediate register	0	0	D	D	D	1	1	0	5-4
MVI M	Move immediate memory	0	0	1	1	0	1	1	0	5-4
LXI B	Load immediate register Pair B & C	0	0	0	0	0	0	0	1	5-5
LXI D	Load immediate register Pair D & E	0	0	0	1	0	0	0	1	5-5
LXI H	Load immediate register Pair H & L	0	0	1	0	0	0	0	1	5-5
STAX B	Store A indirect	0	0	0	0	0	0	1	0	5-6
STAX D	Store A indirect	0	0	0	1	0	0	1	0	5-6
LDAX B	Load A indirect	0	0	0	0	1	0	1	0	5-5
LDAX D	Load A indirect	0	0	0	1	1	0	1	0	5-5
STA	Store A direct	0	0	1	1	0	0	1	0	5-5
LDA	Load A direct	0	0	1	1	1	0	1	0	5-5
SHLD	Store H & L direct	0	0	1	0	0	0	1	0	5-5
LHLD	Load H & L direct	0	0	1	0	1	0	1	0	5-5
XCHG	Exchange D & E. H & L Registers	1	1	1	0	1	0	1	1	5-6
STACK OPS										
PUSH B	Push register Pair B & C on stack	1	1	0	0	0	1	0	1	5-15
PUSH D	Push register Pair D & E on stack	1	1	0	1	0	1	0	1	5-15
PUSH H	Push register Pair H & L on stack	1	1	1	0	0	1	0	1	5-15
PUSH PSW	Push A and Flags on stack	1	1	1	1	0	1	0	1	5-15
POP B	Pop register Pair B & C off stack	1	1	0	0	0	0	0	1	5-15
POP D	Pop register Pair D & E off stack	1	1	0	1	0	0	0	1	5-15
POP H	Pop register Pair H & L off stack	1	1	1	0	0	0	0	1	5-15
POP PSW	Pop A and Flags off stack	1	1	1	1	0	0	0	1	5-15
XTHL	Exchange top of stack. H & L	1	1	1	0	0	0	1	1	5-16
SPHL	H & L to stack pointer	1	1	1	1	1	0	0	1	5-16
LXI SP	Load immediate stack pointer	0	0	1	1	0	0	0	1	5-5
INX SP	Increment stack pointer	0	0	1	1	0	0	1	1	5-9
DCX SP	Decrement stack pointer	0	0	1	1	1	0	1	1	5-9
JUMP										
JMP	Jump unconditional	1	1	0	0	0	0	1	1	5-13
JC	Jump on carry	1	1	0	1	1	0	1	0	5-13
JNC	Jump on no carry	1	1	0	1	0	0	1	0	5-13
JZ	Jump on zero	1	1	0	0	1	0	1	0	5-13
JNZ	Jump on no zero	1	1	0	0	0	0	1	0	5-13
JP	Jump on positive	1	1	1	1	0	0	1	0	5-13
JM	Jump on minus	1	1	1	1	1	0	1	0	5-13
JPE	Jump on parity even	1	1	1	0	1	0	1	0	5-13
JPO	Jump on parity odd	1	1	1	0	0	0	1	0	5-13
PCHL	H & L to program counter	1	1	1	0	1	0	0	1	5-15
CALL										
CALL	Call unconditional	1	1	0	0	1	1	0	1	5-13
CC	Call on carry	1	1	0	1	1	1	0	0	5-14
CNC	Call on no carry	1	1	0	1	0	1	0	0	5-14

Mnemonic	Description	D7	D6	D5	D4	D3	D2	D1	D0	Page
CZ	Call on zero	1	1	0	0	1	1	0	0	5-14
CNZ	Call on no zero	1	1	0	0	0	1	0	0	5-14
CP	Call on positive	1	1	1	1	0	1	0	0	5-14
CM	Call on minus	1	1	1	1	1	1	0	0	5-14
CPE	Call on parity even	1	1	1	0	1	1	0	0	5-14
CPO	Call on parity odd	1	1	1	0	0	1	0	0	5-14
RETURN										
RET	Return	1	1	0	0	1	0	0	1	5-14
RC	Return on carry	1	1	0	1	1	0	0	0	5-14
RNC	Return on no carry	1	1	0	1	0	0	0	0	5-14
RZ	Return on zero	1	1	0	0	1	0	0	0	5-14
RNZ	Return on no zero	1	1	0	0	0	0	0	0	5-14
RP	Return on positive	1	1	1	1	0	0	0	0	5-14
RM	Return on minus	1	1	1	1	1	0	0	0	5-14
RPE	Return on parity even	1	1	1	0	1	0	0	0	5-14
RPO	Return on parity odd	1	1	1	0	0	0	0	0	5-14
RESTART										
RST	Restart	1	1	A	A	A	1	1	1	5-14
INPUT/OUTPUT										
IN	Input	1	1	0	1	1	0	1	1	5-16
OUT	Output	1	1	0	1	0	0	1	1	5-16
INCREMENT AND DECREMENT										
INR r	Increment register	0	0	D	D	D	1	0	0	5-8
DCR r	Decrement register	0	0	D	D	D	1	0	1	5-8
INR M	Increment memory	0	0	1	1	0	1	0	0	5-8
DCR M	Decrement memory	0	0	1	1	0	1	0	1	5-8
INX B	Increment B & C registers	0	0	0	0	0	0	1	1	5-9
INX D	Increment D & E registers	0	0	0	1	0	0	1	1	5-9
INX H	Increment H & L registers	0	0	1	0	0	0	1	1	5-9
DCX B	Decrement B & C	0	0	0	0	1	0	1	1	5-9
DCX D	Decrement D & E	0	0	0	1	1	0	1	1	5-9
DCX H	Decrement H & L	0	0	1	0	1	0	1	1	5-9
ADD										
ADD r	Add register to A	1	0	0	0	0	S	S	S	5-6
ADC r	Add register to A with carry	1	0	0	0	1	S	S	S	5-6
ADD M	Add memory to A	1	0	0	0	0	1	1	0	5-6
ADC M	Add memory to A with carry	1	0	0	0	1	1	1	0	5-7
ADI	Add immediate to A	1	1	0	0	0	1	1	0	5-6
ACI	Add immediate to A with carry	1	1	0	0	1	1	1	0	5-7
DAD B	Add B & C to H & L	0	0	0	0	1	0	0	1	5-9
DAD D	Add D & E to H & L	0	0	0	1	1	0	0	1	5-9
DAD H	Add H & L to H & L	0	0	1	0	1	0	0	1	5-9
DAD SP	Add stack pointer to H & L	0	0	1	1	1	0	0	1	5-9
SUBTRACT										
SUB r	Subtract register from A	1	0	0	1	0	S	S	S	5-7
SBB r	Subtract register from A with borrow	1	0	0	1	1	S	S	S	5-7
SUB M	Subtract memory from A	1	0	0	1	0	1	1	0	5-7
SBB M	Subtract memory from A with borrow	1	0	0	1	1	1	1	0	5-8
SUI	Subtract immediate from A	1	1	0	1	0	1	1	0	5-7

*All mnemonics copyrighted © Intel Corporation 1976.

Courtesy of Intel Corporation.

8085A

8085A INSTRUCTION SET SUMMARY (Cont'd)
Table 5-3

Mnemonic	Description	D7	D6	D5	D4	D3	D2	D1	D0	Page
SBI	Subtract immediate from A with borrow	1	1	0	1	1	1	1	0	5-8
LOGICAL										
ANA r	And register with A	1	0	1	0	0	S	S	S	5-9
XRA r	Exclusive OR register with A	1	0	1	0	1	S	S	S	5-10
ORA r	OR register with A	1	0	1	1	0	S	S	S	5-10
CMP r	Compare register with A	1	0	1	1	1	S	S	S	5-11
ANA M	And memory with A	1	0	1	0	0	1	1	0	5-10
XRA M	Exclusive OR memory with A	1	0	1	0	1	1	1	0	5-10
ORA M	OR memory with A	1	0	1	1	0	1	1	0	5-11
CMP M	Compare memory with A	1	0	1	1	1	0	1	0	5-10
ANI	And immediate with A	1	1	1	0	0	1	1	0	5-10
XRI	Exclusive OR immediate with A	1	1	1	0	1	1	1	0	5-10
ORI	OR immediate with A	1	1	1	1	0	1	1	0	5-11
CPI	Compare immediate with A	1	1	1	1	1	1	1	0	5-11
ROTATE										
RLC	Rotate A left	0	0	0	0	0	1	1	1	5-11
RRC	Rotate A right	0	0	0	0	1	1	1	1	5-12
RAL	Rotate A left through carry	0	0	0	1	0	1	1	1	5-12
RAR	Rotate A right through carry	0	0	0	1	1	1	1	1	5-12
SPECIALS										
CMA	Complement A	0	0	1	0	1	1	1	1	5-12
STC	Set carry	0	0	1	1	0	1	1	1	5-12
CMC	Complement carry	0	0	1	1	1	1	1	1	5-12
DAA	Decimal adjust A	0	0	1	0	0	1	1	1	5-9
CONTROL										
EI	Enable Interrupts	1	1	1	1	1	0	1	1	5-17
DI	Disable Interrupt	1	1	1	1	0	0	1	1	5-17
NOP	No-operation	0	0	0	0	0	0	0	0	5-17
HLT	Halt	0	1	1	1	0	1	1	0	5-17
NEW 8085A INSTRUCTIONS										
RIM	Read Interrupt Mask	0	0	1	0	0	0	0	0	5-17
SIM	Set Interrupt Mask	0	0	1	1	0	0	0	0	5-18

NOTES: 1. DDS or SSS: B 000, C 001, D 010, E 011, H 100, L 101, Memory 110, A 111.

2. Two possible cycle times. (6/12) indicate instruction cycles dependent on condition flags.

*All mnemonics copyrighted © Intel Corporation 1976.

Courtesy of Intel Corporation.

INSTRUCTION SET

The Z80 microprocessr has one of the most powerful and versatile instruction sets available in any 8-bit microprocessor. It includes such unique operations as a block move for fast, efficient data transfers within memory, or between memory and I/O. It also allows operations on any bit in any location in memory.

The following is a summary of the Z80 instruction set which shows the assembly language mnemonic, the operation, the flag status, and gives comments on each instruction. For an explanation of flag notations and symbols for mnemonic tables, see the Symbolic Notations section which follows these tables. The *Z80 CPU Technical Manual* (03-0029-01), the *Programmer's Reference Guide* (03-0012-03), and *Assembly Language Programming Manual* (03-0002-01) contain significantly more details for programming use.

The instructions are divided into the following categories:

□ 8-bit loads

□ 16-bit loads

□ Exchanges, block transfers, and searches

□ 8-bit arithmetic and logic operations

□ General-purpose arithmetic and CPU control

□ 16-bit arithmetic operations

□ Rotates and shifts

□ Bit set, reset, and test operations

□ Jumps

□ Calls, returns, and restarts

□ Input and output operations

A variety of addressing modes are implemented to permit efficient and fast data transfer between various registers, memory locations, and input/output devices. These addressing modes include:

□ Immediate

□ Immediate extended

□ Modified page zero

□ Relative

□ Extended

□ Indexed

□ Register

□ Register indirect

□ Implied

□ Bit

8-BIT LOAD GROUP

Mnemonic	Symbolic Operation	S	Z	Flags H	P/V	N	C	Opcode 76	543	210	Hex	No. of Bytes	No. of M Cycles	No. of T States	Comments
LD r, r'	r ← r'	•	•	X • X	•	•	•	01	r	r'		1	1	4	r, r' Reg.
LD r, n	r ← n	•	•	X • X	•	•	•	00	r	110		2	2	7	000 B
									← n →						001 C
LD r, (HL)	r ← (HL)	•	•	X • X	•	•	•	01	r	110		1	2	7	010 D
LD r, (IX + d)	r ← (IX + d)	•	•	X • X	•	•	•	11	011	101	DD	3	5	19	011 E
								01	r	110					100 H
									← d →						101 L
LD r, (IY + d)	r ← (IY + d)	•	•	X • X	•	•	•	11	111	101	FD	3	5	19	111 A
								01	r	110					
									← d →						
LD (HL), r	(HL) ← r	•	•	X • X	•	•	•	01	110	r		1	2	7	
LD (IX + d), r	(IX + d) ← r	•	•	X • X	•	•	•	11	011	101	DD	3	5	19	
								01	110	r					
									← d →						
LD (IY + d), r	(IY + d) ← r	•	•	X • X	•	•	•	11	111	101	FD	3	5	19	
								01	110	r					
									← d →						
LD (HL), n	(HL) ← n	•	•	X • X	•	•	•	00	110	110	36	2	3	10	
									← n →						
LD (IX + d), n	(IX + d) ← n	•	•	X • X	•	•	•	11	011	101	DD	4	5	19	
								00	110	110	36				
									← d →						
									← n →						

8-BIT LOAD GROUP (Continued)

Mnemonic	Symbolic Operation	S	Z		H		P/V	N	C	76	543	210	Hex	No. of Bytes	No. of M Cycles	No. of T States	Comments
LD (IY + d), n	(IY + d) ← n	•	•	X	•	X	•	•	•	11	111	101	FD	4	5	19	
										00	110	110	36				
										← d →							
										← n →							
LD A, (BC)	A ← (BC)	•	•	X	•	X	•	•	•	00	001	010	0A	1	2	7	
LD A, (DE)	A ← (DE)	•	•	X	•	X	•	•	•	00	011	010	1A	1	2	7	
LD A, (nn)	A ← (nn)	•	•	X	•	X	•	•	•	00	111	010	3A	3	4	13	
										← n →							
										← n →							
LD (BC), A	(BC) ← A	•	•	X	•	X	•	•	•	00	000	010	02	1	2	7	
LD (DE), A	(DE) ← A	•	•	X	•	X	•	•	•	00	010	010	12	1	2	7	
LD (nn), A	(nn) ← A	•	•	X	•	X	•	•	•	00	110	010	32	3	4	13	
										← n →							
										← n →							
LD A, I	A ← I	‡	‡	X	0	X	IFF	0	•	11	101	101	ED	2	2	9	
										01	010	111	57				
LD A, R	A ← R	‡	‡	X	0	X	IFF	0	•	11	101	101	ED	2	2	9	
										01	011	111	5F				
LD I, A	I ← A	•	•	X	•	X	•	•	•	11	101	101	ED	2	2	9	
										01	000	111	47				
LD R, A	R ← A	•	•	X	•	X	•	•	•	11	101	101	ED	2	2	9	
										01	001	111	4F				

NOTE: IFF, the content of the interrupt enable flip-flop, (IFF_2), is copied into the P/V flag.

16-BIT LOAD GROUP

Mnemonic	Symbolic Operation	S	Z		H		P/V	N	C	76	543	210	Hex	No. of Bytes	No. of M Cycles	No. of T States	Comments	
LD dd, nn	dd ← nn	•	•	X	•	X	•	•	•	00	dd0	001		3	3	10	dd	Pair
										← n →							00	BC
										← n →							01	DE
LD IX, nn	IX ← nn	•	•	X	•	X	•	•	•	11	011	101	DD	4	4	14	10	HL
										00	100	001	21				11	SP
										← n →								
										← n →								
LD IY, nn	IY ← nn	•	•	X	•	X	•	•	•	11	111	101	FD	4	4	14		
										00	100	001	21					
										← n →								
										← n →								
LD HL, (nn)	H ← (nn + 1)	•	•	X	•	X	•	•	•	00	101	010	2A	3	5	16		
	L ← (nn)									← n →								
										← n →								
LD dd, (nn)	dd_H ← (nn + 1)	•	•	X	•	X	•	•	•	11	101	101	ED	4	6	20		
	dd_L ← (nn)									01	dd1	011						
										← n →								
										← n →								

NOTE: $(PAIR)_H$, $(PAIR)_L$ refer to high order and low order eight bits of the register pair respectively. e.g., $BC_L = C$, $AF_H = A$.

Z80 CPU

16-BIT LOAD GROUP (Continued)

Mnemonic	Symbolic Operation	S	Z	H	P/V	N	C	76	543	210	Hex	No. of Bytes	No. of M Cycles	No. of T States	Comments
LD IX, (nn)	$IX_H \leftarrow (nn+1)$ $IX_L \leftarrow (nn)$	•	•	X	•	X	•	•	•	11 00 ←n→ ←n→	011 101 101 010	101 010	DD 2A	4	6	20	
LD IY, (nn)	$IY_H \leftarrow (nn+1)$ $IY_L \leftarrow (nn)$	•	•	X	•	X	•	•	•	11 00 ←n→ ←n→	111 101 101 010	101 010	FD 2A	4	6	20	
LD (nn), HL	$(nn+1) \leftarrow H$ $(nn) \leftarrow L$	•	•	X	•	X	•	•	•	00 ←n→ ←n→	100	010	22	3	5	16	
LD (nn), dd	$(nn+1) \leftarrow dd_H$ $(nn) \leftarrow dd_L$	•	•	X	•	X	•	•	•	11 01 ←n→ ←n→	101 dd0	101 011	ED	4	6	20	
LD (nn), IX	$(nn+1) \leftarrow IX_H$ $(nn) \leftarrow IX_L$	•	•	X	•	X	•	•	•	11 00 ←n→ ←n→	011 100	101 010	DD 22	4	6	20	
LD (nn), IY	$(nn+1) \leftarrow IY_H$ $(nn) \leftarrow IY_L$	•	•	X	•	X	•	•	•	11 00 ←n→ ←n→	111 100	101 010	FD 22	4	6	20	
LD SP, HL	$SP \leftarrow HL$	•	•	X	•	X	•	•	•	11	111	001	F9	1	1	6	
LD SP, IX	$SP \leftarrow IX$	•	•	X	•	X	•	•	•	11 11	011 111	101 001	DD F9	2	2	10	
LD SP, IY	$SP \leftarrow IY$	•	•	X	•	X	•	•	•	11 11	111 111	101 001	FD F9	2	2	10	
PUSH qq	$(SP-2) \leftarrow qq_L$ $(SP-1) \leftarrow qq_H$ $SP \rightarrow SP-2$	•	•	X	•	X	•	•	•	11	qq0	101		1	3	11	
PUSH IX	$(SP-2) \leftarrow IX_L$ $(SP-1) \leftarrow IX_H$ $SP \rightarrow SP-2$	•	•	X	•	X	•	•	•	11 11	011 100	101 101	DD E5	2	4	15	
PUSH IY	$(SP-2) \leftarrow IY_L$ $(SP-1) \leftarrow IY_H$ $SP \rightarrow SP-2$	•	•	X	•	X	•	•	•	11 11	111 100	101 101	FD E5	2	4	15	
POP qq	$qq_H \leftarrow (SP+1)$ $qqL \leftarrow (SP)$ $SP \rightarrow SP+2$	•	•	X	•	X	•	•	•	11	qq0	001		1	3	10	
POP IX	$IX_H \leftarrow (SP+1)$ $IX_L \leftarrow (SP)$ $SP \rightarrow SP+2$	•	•	X	•	X	•	•	•	11 11	011 100	101 001	DD E1	2	4	14	
POP IY	$IY_H \leftarrow (SP+1)$ $IY_L \leftarrow (SP)$ $SP \rightarrow SP+2$	•	•	X	•	X	•	•	•	11 11	111 100	101 001	FD E1	2	4	14	

qq	Pair
00	BC
01	DE
10	HL
11	AF

NOTE: (PAIR)$_H$, (PAIR)$_L$ refer to high order and low order eight bits of the register pair respectively, e.g., BC$_L$ = C, AF$_H$ = A.

EXCHANGE, BLOCK TRANSFER, BLOCK SEARCH GROUPS

Mnemonic	Symbolic Operation	S	Z	H	P/V	N	C	76	543	210	Hex	No. of Bytes	No. of M Cycles	No. of T States	Comments
EX DE, HL	DE ↔ HL	•	•	•	•	•	•	11	101	011	EB	1	1	4	
EX AF, AF'	AF ↔ AF'	•	•	•	•	•	•	00	001	000	08	1	1	4	
EXX	BC ↔ BC'	•	•	•	•	•	•	11	011	001	D9	1	1	4	Register bank and auxiliary register bank exchange
	DE ↔ DE'														
	HL ↔ HL'														
EX (SP), HL	H ↔ (SP + 1)	•	•	•	•	•	•	11	100	011	E3	1	5	19	
	L ↔ (SP)														
EX (SP), IX	IX$_H$ ↔ (SP + 1)	•	•	•	•	•	•	11	011	101	DD	2	6	23	
	IX$_L$ ↔ (SP)							11	100	011	E3				
EX (SP), IY	IY$_H$ ↔ (SP + 1)	•	•	•	•	•	•	11	111	101	FD	2	6	23	
	IYL ↔ (SP)							11	100	011	E3				
LDI	(DE) ← (HL)	•	•	0	↕ ①	0	•	11	101	101	ED	2	4	16	Load (HL) into (DE), increment the pointers and decrement the byte counter (BC)
	DE ← DE + 1							10	100	000	A0				
	HL ← HL + 1														
	BC ← BC – 1														
LDIR	(DE) ← (HL)	•	•	0	0 ②	0	•	11	101	101	ED	2	5	21	If BC ≠ 0
	DE ← DE + 1							10	110	000	B0	2	4	16	If BC = 0
	HL ← HL + 1														
	BC ← BC – 1														
	Repeat until														
	BC = 0														
LDD	(DE) ← (HL)	•	•	0	↕ ①	0	•	11	101	101	ED	2	4	16	
	DE ← DE – 1							10	101	000	A8				
	HL ← HL – 1														
	BC ← BC – 1														
LDDR	(DE) ← (HL)	•	•	0	0 ②	0	•	11	101	101	ED	2	5	21	If BC ≠ 0
	DE ← DE – 1							10	111	000	B8	2	4	16	If BC = 0
	HL ← HL – 1														
	BC ← BC – 1														
	Repeat until														
	BC = 0														
CPI	A – (HL)	↕	↕ ③	↕	↕ ①	1	•	11	101	101	ED	2	4	16	
	HL ← HL + 1							10	100	001	A1				
	BC ← BC – 1														

NOTE: ① P/V flag is 0 if the result of BC – 1 = 0, otherwise P/V = 1.
        ② P/V flag is 0 only at completion of instruction.
        ③ Z flag is 1 if A = HL , otherwise Z = 0.

Z80 CPU

EXCHANGE, BLOCK TRANSFER, BLOCK SEARCH GROUPS (Continued)

Mnemonic	Symbolic Operation	S	Z	H	P/V	N	C	76	543	210	Hex	No. of Bytes	No. of M Cycles	No. of T States	Comments
			③		①										
CPIR	A − (HL)	‡	‡	X	‡	1	•	11	101	101	ED	2	5	21	If BC ≠ 0 and A ≠ (HL)
	HL ← HL + 1 BC ← BC − 1 Repeat until A = (HL) or BC = 0							10	110	001	B1	2	4	16	If BC = 0 or A = (HL)
			③		①										
CPD	A − (HL) HL ← HL − 1 BC ← BC − 1	‡	‡	X	‡	1	•	11 10	101 101	101 001	ED A9	2	4	16	
			③		①										
CPDR	A − (HL)	‡	‡	X	‡	1	•	11	101	101	ED	2	5	21	If BC ≠ 0 and A ≠ (HL)
	HL ← HL − 1 BC ← BC − 1 Repeat until A = (HL) or BC = 0							10	111	001	B9	2	4	16	If BC = 0 or A = (HL)

NOTE: ① P/V flag is 0 if the result of BC − 1 = 0, otherwise P/V = 1.
② P/V flag is 0 only at completion of instruction.
③ Z flag is 1 if A = (HL), otherwise Z = 0.

8-BIT ARITHMETIC AND LOGICAL GROUP

Mnemonic	Symbolic Operation	S	Z	H	P/V	N	C	76	543	210	Hex	No. of Bytes	No. of M Cycles	No. of T States	Comments
ADD A, r	A ← A + r	‡	‡	X	‡	X	V	0	‡	10	[000]	r		1	1	4
ADD A, n	A ← A + n	‡	‡	X	‡	X	V	0	‡	11	[000]	110		2	2	7
									← n →							
ADD A, (HL)	A ← A + (HL)	‡	‡	X	‡	X	V	0	‡	10	[000]	110		1	2	7
ADD A, (IX + d)	A ← A + (IX + d)	‡	‡	X	‡	X	V	0	‡	11 10	011 [000]	101 110	DD	3	5	19
									← d →							
ADD A, (IY + d)	A ← A + (IY + d)	‡	‡	X	‡	X	V	0	‡	11 10	111 [000]	101 110	FD	3	5	19
									← d →							
ADC A, s	A ← A + s + CY	‡	‡	X	‡	X	V	0	‡		[001]					
SUB s	A ← A − s	‡	‡	X	‡	X	V	1	‡		[010]					
SBC A, s	A ← A − s − CY	‡	‡	X	‡	X	V	1	‡		[011]					
AND s	A ← A > s	‡	‡	X	1	X	P	0	0		[100]					
OR s	A ← A > s	‡	‡	X	0	X	P	0	0		[110]					
XOR s	A ← A ⊕ s	‡	‡	X	0	X	P	0	0		[101]					
CP s	A − s	‡	‡	X	‡	X	V	1	‡		[111]					

Register table (Comments column, ADD A, r):

r	Reg.
000	B
001	C
010	D
011	E
100	H
101	L
111	A

s is any of r, n, (HL), (IX + d), (IY + d) as shown for ADD instruction. The indicated bits replace the [000] in the ADD set above.

8-BIT ARITHMETIC AND LOGICAL GROUP (Continued)

Mnemonic	Symbolic Operation	S	Z		H		P/V	N	C	76	543	210	Hex	No. of Bytes	No. of M Cycles	No. of T States	Comments
INC r	r ← r+1	‡	‡	X	‡	X	V	0	•	00	r	[100]		1	1	4	
INC (HL)	(HL) ← (HL) + 1	‡	‡	X	‡	X	V	0	•	00	110	[100]		1	3	11	
INC (IX + d)	(IX + d) ← (IX + d) + 1	‡	‡	X	‡	X	V	0	•	11	011	101	DD	3	6	23	
										00	110	[100]					
										← d →							
INC (IY + d)	(IY + d) ← (IY + d) + 1	‡	‡	X	‡	X	V	0	•	11	111	101	FD	3	6	23	
										00	110	[100]					
										← d →							
DEC m	m ← m − 1	‡	‡	X	‡	X	V	1	•			[101]					

NOTE: m is any of r, (HL), (IX + d), (IY + d) as shown for INC. DEC same format and states as INC. Replace [100] with [101] in opcode.

GENERAL-PURPOSE ARITHMETIC AND CPU CONTROL GROUPS

Mnemonic	Symbolic Operation	S	Z		H		P/V	N	C	76	543	210	Hex	No. of Bytes	No. of M Cycles	No. of T States	Comments
DAA	@	‡	‡	X	‡	X	P	•	‡	00	100	111	27	1	1	4	Decimal adjust accumulator.
CPL	A ← $\overline{A}$	•	•	X	1	X	•	1	•	00	101	111	2F	1	1	4	Complement accumulator (one's complement).
NEG	A ← 0 − A	‡	‡	X	‡	X	V	1	‡	11	101	101	ED	2	2	8	Negate acc. (two's complement).
										01	000	100	44				
CCF	CY ← $\overline{CY}$	•	•	X	X	X	•	0	‡	00	111	111	3F	1	1	4	Complement carry flag.
SCF	CY ← 1	•	•	X	0	X	•	0	1	00	110	111	37	1	1	4	Set carry flag.
NOP	No operation	•	•	X	•	X	•	•	•	00	000	000	00	1	1	4	
HALT	CPU halted	•	•	X	•	X	•	•	•	01	110	110	76	1	1	4	
DI ★	IFF ← 0	•	•	X	•	X	•	•	•	11	110	011	F3	1	1	4	
EI ★	IFF ← 1	•	•	X	•	X	•	•	•	11	111	011	FB	1	1	4	
IM 0	Set interrupt mode 0	•	•	X	•	X	•	•	•	11	101	101	ED	2	2	8	
										01	000	110	46				
IM 1	Set interrupt mode 1	•	•	X	•	X	•	•	•	11	101	101	ED	2	2	8	
										01	010	110	56				
IM 2	Set interrupt mode 2	•	•	X	•	X	•	•	•	11	101	101	ED	2	2	8	
										01	011	110	5E				

NOTES: @ converts accumulator content into packed BCD following add or subtract with packed BCD operands.
IFF indicates the interrupt enable flip-flop.
CY indicates the carry flip-flop.
★ indicates interrupts are not sampled at the end of EI or DI.

Z80 CPU

16-BIT ARITHMETIC GROUP

Mnemonic	Symbolic Operation	S	Z	Flags H	P/V	N	C	Opcode 76	543	210	Hex	No. of Bytes	No. of M Cycles	No. of T States	Comments	
ADD HL, ss	HL ← HL + ss	•	•	X	X	X	•	0	‡	00	ssl	001		1	3	11	ss	Reg.
																00	BC
ADC HL, ss	HL ←															01	DE
	HL + ss + CY	‡	‡	X	X	X	V	0	‡	11	101	101	ED	2	4	15	10	HL
										01	ss1	010					11	SP
SBC HL, ss	HL ←															
	HL – ss – CY	‡	‡	X	X	X	V	1	‡	11	101	101	ED	2	4	15		
										01	ss0	010						
ADD IX, pp	IX ← IX + pp	•	•	X	X	X	•	0	‡	11	011	101	DD	2	4	15	pp	Reg.
										01	pp1	001					00	BC
																01	DE
																10	IX
																11	SP
ADD IY, rr	IY ← IY + rr	•	•	X	X	X	•	0	‡	11	111	101	FD	2	4	15	rr	Reg.
										00	rr1	001					00	BC
INC ss	ss ← ss + 1	•	•	X	•	X	•	•	•	00	ss0	011		1	1	6	01	DE
INC IX	IX ← IX + 1	•	•	X	•	X	•	•	•	11	011	101	DD	2	2	10	10	IY
										00	100	011	23				11	SP
INC IY	IY ← IY + 1	•	•	X	•	X	•	•	•	11	111	101	FD	2	2	10		
										00	100	011	23					
DEC ss	ss ← ss – 1	•	•	X	•	X	•	•	•	00	ss1	011		1	1	6		
DEC IX	IX ← IX – 1	•	•	X	•	X	•	•	•	11	011	101	DD	2	2	10		
										00	101	011	2B					
DEC IY	IY ← IY – 1	•	•	X	•	X	•	•	•	11	111	101	FD	2	2	10		
										00	101	011	2B					

ROTATE AND SHIFT GROUP

Mnemonic	Symbolic Operation	S	Z	Flags H	P/V	N	C	Opcode 76	543	210	Hex	No. of Bytes	No. of M Cycles	No. of T States	Comments
RLCA		•	•	X	0	X	•	0	‡	00	000	111	07	1	1	4	Rotate left circular accumulator.
RLA		•	•	X	0	X	•	0	‡	00	010	111	17	1	1	4	Rotate left accumulator.
RRCA		•	•	X	0	X	•	0	‡	00	001	111	0F	1	1	4	Rotate right circular accumulator.
RRA		•	•	X	0	X	•	0	‡	00	011	111	1F	1	1	4	Rotate right accumulator.

ROTATE AND SHIFT GROUP (Continued)

Mnemonic	Symbolic Operation	S	Z	H		P/V	N	C	76	543	210	Hex	No. of Bytes	No. of M Cycles	No. of T States	Comments
RLC r		‡	‡	X	0	X	P	0 • ‡	11	001	011	CB	2	2	8	Rotate left
									00	[000]	r					circular
RLC (HL)		‡	‡	X	0	X	P	0 ‡	11	001	011	CB	2	4	15	register r.
									00	000	110					
RLC (IX + d)	r,(HL),(IX + d),(IY + d)	‡	‡	X	0	X	P	0 ‡	11	011	101	DD	4	6	23	
									11	001	011	CB				
										←d→						
									00	[000]	110					
RLC (IY + d)		‡	‡	X	0	X	P	0 ‡	11	111	101	FD	4	6	23	
									11	001	011	CB				
										←d→						Instruction
									00	[000]	110					format and
RL m	m = r,(HL,(IX + d),(IY + d)	‡	‡	X	0	X	P	0 ‡		[010]						states are as shown for
RRC m	m = r,(HL),(IX + d),(IY + d)	‡	‡	X	0	X	P	0 ‡		[001]						RLCs. To form new opcode
RR m	m = r,(HL),(IX + d),(IY + d)	‡	‡	X	0	X	P	0 ‡		[011]						replace [000] or RLCs with
SLA m	m = r,(HL,(IX + d),(IY + d)	‡	‡	X	0	X	P	0 ‡		[100]						shown code.
SRA m	m = r,(HL,(IX + d),(IY + d)	‡	‡	X	0	X	P	0 ‡		[101]						
SRL m	m = r,(HL),(IX + d),(IY + d)	‡	‡	X	0	X	P	0 ‡		[111]						
RLD		‡	‡	X	0	X	P	0 •	11	101	101	ED	2	5	18	Rotate digit left and right between the accumulator and location (HL).
									01	101	111	6F				
RRD		‡	‡	X	0	X	P	0 •	11	101	101	ED	2	5	18	The content of the upper half of the accumulator is unaffected.
									01	100	111	67				

r	Reg.
000	B
001	C
010	D
011	E
001	H
101	L
111	A

Z80 CPU

BIT SET, RESET AND TEST GROUP

Mnemonic	Symbolic Operation	S	Z	H	P/V	N	C	76	543	210	Hex	No. of Bytes	No. of M Cycles	No. of T States
BIT b, r	$Z \leftarrow r_b$	X	‡	1	X	0	•	11	001	011	CB	2	2	8
								01	b	r				
BIT b, (HL)	$Z \leftarrow (HL)_b$	X	‡	1	X	0	•	11	001	011	CB	2	3	12
								01	b	110				
BIT b, (IX+d)	$Z \leftarrow (IX+d)_b$	X	‡	1	X	0	•	11	011	101	DD	4	5	20
								11	001	011	CB			
									←d→					
								01	b	110				
BIT b, (IY+d)	$Z \leftarrow (IY+d)_b$	X	‡	1	X	0	•	11	111	101	FD	4	5	20
								11	001	011	CB			
									←d→					
								01	b	110				
SET b, r	$r_b \leftarrow 1$	•	•	•	•	•	•	11	001	011	CB	2	2	8
								[11]	b	r				
SET b, (HL)	$(HL)_b \leftarrow 1$	•	•	•	•	•	•	11	001	011	CB	2	4	15
								[11]	b	110				
SET b, (IX+d)	$(IX+d)_b \leftarrow 1$	•	•	•	•	•	•	11	011	101	DD	4	6	23
								11	001	011	CB			
									←d→					
								[11]	b	110				
SET b, (IY+d)	$(IY+d)_b \leftarrow 1$	•	•	•	•	•	•	11	111	101	FD	4	6	23
								11	001	011	CB			
									←d→					
								[11]	b	110				
RES b, m	$m_b \leftarrow 0$ $m \equiv r, (HL),$ $(IX+d), (IY+d)$	•	•	•	•	•	•	[10]						

Comments:

r	Reg.
000	B
001	C
010	D
011	E
100	H
101	L
111	A

b	Bit Tested
000	0
001	1
010	2
011	3
100	4
101	5
110	6
111	7

RES b, m comment: To form new opcode replace [11] of SET b, s with [10]. Flags and time states for SET instruction.

NOTE: The notation m_b indicates location m, bit b (0 to 7).

JUMP GROUP

Mnemonic	Symbolic Operation	S	Z		H		P/V	N	C	76	543	210	Hex	No. of Bytes	No. of M Cycles	No. of T States	Comments
JP nn	PC ← nn	•	•	X	•	X	•	•	•	11	000	011	C3	3	3	10	
										← n →							
										← n →							
JP cc, nn	If condition cc is true PC←nn, otherwise continue	•	•	X	•	X	•	•	•	11	cc	010		3	3	10	
										← n →							
										← n →							
JR e	PC ← PC+e	•	•	X	•	X	•	•	•	00	011	000	18	2	3	12	
										← e-2 →							
JR C, e	If C=0, continue	•	•	X	•	X	•	•	•	00	111	000	38	2	2	7	If condition not met.
										← e-2 →							
	If C=1, PC ← PC+e														2	3	12	If condition is met.
JR NC, e	IF C=1, continue	•	•	X	•	X	•	•	•	00	110	000	30	2	2	7	If condition not met.
										← e-2 →							
	If C=0, PC ← PC+e														2	3	12	If condition is met.
JP Z, e	If Z=0, continue	•	•	X	•	X	•	•	•	00	101	000	28	2	2	7	If condition not met.
										← e-2 →							
	If Z=1, PC ← PC+e														2	3	12	If condition is met.
JR NZ, e	If Z=1, continue	•	•	X	•	X	•	•	•	00	100	000	20	2	2	7	If condition not met.
										← e-2 →							
	If Z=0, PC ← PC+e														2	3	12	If condition is met.
JP (HL)	PC ← HL	•	•	X	•	X	•	•	•	11	101	001	E9	1	1	4	
JP (IX)	PC ← IX	•	•	X	•	X	•	•	•	11	011	101	DD	2	2	8	
										11	101	001	E9				
JP (IY)	PC ← IY	•	•	X	•	X	•	•	•	11	111	101	FD	2	2	8	
										11	101	001	E9				
DJNZ, e	B ← B-1	•	•	X	•	X	•	•	•	00	010	000	10	2	2	8	If B=0
	If B=0, continue									← e-2 →							
	If B≠0, PC ← PC+e														2	3	13	If B≠0.

Comments (condition codes):

cc	Condition
000	NZ (non-zero)
001	Z (zero)
010	NC (non-carry)
011	C (carry)
100	PO (parity odd)
101	PE (parity even)
110	P (sign positive)
111	M (sign negative)

NOTES: e represents the extension in the relative addressing mode.
e is a signal two's complement number in the range < −126, 129 >.
e − 2 in the opcode provides an effective address of pc + e as PC is incremented by 2 prior to the addition of e.

Z80 CPU

CALL AND RETURN GROUP

Mnemonic	Symbolic Operation	S	Z	Flags H		P/VN	C	76	Opcode 543	210	Hex	No. of Bytes	No. of M Cycles	No. of T States	Comments
CALL nn	$(SP-1) \leftarrow PC_H$	•	•	X	•	X	•	11	001	101	CD	3	5	17	
	$(SP-2) \leftarrow PC_L$								←n→						
	$PC \leftarrow nn,$								←n→						
CALL cc,nn	If condition	•	•	X	•	X	•	11	cc	100		3	3	10	If cc is false.
	cc is false								←n→						
	continue,								←n→			3	5	17	If cc is true.
	otherwise same as CALL nn														
RET	$PC_L \leftarrow (SP)$	•	•	X	•	X	•	11	001	001	C9	1	3	10	
	$PC_H \leftarrow (SP+1)$														
RET cc	If condition	•	•	X	•	X	•	11	cc	000		1	1	5	If cc is false.
	cc is false continue,											1	3	11	If cc is true.
	otherwise same as RET														
RETI	Return from interrupt	•	•	X	•	X	•	11 01	101 001	101 101	ED 4D	2	4	14	
RETN[1]	Return from non-maskable interrupt	•	•	X	•	X	•	11 01	101 000	101 101	ED 45	2	4	14	
RST p	$(SP-1) \leftarrow PC_H$	•	•	X	•	X	•	11	t	111		1	3	11	
	$(SP-2) \leftarrow PC_L$														
	$PC_H \leftarrow 0$														
	$PC_L \leftarrow p$														

cc	Condition
000	NZ (non-zero)
001	Z (zero)
010	NC (non-carry)
011	C (carry)
100	PO (parity odd)
101	PE (parity even)
110	P (sign positive)
111	M (sign negative)

t	p
000	00H
001	08H
010	10H
011	18H
100	20H
101	28H
110	30H
111	38H

NOTE: [1]RETN loads $IFF_2 \rightarrow IFF_1$

INPUT AND OUTPUT GROUP

Mnemonic	Symbolic Operation	S	Z	H	P/V	N	C	76	543	210	Hex	No. of Bytes	No. of M Cycles	No. of T States	Comments
IN A, (n)	A ← (n)	•	•	X	• X	•	•	11	011	01	DB	2	3	11	n to $A_0 \sim A_7$
								← n →							Acc. to $A_8 \sim A_{15}$
IN r, (C)	r ← (C)	‡	‡	X	‡ X	P	0 •	11	101	101	ED	2	3	12	C to $A_0 \sim A_7$
	if r = 110 only the flags will be affected							01	r	000					B to $A_8 \sim A_{15}$
INI	(HL) ← (C) ①	X	‡	X	X X	X 1	X	11	101	101	ED	2	4	16	C to $A_0 \sim A_7$
	B ← B − 1							10	100	010	A2				B to $A_8 \sim A_{15}$
	HL ← HL + 1 ②														
INIR	(HL) ← (C) ①	X	1	X	X X	X 1	X	11	101	101	ED	2	5 (If B≠0)	21	C to $A_0 \sim A_7$
	B ← B − 1							10	110	010	B2				B to $A_8 \sim A_{15}$
	HL ← HL + 1											2	4 (If B = 0)	16	
	Repeat until B = 0														
IND	(HL) ← (C) ①	X	‡	X	X X	X 1	X	11	101	101	ED	2	4	16	C to $A_0 \sim A_7$
	B ← B − 1							10	101	010	AA				B to $A_8 \sim A_{15}$
	HL ← HL − 1 ②														
INDR	(HL) ← (C) ①	X	1	X	X X	X 1	X	11	101	101	ED	2	5 (If B≠0)	21	C to $A_0 \sim A_7$
	B ← B − 1							10	111	010	BA				B to $A_8 \sim A_{15}$
	HL ← HL − 1											2	4 (If B = 0)	16	
	Repeat until B = 0														
OUT (n), A	(n) ← A	•	•	X	• X	•	•	11	010	011	D3	2	3	11	n to $A_0 \sim A_7$
								← n →							Acc. to $A_8 \sim A_{15}$
OUT (C), r	(C) ← r	•	•	X	• X	•	•	11	101	101	ED	2	3	12	C to $A_0 \sim A_7$
								01	r	001					B to $A_8 \sim A_{15}$
OUTI	(C) ← (HL) ①	X	‡	X	X X	X 1	X	11	101	101	ED	2	4	16	C to $A_0 \sim A_7$
	B ← B − 1							10	100	011	A3				B to $A_8 \sim A_{15}$
	HL ← HL + 1 ②														
OTIR	(C) ← (HL) ①	X	1	X	X X	X 1	X	11	101	101	ED	2	5 (If B≠0)	21	C to $A_0 \sim A_7$
	B ← B − 1							10	110	011	B3				B to $A_8 \sim A_{15}$
	HL ← HL + 1											2	4 (If B = 0)	16	
	Repeat until B = 0														
OUTD	(C) ← (HL) ①	X	‡	X	X X	X 1	X	11	101	101	ED	2	4	16	C to $A_0 \sim A_7$
	B ← B − 1							10	101	011	AB				B to $A_8 \sim A_{15}$
	HL ← HL − 1														
OTDR	(C) ← (HL) ②	X	1	X	X X	X 1	X	11	101	101	ED	2	5 (If B≠0)	21	C to $A_0 \sim A_7$
	B ← B − 1							10	111	011					B to $A_8 \sim A_{15}$
	HL ← HL − 1											2	4 (If B = 0)	16	
	Repeat until B = 0														

NOTES: ① If the result of B − 1 is zero, the Z flag is set; otherwise it is reset.
② Z flag is set upon instruction completion only.

Z80 CPU

SUMMARY OF FLAG OPERATION

Instructions	D₇ S	Z	H		P/V	N	D₀ C	Comments
ADD A, s; ADC A, s	↕	↕	X	↕	X	V	0	↕	8-bit add or add with carry.
SUB s; SBC A, s; CP s; NEG	↕	↕	X	↕	X	V	1	↕	8-bit subtract, subtract with carry, compare and negate accumulator.
AND s	↕	↕	X	1	X	P	0	0	Logical operation.
OR s, XOR s	↕	↕	X	0	X	P	0	0	Logical operation.
INC s	↕	↕	X	↕	X	V	0	•	8-bit increment.
DEC s	↕	↕	X	↕	X	V	1	•	8-bit decrement.
ADD DD, ss	•	•	X	X	X	•	0	↕	16-bit add.
ADC HL, ss	↕	↕	X	X	X	V	0	↕	16-bit add with carry.
SBC HL, ss	↕	↕	X	X	X	V	1	↕	16-bit subtract with carry.
RLA; RLCA; RRA; RRCA	•	•	X	0	X	•	0	↕	Rotate accumulator.
RL m; RLC m; RR m; RRC m; SLA m; SRA m; SRL m	↕	↕	X	0	X	P	0	↕	Rotate and shift locations.
RLD; RRD	↕	↕	X	0	X	P	0	•	Rotate digit left and right.
DAA	↕	↕	X	↕	X	P	•	↕	Decimal adjust accumulator.
CPL	•	•	X	1	X	•	1	•	Complement accumulator.
SCF	•	•	X	0	X	•	0	1	Set carry.
CCF	•	•	X	X	X	•	0	↕	Complement carry.
IN r (C)	↕	↕	X	0	X	P	0	•	Input register indirect.
INI; IND; OUTI; OUTD	X	↕	X	X	X	X	1	•	Block input and output. Z = 1 if B ≠ 0, otherwise Z = 0.
INIR; INDR; OTIR; OTDR	X	1	X	X	X	X	1	•	Block input and output. Z = 1 if B ≠ 0, otherwise Z = 0.
LDI; LDD	X	X	X	0	X	↕	0	•	Block transfer instructions. P/V = 1 if BC ≠ 0, otherwise P/V = 0.
LDIR; LDDR	X	X	X	0	X	0	0	•	Block transfer instructions. P/V = 1 if BC ≠ 0, otherwise P/V = 0.
CPI; CPIR; CPD; CPDR	X	↕	X	X	X	↕	1	•	Block search instructions. Z = 1 if A = (HL), otherwise Z = 0. P/V = 1 if BC ≠ 0, otherwise P/V = 0.
LD A; I, LD A, R	↕	↕	X	0	X	IFF	0	•	IFF, the content of the interrupt enable flip-flop, (IFF₂), is copied into the P/V flag.
BIT b, s	X	↕	X	1	X	X	0	•	The state of bit b of location s is copied into the Z flag.

SYMBOLIC NOTATION

Symbol	Operation	Symbol	Operation
S	Sign flag. S = 1 if the MSB of the result is 1.	↕	The flag is affected according to the result of the operation.
Z	Zero flag. Z = 1 if the result of the operation is 0.	•	The flag is unchanged by the operation.
P/V	Parity or overflow flag. Parity (P) and overflow (V) share the same flag. Logical operations affect this flag with the parity of the result while arithmetic operations affect this flag with the overflow of the result. If P/V holds parity: P/V = 1 if the result of the operation is even; P/V = 0 if result is odd. If P/V holds overflow, P/V = 1 if the result of the operation produced an overflow. If P/V does not hold overflow, P/V = 0.	0	The flag is reset by the operation.
		1	The flag is set by the operation.
		X	The flag is indeterminate.
		V	P/V flag affected according to the overflow result of the operation.
		P	P/V flag affected according to the parity result of the operation.
H*	Half-carry flag. H = 1 if the add or subtract operation produced a carry into, or borrow from, bit 4 of the accumulator.	r	Any one o the CPU registers A, B, C, D, E, H, L.
		s	Any 8-bit location for all the addressing modes allowed for the particular instruction.
N*	Add/Subtract flag. N = 1 if the previous operation was a subtract.	ss	Any 16-bit location for all the addressing modes allowed for that instruction.
C	Carry/Link flag. C = 1 if the operation produced a carry from the MSB of the operand or result.	ii	Any one of the two index registers IX or IY.
		R	Refresh counter.
		n	8-bit value in range < 0, 255 >.
		nn	16-bit value in range < 0, 65535 >.

*H and N flags are used in conjunction with the decimal adjust instruction (DAA) to properly correct the result into packed BCD format following addition or subtraction using operands with packed BCD format.

INDEX